Strategies for Information Technology Governance

Wim Van Grembergen
University of Antwerp, Belgium

IDEA GROUP PUBLISHING
Hershey • London • Melbourne • Singapore

Acquisitions Editor:	Mehdi Khosrow-Pour
Senior Managing Editor:	Jan Travers
Managing Editor:	Amanda Appicello
Development Editor:	Michele Rossi
Copy Editor:	Ingrid Widitz
Typesetter:	Jennifer Wetzel
Cover Design:	Michelle Waters
Printed at:	Integrated Book Technology

Published in the United States of America by

    Idea Group Publishing (an imprint of Idea Group Inc.)
701 E. Chocolate Avenue, Suite 200
Hershey PA 17033
Tel: 717-533-8845
Fax: 717-533-8661
E-mail: cust@idea-group.com
Web site: http://www.idea-group.com

and in the United Kingdom by

Idea Group Publishing (an imprint of Idea Group Inc.)
3 Henrietta Street
Covent Garden
London WC2E 8LU
Tel: 44 20 7240 0856
Fax: 44 20 7379 3313
Web site: http://www.eurospan.co.uk

Library of Congress Cataloging-in-Publication Data

Strategies for information technology governance / Wim Van Grembergen, editor.
    p. cm.
Includes bibliographical references and index.
  ISBN 1-59140-140-2 -- ISBN 1-59140-141-0 (ebook)
 1. Information technology--Management. 2. Business enterprises--Communication systems--Management. I. Van Grembergen, Wim, 1947-
  HD30.2.S7872 2003
  658.4'038--dc22

                       2003014949

Paperback ISBN 1-59140-284-0

British Cataloguing in Publication Data
A Cataloguing in Publication record for this book is available from the British Library.

Strategies for Information Technology Governance

Table of Contents

Section II: Performance Management as IT Governance Mechanism

Section III: Other IT Governance Mechanisms

Preface

This book, *Strategies for Information Technology Governance*, is aimed at improving the understanding of Information Technology (IT) Governance and its structures, processes and relational mechanisms. As will be defined in this book, IT Governance is the organisational capacity exercised by the Board, executive management and IT management to control the formulation and implementation of IT strategy and in this way ensure the fusion of business and IT. Theoretical models and practices regarding IT Governance will be discussed in the different chapters of this book and attention will be paid to its mechanisms, including IT steering committee structures, Balanced Scorecards, control objectives and management guidelines developed by ISACA, and relational mechanisms such as business/IT job rotation.

This book brings together 14 papers on IT Governance written by academics and practitioners from different countries including Belgium, Canada, Finland, Greece, The Netherlands, Norway, Spain, the United Kingdom and the United States.

The authors of the different chapters have been included in the review process and have reviewed and critiqued the manuscripts of their colleague-authors. I wish to thank the contributors to this book for submitting their chapter(s) and for assisting me in the review process as well.

The overall structure of this book follows a logical sequence: introducing the IT Governance frameworks in Section I; reviewing performance management mechanisms in Section II; presenting other IT Governance mechanisms in Section III; and illustrating how IT Governance can work in practice in Section IV.

Section I: IT Governance Frameworks

This section introduces the IT Governance concepts and consists of three chapters.

Chapter I: *Structures, Processes and Relational Mechanisms for IT Governance* by Wim Van Grembergen, Steven De Haes and Erik Guldentops respectively from the University of Antwerp, the University of Antwerp Management School (Belgium), and the IT Governance Institute (US), defines the IT Governance concepts and overviews the different IT Governance mechanisms. It records and interprets some important existing theories, models and practices on IT Governance. The chapter is based on relevant academic and professional publications and integrates also the main contributions of the other chapters in this book.

Chapter II: *Integration Strategies and Tactics for Information Technology Governance* by Ryan Peterson from the Instituto de Empresa (Spain) has three objectives. First of all, to describe past developments and current challenges complex organisations are facing governing the IT portfolio of IT applications, IT development, IT operations and IT platforms. Secondly, to discuss how organisations can diagnose and design IT governance architecture for future performance improvement and sustained business growth. Finally, to provide a thorough understanding and holistic picture of effective IT governance practices, and to present a new organising logic for IT governance.

Chapter III: *An Emerging Strategy for E-Business IT Governance* by Nandish Patel from Brunel University (UK) develops a framework for global e-business IT governance. This framework is based on fundamental re-directions in global e-business IT governance thinking and it applies to companies that seek to integrate Internet, intranet and World Wide Web technologies into their business activities in some form of an e-business model. The framework explains and elaborates e-business strategies for coping with emergent organisations and planned aspects of IT. The basic premise of the proposed framework is that organisation, especially virtual organisation, is both planned and emergent, diverging from the dominant premise of central control in IT governance.

Section II: Performance Management as IT Governance Mechanism

Section II: *Performance Management as IT Governance Mechanism* reviews IT governance mechanisms including Balanced Scorecards, business-IT alignment maturity assessment models, ROI measurement and technical IT measurements. This part consists of six chapters.

Chapter IV: *Assessing Business-IT Alignment Maturity* by Jerry Luftman from Stevens Institute of Technology (USA) discusses an approach for assessing the maturity of the business-IT alignment. The proposed strategic alignment maturity assessment approach provides a vehicle to evaluate where an organisation is and where it needs to go to attain and sustain business-IT alignment. The careful assessment of a firm's alignment maturity is an important step in identifying the specific actions necessary to ensure IT is being used to appropriately enable or drive the business strategy.

Chapter V: *Linking the IT Balanced Scorecard to the Business Objectives at a Major Canadian Financial Group* by Wim Van Grembergen, Ronald Saull and Steven De Haes respectively from the University of Antwerp (Belgium), Great-West Life, Londen

Life, Investors Group (Canada), and the University of Antwerp Management School (Belgium) illustrates how the Balanced Scorecard concepts can be used to support the business-IT fusion. The development and implementation of an IT Balanced Scorecard within this financial group is described and discussed. An IT Balanced Scorecard maturity model is developed and used to determine the maturity level of the scorecard under review. An important conclusion is that an IT Balanced Scorecard must go beyond the operational level and must be integrated across the enterprise in order to generate business value. This can be realised through establishing a linkage between the business Balanced Scorecard and different levels of IT scorecards.

Chapter VI: *Measuring and Managing E-Business Initiatives through the Balanced Scorecard* by Wim Van Grembergen and Isabelle Amelinckx both from the University of Antwerp (Belgium), applies the Balanced Scorecard concepts to e-business projects. A generic e-business scorecard is developed and presented as a measuring and management instrument. The proposed e-business scorecard consists of four perspectives: the Customer Perspective representing the evaluation of the consumer and business clients, the Operational Perspective focusing on the business and IT processes, the Future Perspective showing the human and technology resources needed to deliver the e-business application, and the Contribution Perspective capturing the e-business benefits. It is argued that a monitoring instrument such as the proposed e-business scorecard is a must when building, implementing and maintaining an e-business system because these initiatives are often too technically management and are often initiated without a clear business case.

Chapter VII: *A View on Knowledge Management: Utilizing a Balanced Scorecard Methodology for Analyzing Knowledge Metrics* by Alea Fairchild from Vesalius College/Vrije Universiteit Brussel (VUB) (Belgium) addresses the problem of developing measurement models for Knowledge Management metrics and discusses what current Knowledge Management metrics are in use, and examines their sustainability and soundness in assessing knowledge utilisation and retention of generating revenue. The chapter also discusses the use of the Balanced Scorecard approach to determine a business-oriented relationship between strategic Knowledge Management usage and IT strategy and implementation.

Chapter VIII: *Measuring ROI in E-Commerce Applications: Analysis to Action* by Manuel Mogollon and Mahesh Raisinghani respectively from Nortel Networks (US) and the University of Dallas (USA) focuses on measuring the Return on Investment (ROI) as a key element of the IT Governance process. The research in this chapter aims to provide an overview of how to calculate the ROI for e-commerce applications so that this information, and the attached ROI Calculator Tool Template, can be used by organisations to reduce time in preparing the ROI for a project

Chapter IX: *Technical Issues Related to IT Governance Tactics: Product Metrics, Measurements and Process Control* by Michalis Xenos from the Hellenic Open University (Greece) deals with some technical aspects of the strategies for IT Governance and aims at introducing the reader to software metrics that are used to provide knowledge about different elements of IT projects. Internal metrics are presented that can be applied prior to the release of IT products to provide indications relating to quality characteristics, and external metrics are introduced that can be applied after IT product delivery to give information about user perception of product quality. The chapter also analyzes the correlation between internal and external metrics and discusses how these metrics can be combined in a measurement program.

Section III: Other IT Governance Mechanisms

Section III: *Other IT Governance Mechanisms* describes other mechanisms including roles and responsibilities within the IT organisation, the control objectives and management guidelines of COBIT, and the IT outsourcing solution. This part consists of three chapters.

Chapter X: *Managing IT Functions* by Petter Gottschalk from the Norwegian School of Management (Norway) discusses imperatives for IT functions, organisation of IT functions, roles of IT functions, roles of chief information officers (CIOs) and key issues in IT management. A survey conducted in Norway revealed that CIOs find the role of entrepreneur most important and the role of liaison least important. This survey also revealed that "Improving links between information systems strategy and business strategy" was ranked as most important key issue in IT management in Norway.

Chapter XI: *Governing Information Technology through COBIT* by Erik Guldentops from the IT Governance Institute (USA) reviews the COBIT framework that incorporates material on IT Governance. COBIT presents an international and generally accepted IT control framework enabling organisations to implement an IT Governance structure throughout the enterprise. Its management guidelines component consists of maturity models, critical success factors, key goal indicators and key performance indicators for 34 identified IT processes. This structure delivers a significantly improved framework responding for management's need for control and measurability of IT by providing means to assess and measure the organisation's IT environment against COBIT's IT processes.

Chapter XII: *Governance in IT Outsourcing Partnerships* by Erik Beulen from Tilburg University (The Netherlands) is based on 11 international IT outsourcing partnerships, five expert interviews and on literature. Three dimensions are described in a descriptive IT outsourcing partnership governance framework: the outsourcing organisation, the maintenance of the relationship, and the IT supplier. In this framework, 11 governance factors are defined including the existence of a clear IT strategy at the outsourcing organisation, a mutual trust between the outsourcing organisation and the IT supplier, and an adequate contract and account management. Furthermore, the chapter focuses on the IT outsourcing contract.

Section IV: IT Governance in Action

Section IV: *IT Governance in Action* describes the application of IT Governance structures in respectively an enterprise and in the health care industry. Section IV includes two chapters.

Chapter XIII: *The Evolution of IT Governance at NB Power* by Joanne Callahan, Cassio Bastos and Dwayne Keyes, from New Brunswick Power Corporation (Canada) describes the IT Governance framework that NB Power has implemented. Through IT Governance the organisation was able to address the results of a diagnostic study on their internal IT service provider who was attempting to respond to a seemingly endless list of requests for IT support. Now, after four years, factors critical to the success of implementing an IT Governance framework are evident. The IT Governance framework is still evolving, but the organisation is now well positioned to take advantage of its IT investment.

Chapter XIV: *Governance Structures for IT in the Health Care Industry* by Reima Suomi and Jarmo Tähkäpää from the Turku School of Economics and Business Administration (Finland) discusses the role of IT in the health care industry and focuses on the question of which governance structures are best for managing IT within this industry. Two Finnish cases are described — a small health care federation of municipalities and a medium-sized health care unit — to illustrate internal and external governance structures. It is shown that internal governance structures such as developing a comprehensive business strategy are essential parts of IT Governance and that outsourcing activities suggest that there is a need for developing and managing external governance structures.

Wim Van Grembergen, PhD
Sint-Pauwels (Belgium)
April 2003

Acknowledgments

I would like to thank all involved in the collation and review process of this book on IT Governance, without whose support the project could not have been satisfactorily completed. I would like to begin with my colleague, Mehdi Khosrow-Pour, who took a risk inviting me to edit a book on the relatively new concept of IT Governance. A special word of thanks to Jan Travers, Amanda Appicello and Michele Rossi of Idea Group Publishing who both supported me in managing this project and guided me via their e-mails when they thought I had forgotten about the book.

A very special acknowledgment goes to all the authors for their insights and excellent contributions to this book and to the understanding of IT Governance and its structures, processes and relational mechanisms. I also want to thank the authors for assisting me in the review process.

I appreciated support provided for this project by the Business Faculty of the University of Antwerp and the University Antwerp Management School. I also would like to thank my undergraduate and executive students who provided me with many ideas on the subject of IT Governance and the related IT Balanced Scorecard concepts.

A special note of thanks to Isabelle Amelinckx and Steven De Haes who both delivered research support and kept me focused on IT and E-business governance. A further special note of thanks goes to Erik Guldentops for sharing his ideas on IT Governance and for the many stimulating discussions on this subject and IT-audit related issues.

Finally, last but not least, I would like to thank Hilde, Astrid and Helen who always support and help me with every project including this book.

Wim Van Grembergen, PhD
Sint-Pauwels (Belgium)
April 2003

SECTION I:

IT GOVERNANCE FRAMEWORKS

Chapter I

Structures, Processes and Relational Mechanisms for IT Governance

Wim Van Grembergen
University of Antwerp, Belgium

Steven De Haes
University of Antwerp Management School, Belgium

Erik Guldentops
IT Governance Institute, USA

ABSTRACT

In many organisations, Information Technology (IT) has become crucial in the support, the sustainability and the growth of the business. This pervasive use of technology has created a critical dependency on IT that calls for a specific focus on IT Governance. IT Governance consists of the leadership and organisational structures and processes that ensure that the organisation's IT sustains and extends the organisation's strategy and objectives. This introductory chapter records and interprets some important existing theories, models and practises in the IT Governance domain and aims to contribute to the understanding of IT Governance and its structures, processes and relational mechanisms.

INTRODUCTION

Information Technology (IT) has become **pervasive** in current dynamic and often turbulent business environments. While in the past, business executives could delegate, ignore or avoid IT decisions, this is now impossible in most sectors and industries (Peterson, 2003; Duffy, 2002; Van Der Zee & De Jong, 1999). To emphasise this pervasiveness, Broadbent and Weill (1998) refer to three layers of the 'new infrastructure': local IT for business processes, firm IT infrastructure and public IT infrastructures (Figure 1).

The *Public Infrastructure* (Figure 1) is the foundation of the *New Infrastructure*, which is in turn linked to external industry infrastructures such as Internet, EDI networks, etc. This enables the business to communicate and do business with customers, suppliers, partners, etc. Together with the *Firm Information Technology Infrastructure*, such as e-mail, customer databases, etc., these infrastructures make up the *New Infrastructure*. The *New Infrastructure*, plus the local IT needed to perform business processes, can be defined as the *Firm Information Technology Portfolio*.

The *Information Technology Portfolio* not only has the potential to support existing business strategies, but also to shape new strategies (Henderson, Venkatraman, & Oldach, 1993; Henderson & Venkatraman, 1993; Guldentops, 2003). In this mindset, IT becomes not only a success factor for survival and prosperity, but also an **opportunity** to differentiate and to achieve competitive advantage. IT also offers a means for increasing productivity. Leveraging IT successfully to transform the enterprise and create products and services with added value has become a universal business competency (Guldentops, 2003). In this viewpoint, the IT department moves from a commodity service provider to a strategic partner, as illustrated by Venkatraman (1999) (Table 1).

Figure 1. The New Infrastructure

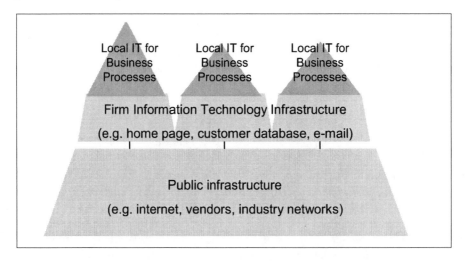

Broadbent, M. & Weill, P. (1998). Leveraging the New Infrastructure – How Market Leaders Capitalize on Information Technology. *Harvard Business School Press.*

Table 1. IT as Service Provider or as Strategic Partner

Service provider	Strategic partner
• IT is for efficiency	• IT for business growth
• Budgets are driven by external benchmarks	• Budgets are driven by business strategy
• IT is separable from the business	• IT is inseparable from the business
• IT is seen as an expense to control	• IT is seen as an investment to manage
• IT managers are technical experts	• IT managers are business problem solvers

Venkatraman, N. (1999). Valuing the IS Contribution to the Business. *Computer Sciences Corporation.*

The dependency on IT becomes even more imperative in our **knowledge-based economy**, where organisations are using technology in managing, developing and communicating intangible assets such as information and knowledge (Patel, 2003). Corporate success can of course only be attained when information and knowledge, very often provided and sustained by technology, is secure, accurate, and reliable, and provided to the right person, at the right time, at the right place (ITGI, 2000; Kakabadse & Kakabadse, 2001).

This major IT dependency also implies a huge vulnerability that is inherently present in certain complex IT environments (ITGI, 2001; Duffy, 2002). System and network downtime has become far too costly for any organisation in these days of doing business globally around the clock. Take for example the impact of downtime in the banking sector or in a medical environment. The **risk factor** is accompanied by a wide spectrum of external threats, such as errors and omissions, abuse, cybercrime and fraud.

Information Technology often entails large capital investments in organisations while companies are faced with multiple shareholders that are demanding the creation of business value through these investments. The question of the '**productivity paradox**', why Information Technologies have not provided a measurable value to the business world, has puzzled many practitioners and researchers (Kakabadse & Kakabadse, 2001; ITGI, 2000, 2001; Lie, 2001; Henderson & Venkatraman, 1993; Duffy, 2002; Strassman, 1990; Brynjolfsson, 1993; Brynjolfsson & Hitt, 1998).

All the issues described above point out that the critical dependency on IT calls for a specific focus on IT Governance. This is needed to ensure that the investments in IT will generate the required business value and that risks associated with IT are mitigated. This chapter records and interprets some important existing theories, models and practises on IT Governance and its structures, processes and relational mechanisms. The chapter is based on relevant academic and professional publications and integrates also the main contributions of the other chapters in this book (whenever the text references to one of the other chapters, the reference is printed in bold).

The first section provides a definition of IT Governance and draws a link with the Corporate Governance principles. The second part elaborates on the core issues in the IT Governance domain: strategic alignment, value delivery, risk management and performance management. The third part delivers an overview of some important structures, processes and relational mechanisms that can be helpful when designing and implementing an IT Governance framework. The final section describes a model for assessing and diagnosing IT Governance implementations.

IT GOVERNANCE AND
CORPORATE GOVERNANCE

IT Governance Definitions

IT, and its use in business environments, has experienced a fundamental transformation in the past decades. Since the introduction of IT in organisations, academics and practitioners conducted research and developed theories and best practises in this emerging knowledge domain (**Peterson**, 2003). This resulted in a variety of IT Governance definitions, some of which are formulated in Table 2.

Although the definitions in Table 2 differ on some aspects, they are all mainly focused to the same issues, such as the link between business and IT. The definition of the IT Governance Institute (ITGI), however, also explicitly states that IT Governance is an integral part of enterprise governance, which is in our opinion a very important premise. The IT Governance definition of ITGI will therefore be used as the reference in this chapter, even though it should be recognised that the link with enterprise governance is implicitly present in Van Grembergen's definitions as well.

IT Governance vs. IT Management

An important (implicit) common concern in the definitions of Table 2 is certainly the link of IT with the present and future business objectives. This goes back to the not always that clear difference between IT Governance and IT Management, which is visualised in Figure 2. IT Management is focused on the internal effective supply of IT services and products and the management of present IT operations. IT Governance in turn is much broader, and concentrates on performing and transforming IT to meet present and future demands of the business (internal focus) and the business' customers (external focus) (**Peterson**, 2003). "This does not undermine the importance and complexity of IT management, …, but whereas elements of IT Management and the supply of (commodity) IT services and products can be commissioned to an external provider, IT Governance is organisation specific, and direction and control over IT can not be delegated to the market" (**Peterson**, 2003).

IT Governance vs. Corporate Governance and the Board

The definition of IT Governance as proposed by the IT Governance Institute (Table 2) expresses that "IT Governance is the responsibility of the Board and Executive Management and that IT Governance should be an integral part of enterprise governance." How can we explain this relationship between IT Governance, Corporate Governance (or Enterprise Governance) and the Board?

Enterprise Governance is the system by which entities are directed and controlled. The business dependency on information technology has made it so that the enterprise governance issues cannot be solved without considering Information Technology. As shown in the first part of Figure 3, enterprise governance should therefore drive and set IT Governance. Information Technology in its turn can influence strategic opportunities as outlined by the enterprise and can provide critical input to strategic plans. In this way, IT Governance enables the enterprise to take full advantage of its information, and can

Table 2. Definitions of IT Governance

- The organisational capacity to control the formulation and implementation of IT strategy and guide to proper direction for the purpose of achieving competitive advantages for the corporation
  The Ministry of International Trade and Industry (1999)

- IT Governance is the responsibility of the Board of Directors and executive management. It is an integral part of enterprise governance and consists of the leadership and organisational structures and processes that ensure that the organisation's IT sustains and extends the organisation's strategy and objectives.
  IT Governance Institute (2001)

- IT Governance is the organisational capacity exercised by the Board, executive management and IT management to control the formulation and implementation of IT strategy and in this way ensure the fusion of business and IT.
  Van Grembergen (2002)

---

be seen as a driver for enterprise governance. Looking at this interplay in more depth (second part of Figure 3), enterprise activities require information from IT activities to meet business objectives, and IT must be aligned with enterprise activities to take full advantage of its information (ITGI, 2000). IT Governance and Enterprise Governance can therefore not be considered as pure distinct disciplines, and IT Governance needs to be integrated into the overall enterprise governance structure, as denoted by several authors and entities (**Guldentops**, 2003; ITGI, 2001; **Peterson**, 2003; Duffy, 2002).

The close relationship between corporate and IT Governance can also be derived from Shleifer and Vishny's definition of Corporate Governance (1982): Corporate Governance "deals with the ways in which suppliers of finance assure themselves of getting a return on investment." According to Shleifer and Vishny (1997), typical Corporate

Figure 2. IT Governance and IT Management

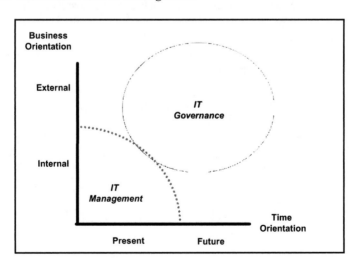

Peterson (2003). Information Strategies and Tactics for Information Technology Governance. In W. Van Grembergen (Ed.), Strategies for Information Technology Governance. Hershey, PA: Idea Group Publishing.

Figure 3. Enterprise Governance and IT Governance

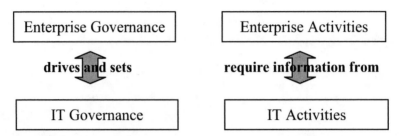

ITGI (2000). CobiT: Governance, Control and Audit for Information and Related Technology. *Available online: www.itgi.org.*

Governance questions are: (1) How do suppliers of finance get managers to return some of the profits to them? (2) How do suppliers of finance make sure that managers do not steal the capital they supply or invest it in bad projects? (3) How do suppliers of finance control managers? The business dependency on IT means that the Corporate Governance issues cannot be solved without considering IT. To make sure that the Corporate Governance matters are covered, IT needs to be governed properly first. This relationship can be made more eloquent by translating the Corporate Governance questions into specific IT Governance questions (Table 3) which discloses that Corporate Governance issues cannot be addressed without considering IT Governance issues.

As IT Governance becomes an integral part of Corporate Governance, it is of course a responsibility of the Board of Directors. The composition of the Board varies widely from organisation to organisation, but generally involves a mix of executive directors (those who are employed directly by the business) and non-executive or 'independent' directors (those who are appointed from outside the business). There are also important differences between countries regarding the role, composition and modus operandi of the Board (Duffy, 2002). These differences naturally lead to variations in expectations, emphasis, etc., but the fundamental responsibilities of the Board do not change and attention should be paid to the close link between technology management and the achievement of business goals (Duffy, 2002). Moreover, market analysts state that investors are willing to pay more for the shares of a well-governed company. Although

Table 3. IT Governance and Corporate Governance Questions

Corporate Governance questions	⇨	IT Governance questions
How do suppliers of finance get managers to return some of the profits to them?	⇨	How does top management get their CIO and IT organisation to return some business value to them?
How do suppliers of finance make sure that managers do not steal the capital they supply or invest it in bad projects?	⇨	How does top management make sure that their CIO and IT organisations do not steal the capital they supply or invest in bad projects?
How do suppliers of finance control managers?	⇨	How does top management control their CIO and IT organisation?

Adapted from: Shleifer, A. & Vishny, W. (1997). A survey on corporate governance. The Journal of Finance, 52*(2).*

hypothetical premiums are difficult to measure, there is little question that good governance makes a difference to corporate value (ITGI, 2002; Duffy, 2002).

STRATEGIC ALIGNMENT AND THE ACHIEVEMENT OF BUSINESS VALUE

The definitions in Table 2 implicitly or explicitly underline that an important aspect of IT Governance is the alignment of Information Technology with the business, often referred to as strategic alignment. Strategic alignment is an important driving force to achieve business value through investments in IT (ITGI, 2001; **Guldentops**, 2003). These two elements of IT Governance — strategic alignment and the achievement of business value through IT — will be discussed in more detail in the following paragraphs. Additionally, two related elements — risk management and performance management — will be described.

Business/IT Alignment: The Strategic Alignment Model (SAM)

What do we exactly mean by strategic alignment between the business and IT? Duffy (2002) formulated the following definition: "the process and goal of achieving competitive advantage through developing and sustaining a symbiotic relationship between business and IT." The idea behind strategic alignment is very comprehensive, but the question is how organisations can achieve this ultimate goal. Henderson and Venkatraman (1993) developed a Strategic Alignment Model to conceptualise and direct the area of strategic management of Information Technology (Figure 4). They were the first to describe in a clear way the interrelationship between business strategies and IT strategies in their well-known Strategic Alignment Model (SAM) (Smaczny, 2001). Many authors used this model for further research, including Luftman and Brier (1999), Burn and Szeto (2000) and Smackzny (2001).

The concept in Figure 4 is based on two building blocks: *strategic fit* and *functional integration*. *Strategic fit* recognises that the IT strategy should be articulated in terms of an external domain — how the firm is positioned in the IT marketplace — and an internal domain — how the IT infrastructure should be configured and managed. The position of an organisation in the IT marketplace (external IT domain) involves three decisions: (1) Information Technology scope (those specific information technologies, such as local and wide area networks, that support business strategy initiatives or could shape new business strategy initiatives for the firm), (2) systemic competencies (those attributes of IT strategy, e.g., cost-performance levels and flexibility, that could contribute positively to the creation of new business strategies or better support of existing business strategy), (3) IT Governance (selection and use of mechanisms, e.g., strategic alliances, for obtaining the required IT competencies). The internal IT domain must address three components: (1) IT architecture (choices that define the portfolio of applications, the configurations of hardware, software and communications, and the data architecture that collectively define the technical infrastructure), (2) IT processes (choices that define the work processes central to the operations of the IT infrastructure,

Figure 4. Strategic Alignment Model

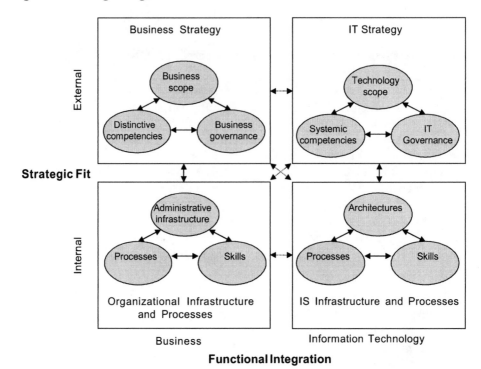

Henserson, J. C. & Venkatraman, N. (1993). Strategic alignment: Leveraging Information Technology for transforming organizations. IBM Systems Journal, 32*(1).*

e.g., systems development maintenance), (3) IT skills (choices pertaining to the acquisition, training and development of the knowledge and capabilities of the individuals required to effectively manage and operate the IT infrastructure). Henderson and Venkatraman (1993) argue that the external and the internal domains are equally important, but that managers traditionally think of IT strategy in terms of the internal domain, since historically IT is viewed as a support function less essential to the business. Relating this to the difference between IT Governance and IT management as referred to in Figure 1, the historical internal view coincides with the IT management perspective, which is focused on the internal domain (while the IT Governance perspective is focused on both the internal and the external domains).

Strategic fit is equally relevant within the business domain, as is also illustrated in Figure 3: the business strategy should take as well the internal as the external domain into account. The attributes are similar, but focussed to business: business scope (choices regarding the product-market offerings in the output market), distinctive competencies (those attributes that contribute to a competitive advantage), business governance (make-vs-buy decisions, inter-company relationships), administrative architectures (roles, responsibilities, authority), business processes (that support and shape the firm's ability to execute business strategies) and business skills (required to execute a given strategy).

In the *functional integration* dimension of the Strategic Alignment model, the authors propose two types of integration which consider how choices made in the IT domain enhance or threaten those made in the business domain and vice versa. *Strategic integration* is the link between business strategy and IT strategy reflecting the external components, which is as important as IT and for many companies has emerged as a source of strategic advantage. The second type, *operational integration*, covers the internal domain and deals with the link between organisational infrastructure and processes, and IT infrastructure and process. This emphasises the importance of internal coherence between the requirements and expectations of the business and the capability of IT to deliver against it.

An important premise of the Strategic Alignment model is that effective governance of IT requires a balance among the choices made in all the four domains of Figure 4. Henderson and Venkatraman (1993) describe two cross-domain relationships in which business strategy plays the role of driver, and two relationships where IT strategy is the enabler (Figure 5). The *strategic execution* perspective is probably the most widely understood, as it is the classic, hierarchical view of strategic management. The perspective starts from the premise that business strategy is articulated and that this strategy is the driver for the choices in organisational design and the design in IT infrastructure. The *technology transformation* perspective also starts from an existing business strategy, but focuses on the implementation of this strategy through appropriate IT strategy and the articulation of the required IT infrastructure and processes. The *competitive potential* perspective allows the adaptation of business strategy through emerging IT capabilities. Starting from the IT strategy, the best set of strategic options

Figure 5. Strategic Alignment Domains

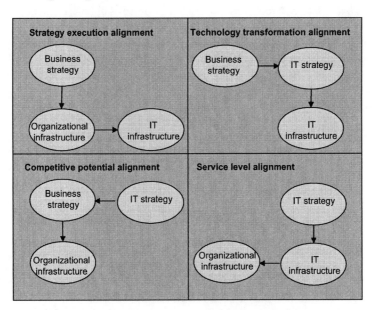

Henserson, J. C. & Venkatraman, N. (1993). Strategic alignment: Leveraging Information Technology for transforming organizations. IBM Systems Journal, 32(1).

for business strategy and a corresponding set of decisions regarding organisational infrastructure and processes are determined. The *service level* perspective focuses on how to build a world-class IT service organisation. This requires an understanding of the external dimensions of IT strategy with the corresponding internal design of the IT infrastructure and processes.

Relevance of the Strategic Alignment Model and Its Relationship to Other Mechanisms

Henderson and Venkatraman (1993) stress that alignment is not a one-point-in-time action. The challenge is to ensure the continual assessment of the trends across the four domains and to evolve from one perspective to another based on shifts in the business environment, both internal and external.

Although the Strategic Alignment model clearly recognizes the need for continual alignment, it does not provide a practical framework to implement this (Van Der Zee & De Jong, 1999). In that case, the question of how to realize strategic alignment is still not solved. Van Der Zee and De Jong (1999) propose the Balanced Scorecard as an implementation solution (see next section).

Another approach for the practical implementation of strategic alignment is provided by Luftman (2000) and Luftman and Brier (1999), who state that achieving alignment in environments of dynamic business strategies and continuously evolving technologies is very hard to accomplish. According to them, strategic alignment should be viewed as a process, and they propose a six-step approach (Table 4) that incorporates organisational assessment using a strategic alignment based on the Henderson and Venkatraman model (Luftman & Brier, 1999).

Guldentops (2003) also promotes some pragmatic practises to achieve alignment, and makes a distinction between vertical and horizontal alignment (Figure 6). According to this author, there are two types of practises, re-enforcing the point that alignment is not only needed at the strategic level but also at the operational level. Vertical alignment is primarily driven by repeatedly communicating an integrated Business and IT strategy down into the organisation, and translating it at each organisational layer into the language, responsibilities, values and challenges at that level. Furthermore, this 'cascad-

Table 4. Six-Step Process for Alignment

Set the goals and establish a team
Understand the business-IT linkage
Analyse and prioritise gaps
Specify the actions (project management)
Choose and evaluate success criteria
Sustain alignment

Luftman, J. & Brier, T. (1999). Achieving and sustaining business-IT alignment. California Management Review, 42*(1), 109-122.*

Figure 6. Vertical and Horizontal Alignment Practises

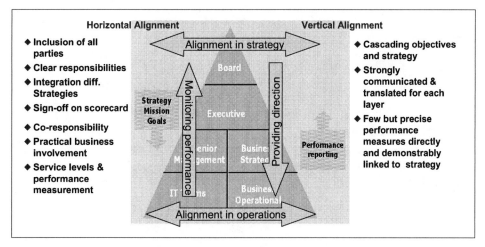

Guldentops, E. (2003). *IT Governance: Part and parcel of Corporate Governance.* CIO Summit, European Financial Management & Marketing (EFMA) Conference, Brussels.

ing down' of the strategic objectives should be clearly linked to performance measures that are reported upwards. Horizontal alignment is primarily driven by cooperation between Business and IT on integrating the strategy, on developing and agreeing on performance measures (e.g., SLAs and IT BSC) and on sharing responsibilities (e.g., IT project co-responsibility) (Guldentops, 2003).

Alignment Practise: Success Factors and Inhibitors

A study of Burn and Szeto (2000) revealed that only 50% of the business managers and 60% of IT managers indicated that the matching of business and IT strategies in their companies was either successful or highly successful. In this study, the key success factors for alignment were identified as 'top management selections of appropriate alignment approach to accomplish business objectives' and 'matching the internal IT with external market'.

Broadbent and Weill (1998) described different difficulties (barriers) that organisations have experienced in aligning business with IT. The *expression barriers* arise from the organisation's strategic context and from senior management behaviour, including lack of direction in business strategy, changing strategic intents, etc. This results in insufficient understanding of and commitment to the organisation's strategic focus by operational management. *Specification barriers* arise from the circumstances of the organisation's IT strategy (such as lack of IT involvement in strategy development, business and IT management conducting two independent monologues, etc.), which ends up in a situation where business and IT strategies are set in isolation and not adequately related. The nature of the organisation's current IT portfolio creates *implementation barriers*, which arise when there are technical, political, or financial constraints (e.g., difficulties in integrating legacy systems) on the current infrastructure.

Table 5. Enablers — Inhibitors of Strategic Alignment

ENABLERS	INHIBITORS
Senior executive support for IT	IT/business lack close relationships
IT involved in strategy development	IT does not prioritise well
IT understands the business	IT fails to meet commitments
Business-IT partnerships	IT does not understand the business
Well-prioritised IT projects	Senior executives do not support IT
IT demonstrates leadership	IT management lack leadership

Luftman, J. & Brier, T. (1999). Achieving and sustaining Business-IT alignment. California Management Review, 42*(1), 109-122.*

Luftman (2000) and Luftman and Brier (1999) have also identified some enablers and inhibitors (Table 5) that help and hinder this alignment process. These points for attention should be closely monitored by management in their effort of aligning the business and IT.

Maturity Models for Strategic Alignment

Insight into the key success factors, barriers, enablers and inhibitors can be very helpful when an organisation strives for a more mature strategic alignment process. To be able to measure its alignment maturity, organisations can use a maturity model (Figure 7). This is a method of scoring that enables the organisation to grade itself from non-existent (0) to optimised (5). This tool offers an easy-to-understand way to determine the "as-is" and the "to-be" (according to enterprise strategy) position, and enables the organisation to benchmark itself against best practises and standard guidelines. In this way, gaps can be identified and specific actions can be defined to move towards the desired level of strategic alignment maturity (ITGI, 2000, 2001; Guldentops, 2003).

Good examples of strategic alignment maturity models are developed by Luftman (2000), Duffy (2002) and the IT Governance Institute (ITGI, 2000). Each of these models uses criteria composed of a variety of attributes to build different levels of maturity.

Luftman (1993) defines five maturity levels using the criteria and attributes described in the first two columns of Table 6. The last two columns of Table 6 indicate the characteristics or values of each attribute to obtain a level 1 or level 5 of the maturity model. When doing this maturity assessment, it is important to comply with the basic principles of maturity measurement: One can only move to a higher maturity when all conditions described in a certain maturity level are fulfilled. This implies that, in order to obtain a maturity level 5, all attributes must have the values described in the last column of Table 6.

Duffy (2002) developed a similar maturity model (Table 7) which is composed of four maturity levels. Although this maturity model differs from the previous example, it aspires to the same goal, i.e., providing a tool to help management in their journey to alignment

Figure 7. Generic Maturity Model (CobiT's Framework)

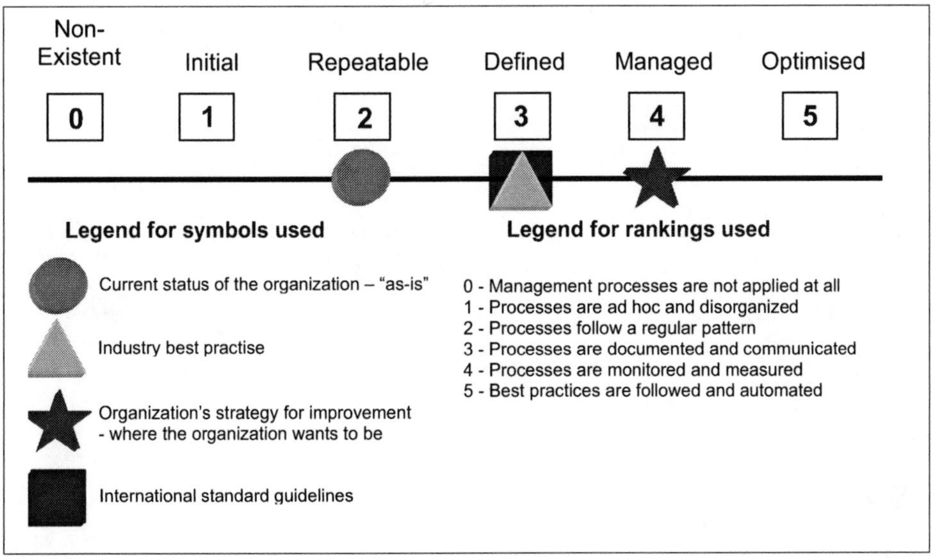

ITGI (2000). CobiT: Governance, control and audit for information and related technology. *Available online: www.itgi.org.*

between the business and IT. This maturity model states that in level one, there is a fundamental disconnect between the technology executive and the rest of corporate management. A maturity level of four (the highest level in this model), however, implies that IT and business are inextricably entwined and there is only one single strategy that incorporates both business and IT.

The third example of an alignment maturity model is provided by the IT Governance Institute (ITGI, 2000). One of the products developed by ITGI is the open standard CobiT (Control Objectives for IT and related Technologies). The CobiT Framework identifies 34 IT processes within an IT environment. For each process, it provides a high-level control statement and between three and thirty detailed control objectives. With CobiT third edition, a management layer was added — called Management Guidelines — providing critical success factors, key performance indicators and maturity models for each of the processes. The first process identified by CobiT is 'define a strategic Information Technology plan'. As this process "satisfies the business requirement to strike an optimum balance of Information Technology opportunities and IT business requirements" (ITGI, 2000), this process plays a very important role in strategic alignment. In the maturity model for this process (Table 8), maturity level one entails that the need for IT strategic planning is known by IT management, but there is no structured decision process in place. To achieve the highest maturity level in this model, IT strategic planning should at least be a documented and a living process, continuously be considered in business goal setting and result in discernable business value through investments in IT.

As already mentioned, maturity models can be a very comprehensive tool to benchmark the organisation through time or against other organisations (in specific

Table 6. Strategic Alignment Maturity Levels (Luftman)

Criteria	Attribute	Characteristics level 1	Characteristics level 5
Communications	Understanding of business by IT	Minimum	Pervasive
	Understanding of IT by business		
	Inter/intra-organisational learning	Minimum	Pervasive
	Protocol rigidity		
	Knowledge sharing	Casual, ad-hoc	Strong and structured
	Liaison(s) breadth/effectiveness		
		Command and control	Informal
		Ad-hoc	Extra-enterprise
		None or ad-hoc	Extra-enterprise
Competency/value measurement	IT metrics	Technical, not related to business	Extended to external partners
			Extended to external partners
	Business metrics	Ad-hoc, not related to IT	Business, partner, & IT metrics
			Extended to external partners
	Balanced metrics	Ad-hoc unlinked	Routinely performed with partners
	Service Level Agreements	Sporadically present	Routinely performed
			Routinely performed
	Benchmarking	Not generally practised	
	Formal assessments/reviews	None	
	Continuous improvement	None	
Governance	Business strategic planning	Ad-hoc	Integrated across, external
	IT strategic planning	Ad-hoc	Integrated across, external
	Reporting/organization structure	Central/decentral, CIO report to CFO	CIO reports to CEO, federated
	Budgetary control	Cost center, erratic spending	Investment center, profit center
			Business value
	IT investment management		Partnership
	Steering committee(s)	Cost based, erratic spending	Value added partner
	Prioritization process	Not formal/regular	
		Reactive	
Partnership	Business perception of IT value	IT perceived as a cost of business	IT co-adapts with business
	Role of IT in strategic business planning	No seat at the business table	Co-adaptive with business
	Shared goals, risks, rewards/penalties	IT takes risk with little reward	Risks & rewards shared
	IT program management	Ad-hoc	
	Relationship/trust style	Conflict/minimum	Continuous improvement
	Business sponsor/champion	None	Valued partnership
			At the CEO level
Scope and architecture	Traditional enabler/driver, external	Traditional (e.g. accounting, email)	External scope, business strategy driver/enabler
	Standards articulation	None or ad-hoc	Inter-enterprise standards
	Architectural integration	No formal integration	Evolve with partners
	Functional organization		Integrated
	Enterprise		
			Standard enterprise architecture
	Inter-enterprise		With all partners
	Architectural transparency, flexibility	None	Across the infrastructure
Skills	Innovation, entrepreneurship	Discouraged	The norm
	Locus of Power	In the business	All executives, including CIO
			Relationship based
	Management style	Command and control	High, focused
	Change readiness	Resistant to change	Across the enterprise
	Career crossover	None	Across the enterprise
	Education, cross-training	None	Effective program for hiring
	Attract and retain best-talent	No program	and retaining

Luftman, J. (2000). Assessing business-IT alignment maturity. Communications of AIS, 4.

Table 7. Strategic Alignment Maturity Model (Duffy)

Maturity Level One: "Uneasy Alliance"	Maturity Level Two: "Supplier/Consumer Relationship"
In this stage, there is a fundamental disconnect between the technology executive and the rest of corporate management. IT responds to business demands with little understanding of how the technology can contribute to value. IT is viewed primarily as something to make the company more efficient. Business units have little understanding of technology and prefer to hold the IT organisation accountable for the success and/or failure of any IT-related project.	If IT has a strategic plan it is developed in response to the corporate strategy. IT is probably viewed as a cost center and there is little appreciation for the value that IT contributes to corporate success. In this stage, IT is still not viewed as a strategic tool and IT executives are unlikely to be involved in developing corporate strategy.
Maturity Level Three: "Co-dependence/Grudging Respect"	Maturity Level Four: "United we succeed, divided we fail"
In this stage, the business is dependent on IT and there are early signs of recognition that it is a strategic tool. CIOs are becoming more knowledgeable about cross-functional business processes because of ERP, CRM, etc. The Internet and interest in e-business forces some level of IT/business alignment. CEO's begin to recognize that IT is a competitive tool.	In this stage, IT and business are inextricably entwined. Business executives have less time to prove they can deliver. Business cannot continue without IT and IT has little real value if it is not to support the corporate strategy. There is only a single strategy and it incorporates both IT and business. Whether the business is a pure play Internet company, or a "bricks 'n clicks" company, IT and business move in lockstep.

Duffy, J. (2002). IT/Business alignment: Is it an option or is it mandatory? *IDC document #26831.*

sectors and geographies, and from specific sizes). To be able to benchmark against other organisations, ISACA (Information Systems Audit and Control Association) conducted a maturity survey in 2002, asking the respondents to assign a maturity score for 15 of the 34 IT processes identified in CobiT. To establish this self-assessment, respondents were asked to use the maturity models that are described within CobiT for each process, as the one for 'IT strategic planning' in Table 8. The main conclusion of the survey is that, on average, the maturity of enterprises in controlling the 15 identified CobiT IT processes fluctuates between 2.0 (repeatable but intuitive) and 2.5. The average maturity score for IT strategic planning was also situated in this range. Filtering the results by geography, size or industry revealed that global working organisations, large organisations and financial institutions attain on average higher maturity levels for their IT processes, mostly within the bracket of 2.5 and 3.0 (defined process) (Guldentops, Van Grembergen, & De Haes, 2002).

Business Value through IT

Although strategic alignment is complex, multifaceted and - perhaps- never completely achieved, it remains a worthwhile ambition because there is a real concern about the value of the IT investment, i.e., the creation of business value (ITGI, 2001; Broadbent & Weill, 1998). "The value that IT adds to the business is a function of the degree to which the IT organisation is aligned with the business and meets the expectations of the business" (ITGI, 2001). The question is how investments in IT will results in measurable value for the entire business. The basic principles of IT value are delivery on time, within budget and with the benefits that were promised (ITGI, 2001; **Guldentops**, 2003). "In business terms, this is often translated into: competitive advantage, elapsed time for order/service fulfilment, customer satisfaction, customer wait time, employee productiv-

Table 8. Maturity Model for IT Strategic Planning

0   Non-existent IT strategic planning is not performed. There is no management awareness that IT strategic planning is needed to support business goals. **1   Initial/Ad Hoc** The need for IT strategic planning is known by IT management, but there is no structured decision process in place. IT strategic planning is performed on an as needed basis in response to a specific business requirement and results are therefore sporadic and inconsistent. IT strategic planning is occasionally discussed at IT management meetings, but not at business management meetings. The alignment of business requirements, applications and technology takes place reactively, driven by vendor offerings, rather than by an organisation-wide strategy. The strategic risk position is identified informally on a project-by-project basis. **2   Repeatable but Intuitive** IT strategic planning is understood by IT management, but is not documented. IT strategic planning is performed by IT management, but only shared with business management on an as needed basis. Updating of the IT strategic plan occurs only in response to requests by management and there is no proactive process for identifying those IT and business developments that require updates to the plan. Strategic decisions are driven on a project-by-project basis, without consistency with an overall organisation strategy. The risks and user benefits of major strategic decisions are being recognised, but their definition is intuitive. **3   Defined Process** A policy defines when and how to perform IT strategic planning. IT strategic planning follows a structured approach, which is documented and known to all staff. The IT planning process is reasonably sound and ensures that appropriate planning is likely to be performed. However, discretion is given to individual managers with respect to implementation of the process and there are no procedures to examine the process on a regular basis. The overall IT strategy includes a consistent definition of risks that the organisation is willing to take as an innovator or follower. The IT financial, technical and human resources strategies increasingly drive the acquisition of new products and technologies. **4   Managed and Measurable** IT strategic planning is standard practice and exceptions would be noticed by management. IT strategic planning is a defined management function with senior level responsibilities. With respect to the IT strategic planning process, management is able to monitor it, make informed decisions based on it and measure its effectiveness. Both short-range and long-range IT planning occurs and is cascaded down into the organisation, with updates done as needed. The IT strategy and organisation-wide strategy are increasingly becoming more coordinated by addressing business processes and value-added capabilities and by leveraging the use of applications and technologies through business process re-engineering. There is a well-defined process for balancing the internal and external resources required in system development and operations. Benchmarking against industry norms and competitors is becoming increasingly formalised. **5   Optimised** IT strategic planning is a documented, living process, is continuously considered in business goal setting and results in discernable business value through investments in IT. Risk and value added considerations are continuously updated in the IT strategic planning process. There is an IT strategic planning function that is integral to the business planning function. Realistic long-range IT plans are developed and constantly being updated to reflect changing technology and business-related developments. Short-range IT plans contain project task milestones and deliverables, which are continuously monitored and updated, as changes occur. Benchmarking against well-understood and reliable industry norms is a well-defined process and is integrated with the strategy formulation process. The IT organisation identifies and leverages new technology developments to drive the creation of new business capabilities and improve the competitive advantage of the organisation.

ITGI (2000). CobiT: Governance, control and audit for information and related technology. *Available online: www.itgi.org.*

ity and profitability. Several of these items are either subjective or difficult to measure, something all stakeholders need to be aware of" (ITGI, 2001).

Different levels of management and users will perceive the value of IT differently. Broadbent and Weill (1998) refer in this context to the 'business value hierarchy'. This hierarchy is composed of four layers: firm-wide IT infrastructure business value, business unit IT applications business value, business unit operational business value and business unit financial business value (Figure 8).

Very successful investments in Information Technology will have a positive impact on all those levels of the business value hierarchy. Less successful investments will not

Figure 8. Business Value Hierarchy

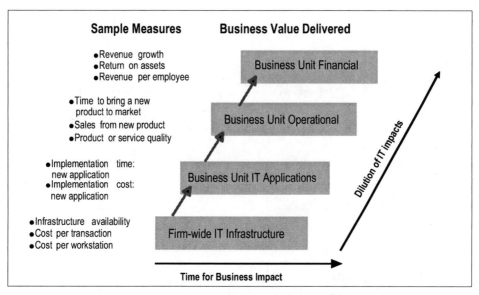

Broadbent, M. & Weill, P. (1998). Leveraging the new infrastructure – How market leaders capitalize on Information Technology. *Harvard Business School Press.*

be strong enough to impact the higher levels and will only have an influence on the lower levels. The higher one goes in the measurement hierarchy, the more dilution will occur by factors such as pricing decisions and competitor's moves. This also means that measuring the impact of an IT investment is much easier at the bottom of the hierarchy than at the top, where many factors dilute the effect. (Broadbent & Weill, 1998; ITGI, 2001).

"The first level of business value is provided by firm-wide Information Technology infrastructure, with measures such as infrastructure availability (e.g., percentage of downtime), and cost per transaction and workstation. The second level of business value is provided by business-unit Information Technology performance of the business, with measures such as time and cost to implement new applications. The third level is provided by the operational performance of the business, with measures such as quality and time to market for new products. The top and most important level is the financial performance of the firm, with measures such as return on assets (ROA) and revenue growth. Investments in Information Technology are made at the bottom two levels in the hierarchy by both information systems departments and line managers. Measuring Information Technology investments at the bottom two levels and performance at all four levels is key to assessing business value. Then we can track the impact of Information Technology investments up this hierarchy of business value, providing solid evidence and insight on how value is or is not created" (Broadbent & Weill, 1998).

Figure 9. Management Practises that Lead to IT-Enabled Business Value

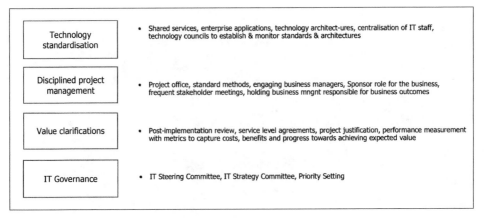

Based on: Weill, P. (2002). Research Briefing. MIT Sloan, 2, *nr. 2C.*

At the top of the hierarchy, the financial measures are typically lagging measures of business value. This means that they only focus on past performance of the enterprise. An indication or prediction of future business value can be obtained by looking at the measures of operational performance, which are leading indicators of business value. The measures of IT performance and Information Technology infrastructure performance track in their turn the efficiency of using IT assets (Broadbent & Weill, 1998).

To be successful, an organisation also needs to be aware that a different strategic context requires different indicators of value. A commercial enterprise, for example, will have different value drivers/indicators compared to a governmental institution (ITGI, 2001; Broadbent & Weil, 1998; Luftman, 2000).

But how can business value now pragmatically be achieved through IT? Weill (2002) identified some emerging management practises that lead to IT-enabled business value. Implementing these practises implies the use of a number of mechanisms, as shown in Figure 9.

Alignment, Value Delivery, Risk Management, Performance Management

We have now studied two important elements of IT Governance: value delivery (which is the end goal) and strategic alignment (which is the means). The IT Governance Institute (ITGI, 2001) introduces two related IT Governance elements — risk management and performance management — and links them all together as follows: "Fundamentally, IT Governance is concerned about two things: that IT delivers value to the business and that IT risks are mitigated. The first is driven by strategic alignment of IT with the business. The second is driven by embedding accountability into the enterprise. Both need measurement, for example, by a Balanced Scorecard. This leads to the four main focus areas for IT Governance, all driven by stakeholder value. Two of them are

Figure 10. Alignment, Value Delivery, Risk Management and Performance Management

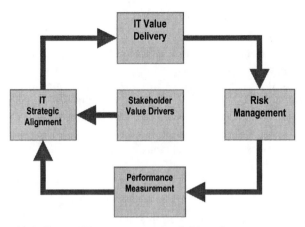

ITGI (2001). Board briefing on IT governance. *Available online: www.itgi.org.*

outcomes: value delivery and risk mitigation. Two of them are drivers: strategic alignment and performance measurements." These relationships can be visualised as illustrated in Figure 10.

This relationship introduces two associated elements (risk management and performance measurement) that are not directly referred to in the definitions of Table 2, but that play an important role in the governance of IT. The relevance of a performance measurement system, such as the Balanced Scorecard, was already mentioned in the section on IT strategic alignment, where it was identified as a mechanism to achieve strategic alignment. The Balanced Scorecard concept is discussed in more detail in the following section. "Risk management concerns itself with safeguarding assets and preparing for disaster. Risk management establishes IT security to protect assets and enable business recovery from it failures. It ensures privacy for users and builds resilience into systems. Risk management knows the importance of establishing trust in the enterprise's services and among its partners. It manages internal and external threats — internal from misuse and errors and external from deliberate attacks, market volatility and the pace of change" (Guldentops, 2002). Effective risk management begins with a clear understanding of the organisation's appetite for risk and the risk exposure. Depending on the type of risk and its significance to the business, management can walk different paths to manage this risk. The risk can be mitigated by, e.g., acquiring and deploying security technology to protect the IT infrastructure. Other possibilities are the transfer of risk, i.e., sharing the risk with partners or transferring to insurance to cover, and the acceptance of risk, i.e., formally acknowledging that the risks exists and monitoring it (ITGI, 2000, 2001). While value delivery (addressed in the previous section) is focused on the *creation of business value,* risk management is focused on the *preservation of business value.*

IT GOVERNANCE STRUCTURES, PROCESSES AND RELATIONAL MECHANISMS

We now have a better understanding of what IT Governance is. The question now arises of how enterprises can pragmatically implement an IT Governance structure. The decision to implement an IT Governance framework can sometimes be initiated by a specific issue or major critical problems. This was, for example, the case at NB Power in Canada, where the decision to implement an IT Governance framework was taken at a time when the Y2K problem required a lot of attention, a major SAP implementation project was running and an endless list of requests for IT support needed to be managed urgently (**Callahan & Keyes**, 2003).

A Holistic Approach

An IT Governance framework can be deployed using a mixture of various structures, processes and relational mechanisms. When designing IT Governance, it is important to recognise that it is contingent upon a variety of sometimes conflicting internal and external factors. Determining the right mechanisms is therefore a complex endeavour and it should be recognised that what strategically works for one company does not necessarily work for another (**Patel**, 2003), even if they work in the same industry sector. A good example of the latter is given by **Suomi and Tähkäkää** (2003), who revealed that the differences in public and private health care have an impact on the appropriate (IT) governance structure to follow. Although working in the same sector, the difference between the public versus private environment (e.g., private sector organisations are typically more flexible in terms of budget allocation, personnel decisions and organisational procedures, while public organisations are more characterized by rigid procedures, structured decision making, dependency on politics, etc.) has a great impact on the IT Governance Framework to follow and its outcomes. The analogous conclusion is made by Ribbers, Peterson and Parker (2002), who point out that environmental contingencies will impact the outcomes of the IT Governance processes (Figure 11).

However, it is not because IT Governance is a complex matter that it should be separated from the overall governance responsibilities. Dividing a complex problem into smaller pieces and solving each problem separately does not always solve the complete problem (**Peterson**, 2003). A holistic approach towards IT Governance acknowledges its complex and dynamic nature, consisting of a set of interdependent subsystems that deliver a powerful whole (Samamurthy & Smud, 1999; **Peterson**, 2003; **Patel**, 2003; Duffy, 2002). Moreover, taking the context of hypercompetition and fluctuating economic conditions into account, IT Governance within an organisation cannot be a static model. It should address both the current and emerging requirements and thus be able to continuously evolve (**Patel**, 2003).

Structures, Processes and Relational Mechanisms

To be able to place IT Governance structures, processes and relational mechanisms in a comprehensible relationship to each other, we propose the framework displayed in Table 9, which is based on Peterson's framework (**Peterson**, 1996). Structures involve

Figure 11. IT Governance Contingencies

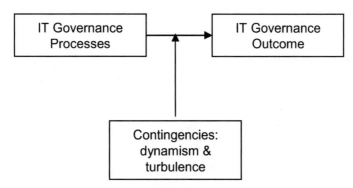

Ribbers, P. M. A., Peterson, R. R., & Parker, M. M. (2002). Designing information technology governance processes: Diagnosing contemporary practises and competing theories. In Proceedings of the 35th Hawaii International Conference on System Sciences (HICCS), CD-ROM, Maui.

the existence of responsible functions such as IT executives and accounts, and a diversity of IT committees. Processes refer to strategic IT decision-making and monitoring. The relational mechanisms include business/IT participation and partnerships, strategic dialogue and shared learning.

Table 9 provides a rich overview of mechanisms that can support IT Governance. The paragraphs below will discuss in more detail some of these mechanisms — with primarily a focus on the IT Governance structures and processes, respectively the roles and responsibilities of the major participants, the IT strategy committee, IT steering committees, the IT organisation structure, the Balanced Scorecard (BSC), the Strategic Information Systems Planning (SISP), COBIT's framework and ITIL, Service level agreements (SLA), and Information Economics. An overall IT Governance maturity model will be presented in the next session, and the strategic alignment model (SAM) and the business/IT alignment model are already covered in the preceding section.

Roles and Responsibilities

Clear and unambiguous definitions of the roles and the responsibilities of the involved parties are a crucial prerequisite for an effective IT Governance framework. It is the role of the Board and Executive management to communicate these roles and responsibilities and to make sure that they are clearly understood throughout the whole organisation (ITGI, 2001; Duffy, 2002). The Board as well as the business and IT management have to play an important role in assuring the governance of IT. The CIO is certainly not the only and primary stakeholder in the process. "IT Governance effectiveness is only partially dependent on the CIO and other IT executives, and should be viewed as a shared responsibility and enterprise-wide commitment towards sustaining and maximising IT business value" (**Peterson**, 2003). The CEO has singular responsibility for carrying out the strategic plans and policies that have been established by the Board, and the CEO should ensure that the CIO is included and accepted in the senior-

Table 9. Structures, Processes and Relational Mechanisms for IT Governance

Integration strategy	Structures	Processes		Relational mechanisms
Tactics	IT Executives & accounts	Strategic IT decision -making	Stakeholder participation	Strategic dialogue
	Committees & councils	Strategic IT monitoring	Business-IT partnerships	Shared learning
Mechanisms	- roles and responsibilities - IT strategy committee - IT steering committee - IT organisation structure - CIO on Board - project steering committees - e-business advisory board - e-business task force	-Balanced (IT) scorecards -Strategic Information Systems Planning - COBIT and ITIL - Service Level Agreements -Information economics - Strategic Alignment Model - Business/IT alignment models - IT Governance maturity models	-Active participation by principle stakeholders -Collaboration between principle stakeholders -Partnership rewards and incentives -Business/IT co-location	-Shared understanding of business/IT objectives -Active conflict resolution ('non-avoidance') -Cross-functional business/IT training -Cross-functional business/IT job rotation

Based on: Peterson (2003). Information strategies and tactics for Information Technology *governance. In W. Van Grembergen (Ed.),* Strategies for Information Technology Governance. *Hershey, PA: Idea Group Publishing.*

level decision-making process (Duffy, 2002). The CIO and the CEO should report on a regular basis to the Board, and the Board in its turn has to play the role of independent overseer of business performance and compliance (Duffy, 2002). The Board members should keep their knowledge up-to-date of current business models, management techniques, technologies, and of course the potential risks and benefits associated with each of them. This enables them to ask the right questions (ITGI, 2001; Duffy, 2002). The establishment of an IT Strategy Committee (cf., infra) at Board level can be a very helpful mechanism to achieve these goals. In the Appendix, a more detailed description is provided of the responsibilities of the CEO, the CIO and the Board, as proposed by IDC (Duffy, 2002).

IT Strategy Committee and IT Steering Committees

As mentioned earlier in this chapter, IT Governance should be an integral part of enterprise governance, and in this way it is a concern of the Board of Directors that is responsible for governing the enterprise. Many Boards carry out their governance duties through committees that oversee critical areas such as audit, compensation and acquisitions (COSO, 1992). Taking the criticality of IT into account, IT should be managed with the same commitment and accuracy, and the set-up of an IT committee at Board level — the IT Strategy Committee — can be an important mechanism to achieve this goal. The IT Strategy Committee, composed of Board and non-Board members, should assist the Board in governing and overseeing the enterprise's IT-related matters. The Committee

Table 10. Authority and Membership of IT Strategy/Steering Committee

	IT Strategy Committee	IT Steering Committee
Authority	• Advises the Board and Management on IT strategy • Is delegated by the Board to provide input to the strategy and prepare its approval • Focuses on current and future strategic IT issues	• Assists the Executive in the delivery of the IT strategy • Oversees day-to-day management of IT service delivery and IT projects • Focuses on implementation
Membership	• Board members and (specialist) non-Board members	• Sponsoring executive • Business executive (key users) • CIO • Key advisors as required (IT, audit, legal, finance)

ITGI (2002). IT Strategy Committee. *Available online: www.itgi.org.*

should ensure that IT is a regular item on the Board's agenda and that it is addressed in a structured manner. In addition, the Committee must ensure that the Board has the information it needs to achieve the ultimate objectives of IT Governance (ITGI, 2001, 2003; COSO, 1992; **Callahan & Keyes**, 2003).

The IT Strategy Committee should of course work in close partnership with the other Board committees and management (committees) to provide input to, review and amend, the aligned corporate and IT strategies (ITGI, 2002; Duffy, 2002). The detailed implementation of the IT strategy will be the responsibility of Executive Management, assisted by one or more IT "Steering" Committees. Typically, such a Steering Committee has the specific responsibility for overseeing a major project or managing IT priorities, IT costs, IT resource allocation, etc. While the IT Strategy Committee operates at Board level, the IT Steering Committee is situated at Executive level, which of course implies that these committees have different membership and a different authority (Table 10) (ITGI, 2002).

Luftman and Brier (1999) provide a list of Critical Success Factors for sustaining a Steering Committee (Table 11). In practise, the terminology used and roles and responsibilities described to define these Strategy and/or Steering Committees can vary a lot. Most important is that the concepts and rationale of these mechanisms is applied and customised to the specific organisational environment (**Callahan & Keyes**, 2003; ITGI, 2002).

IT Organisation Structure

The possibility of effective governance over IT is of course also determined by the way the IT function is organised and where the IT decision-making authority is located in the organisation. Regarding the former, it should however be noted that "given the widespread proliferation and infusion of IT in organisations, involving, e.g., technical platforms, shared IT services centres, and local business-embedded applications, the notion of a single homogenous IT function is obsolete" (**Peterson**, 2003). A lot of research has been performed with regard to the location of the decision-making authority (e.g., Zambamurthy & Smud, 1999; the Ministry on International Trade and Industry,

Table 11. Critical Success Factors for Sustaining Steering Committees

Bureaucracy	Focus on reduction/elimination to expedite opportunities to leverage IT
Career Building	Opportunities for participants to learn and expand responsibilities
Communication	Primary vehicle for IT and business discussions and sharing knowledge across parts of the organisation
Complex Decisions	Do not get involved in 'mundane areas'
Influence/Empowerment	Authority to have decisions carried out
Low hanging fruit/Quick hits	Immediate changes carried out when appropriate
Marketing	Vehicle for 'selling' the value of IT to the business
Objective Measurement	Formal assessment and review of IT's business contributions
Ownership	Responsible/accountable for the decisions made
Priorities	Primary vehicle for selecting what is done, and how much resources to allocate
Relationships	Partnerships of business and IT
Right Participants	Cooperative, committed, respected team members with knowledge of business and IT
Share risks	Equal accountability, recognition, responsibility, rewards, and uncertainty
Structure, facilitator	Processes and leadership to ensure the right focus

Luftman, J. & Brier, T. (1999). Achieving and sustaining business-IT alignment. California Management Review, 42*(1), 109-122.*

1999; **Peterson**, 2003; **Gottschalk**, 2003) and several models of modes are developed, such as centralised, decentralised and federal. The adoption of a particular mode is influenced by different determinants, such as history, size, economies of scale, Corporate Governance model, business strategy and absorptive capacity (i.e., the ability of employees to develop relevant knowledge, recognise valuable external information, make appropriate decisions, etc.) (**Peterson**, 2003; Zambamurthy & Smud, 1999). **Peterson** (2003) summarised the empirical findings of several authors, which determine the choice for a centralised or a decentralised organisation, as shown in Table 12.

However, studies indicate that a federal structure (i.e., a hybrid design of centralised infrastructure control and decentralised application control), is the dominant model in many contemporary enterprises. This model tries to achieve the 'best of both worlds', i.e., efficiency and standardisation under centralisation, and effectiveness and flexibility under decentralisation (**Peterson**, 2003; Ribbers, Peterson, & Parker, 2002).

Balanced Scorecards

Kaplan and Norton (1992) have introduced the Balanced Scorecard (BSC) at enterprise level. Their fundamental premise is that the evaluation of a firm should not be

Table 12. Determinants of Centralised/Decentralised IT Organisation

	Centralized	**Decentralized**
Business strategy	Cost focus	Innovation focus
Business governance	Centralised	Decentralised
Organisation size	Small	Large
Information intensity	Low	High
Environment stability	High	Low
Business competency	Low	High

Peterson (2003). Information strategies and tactics for Information Technology Governance. In W. Van Grembergen (Ed.), Strategies for Information Technology Governance. *Hershey, PA: Idea Group Publishing.*

restricted to a traditional financial evaluation but should be supplemented with measures concerning customer satisfaction, internal processes and the ability to innovate. Results achieved within these additional perspective areas should assure future financial results and drive the organisation towards its strategic goals while keeping all four perspectives in balance. For this balanced measurement framework, they proposed a three-layer structure for each of these four perspectives: mission, objectives and measures from which targets would be set and initiatives created (Kaplan & Norton, 1992, 1993, 1996a, 1996b). This Balanced Scorecard has been applied in the IT function and its processes (Gold, 1994; Willcocks, 2002; Van Grembergen & Saull, 2001; Van Grembergen & Van Bruggen, 1997). Recognising that IT is an internal service provider, the proposed perspectives of the Balanced Scorecard should be changed accordingly, with the following perspectives: corporate contribution, customer (user) orientation, operational excellence, and future orientation. By using a "cascade or waterfall of Balanced Scorecards," a method for business and IT fusion and control mechanisms are provided to top management. To achieve this, the IT Development Balanced Scorecard and the IT Operational Balanced Scorecard are defined as enablers for the Strategic Balanced Scorecard that is in turn the enabler of the Business Balanced Scorecard. This relationship is shown in Figure 12.

Linking the business BSC and the IT BSCs is a supportive mechanism for IT Governance. Van Der Zee and De Jong (1999) argue that the Balanced Scorecard technique is uniquely placed to address two main problems in business and IT management. The first problem is the time lag between business and IT planning process. The second is the lack of common 'language' between business and IT management. When using the BSC concepts in this way, it becomes an alignment method: business goals and the drivers of business success are identified, including specific IT drivers.

A major Canadian Financial group, who implemented the IT Balanced Scorecard, accomplished more alignment through the Balanced Scorecard by establishing cause-and-effect relationships between the different domains of the scorecard. This is visualised in Figure 13: building the foundation for delivery and continuous learning and growth

Figure 12.   Cascade of Balanced Scorecards

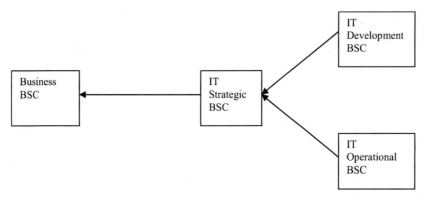

Van Grembergen, W. & Saull, R. (2001). Aligning business and Information Technology through the balanced scorecard at a major Canadian financial group: Its dtatus measured with and IT BSC Maturity Model. In Proceedings of the 34th Hawaii International Conference on System Sciences (HICCS), CD-ROM, Maui.

(future orientation perspective) is an enabler for carrying out the roles of the IT division's mission (operational excellence perspective) that is in turn an enabler for measuring up to business expectations (customer expectations perspective), that eventually must lead to ensuring effective IT Governance (corporate contribution perspective) (Van Grembergen and Saull, 2001; **Van Grembergen, Saull & De Haes**, 2003).

Strategic Information Systems Planning

According to Earl (1993) Strategic Information Systems Planning (SISP) has four components: aligning IT with business goals, exploiting IT for competitive advantage, directing efficient and effective management of IT resources, and developing technology policies and architectures. A broad variety of governance mechanisms for the two high level components — alignment and competitive advantage — have been developed and are used by organisations to achieve the business/IT fusion: Business Systems Planning (Rockart, 2001), Critical Success Factors (Rockart, 1979), the competitive forces model and the value chain models of Porter (1980, 1985) and the Business Process Reengineering approach (Hammer & Champy, 1993; Van Grembergen et al., 1997). Recently, Porter adapted his models to the e-business phenomenon in his "Strategy and the Internet" article (Porter, 2001) concluding that "the internet per se will rarely be a competitive advantage" and "many of the companies that succeed will be ones that use the internet as a complement to traditional ways of competing, not those that set their internet initiatives apart from their established operations."

CobiT and ITIL

As already explained, CobiT provides for 34 IT processes their corresponding high-level control objectives and management guidelines, including their maturity models and their scorecards in the form of key goal indicators and key performance indicators. As

Figure 13. Alignment through the IT Balanced Scorecard

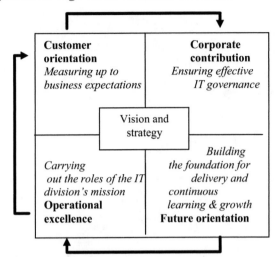

Van Grembergen, W., Saull, R., & De Haes, S. (2003). Linking the IT Balanced Scorecard to the business objectives at a major Canadian financial group. In W. Van Grembergen (Ed.), Strategies for Information Technology Governance.

illustrated in other sections of this chapter, the maturity models and scorecards enable organisations to implement an IT Governance structure (**Guldentops**, 2003).

The CobiT control objectives also can help to support IT Governance within an enterprise. The control objectives of the "Assist and advise IT customers" process, e.g., consist of establishing a help desk, registration of the customer queries, customer query escalation, monitoring of clearance, and trend analysis and reporting (ITGI, 2000). These high-level control objectives can be implemented through the use of the IT Infrastructure Library (ITIL) of Central Computer and Telecommunications Agency (UK). Its help desk module (CCTA, 1998), e.g., complements and provides details on the help desk process including the planning, implementation, post-implementation, benefits and costs, and tools. So, CobiT tells *what* is to be done and ITIL explains in detail *how* it is to be done.

Service Level Agreements

In a maturing IT Governance environment, Service Level Agreements (SLAs) and their supporting Service Level Management (SLM) process need to play an important role. The functions of SLAs are (1) the definition of what levels of service are acceptable by users and are attainable by the service provider and (2) the definition of mutually acceptable and agreed upon set of indicators of the quality of service. The SLM process includes the definition of a SLA framework, establishing SLAs including levels of service and their corresponding metrics, monitoring and reporting on the achieved services and problems encountered, reviewing SLAs, and establishing improvement programs. The major governance challenges are that the service levels are to be expressed in business terms and that the right SLM/SLA process has to be put in place (Hiles, 2000).

Figure 14. Information Economics

Traditional ROI (+)		
+ value linking (+) + value acceleration (+)	+ value restructuring (+) + innovation (+)	
= Adjusted ROI	**+ Business Value**	**+ IT Value**
	■ Strategic match (+) ■ Competitive advantage (+) ■ Competitive response (+) ■ Management information (+) ■ Service and quality (+) ■ Environmental quality (+) ■ Empowerment (+) ■ Cycle time (+) ■ Mass customization (+)	■ Strategic IT architecture (+)
	- Business	**- IT Risk**
	■ Business strategy risk (-) ■ Business organization risk (-)	■ IT Strategy risk (-) ■ Definitional uncertainty (-) ■ Technical risk (-) ■ IT service delivery risk (-)
= VALUE (business contribution)		

Van Grembergen, W. & Van Bruggen, R. (1997). Measuring and improving corporate information technology through the balanced scorecard technique. In Proceedings of the European Conference on the Evaluation of Information Technology, Delft, The Netherlands.

Information Economics

The information economics method developed by Benson and Parker (Parker, 1996) can be used as an alignment technique whereby both business and IT people score IT projects and in this way prioritise and select projects. It departs from the Return on Investment (ROI) of a project and different non-tangibles such as "strategic match of the project" (business evaluation) and "match with the strategic IT architecture" (IT evaluation). In essence, information economics is a scoring technique resulting in a weighted total score based on the scores for the ROI and the non-tangibles (Figure 14). Typically scores from 0 to 5 are attributed whereby 0 means no contribution and 5 refers to a high contribution; the values obtain a positive score and the risks a negative score.

Relational Mechanisms: Effective Communication and Knowledge Sharing

Another prior mechanism for IT Governance is an effective two-way communication and a good participation/collaboration relationship between the business and the IT department, because often there is little business awareness on the part of IT or little IT appreciation on the part of the business. Ensuring ongoing knowledge sharing across departments and organisations is paramount for attaining and sustaining business-IT alignment (Luftman, 2000; Broadbent & Weill, 1998; Henderson, Venkatraman & Oldach, 1993; **Callahan & Keyes**, 2003). It is important to facilitate the sharing and the management of knowledge by using mechanisms such as career cross-over (IT staff working in the business unit; business staff working in IT), continuous education, cross-training, etc. (Luftman & Brier, 1999; Luftman, 2000). To support a Knowledge Management initiative in the organisation, the Balanced Scorecard framework can be extended in terms of its perspectives to cover specific Knowledge Management metrics, as described by **Fairchild** (2003).

IT GOVERNANCE DIAGNOSIS AND ASSESSMENT

To implement and improve an IT Governance framework, organisations need to have a self-diagnosing tool to be able to assess IT Governance effectiveness and to identify opportunities for improvement (ITGI, 2001; **Peterson**, 2003).

An easy to understand method to self-asses and benchmark the IT Governance performance is the use of maturity models. The basic principles of maturity models are already addressed in the section on strategic alignment. The IT Governance Institute (2001) recently developed a detailed IT Governance maturity model, which identifies six (from 0 to 5) levels of maturity, from 'non-existent' to 'optimised' (ITGI, 2001).

According to this model, organisations that are situated in level zero are characterised by a complete lack of any recognisable IT Governance process. To move up to level one, the organisation at least needs to recognise the importance of addressing IT Governance issues. Maturity level five at least implies an advanced and forward-looking understanding of IT Governance issues and solutions, supported by an established framework and best practises of structures, processes and relational mechanisms. As mentioned before, this maturity model provides a comprehensive tool for determining the 'as-is' and the 'to-be' position. It should be recognised that the desired 'to-be' position should be identified in function of the context where one operates in (industry, geography, size, etc.) and of the enterprise strategy. When the 'as-is' and 'to-be' positions are known, gaps can be determined, projects defined and specific actions be taken.

CONCLUSION

This introductory chapter to *Strategies for Information Technology Governance* described relevant structures, processes and relational mechanisms for IT Governance. At the same time, this chapter introduced the main contributions of the remaining chapters in this book

A major conclusion is that governing the enterprise's Information Technology is becoming more and more important in our knowledge-based and complex society. Key elements in IT Governance are the alignment of the business and IT that must lead to the achievement of business value through IT. These high level goals of IT Governance can be achieved by acknowledging IT Governance as a part of Corporate Governance and by setting up an IT Governance framework and its corresponding best practises. Such a framework and practises should be composed of a variety of structures, processes and

Table 13. IT Governance Maturity Model

---

0   Non-existent

There is a complete lack of any recognisable IT Governance process. The organisation has not even recognised that there is an issue to be addressed and hence there is no communication about the issue.

Governance, such as it is, is predominantly centralised within the IT organisation, and IT budgets and decisions are made centrally. Business unit input is informal and done on a project basis. In some cases, a steering committee may be in place to help make resource decisions.

1   Initial /Ad Hoc

The organisation has recognised that IT Governance issues exist and need to be addressed. There are, however, no standardised review processes, but instead management considers IT management issues on an individual or case-by-case basis. Management's approach is unstructured and there is inconsistent communication on issues and approaches to address the problems that arise. Although it is recognised that the performance of the IT function ought to be measured, there are no proper metrics in place -- reviews are based on individual managers' requests. IT monitoring is implemented only reactively to an incident that has caused some loss or embarrassment to the organisation.

Governance is difficult to initiate and the central IT organisation and business units may even have an adversarial relationship. The organisation is trying to increase trust between IT and the business and there are normally periodic joint meetings to review operational issues and new projects. Upper management is involved only when there are major problems or successes.

2   Repeatable but Intuitive

There is awareness of IT Governance objectives, and practices are developed and applied by individual managers. IT Governance activities are becoming established within the organisation's change management process, with active senior management involvement and oversight. Selected IT processes have been identified for improvement that would impact key business processes. IT management is beginning to define standards for processes and technical architectures. Management has identified basic IT Governance measurements, assessment methods and techniques, but the process has not been adopted across the organisation. There is no formal training and communication on governance standards and responsibilities are left to the individual.

An IT steering committee has begun to formalise and establish its roles and responsibilities. There is a draft governance charter (e.g., participants, roles, responsibilities, delegated powers, retained powers, shared resources and policy). Small and pilot governance projects are initiated to see what works and what does not. General guidelines are emerging for standards and architecture that make sense for the enterprise and a dialogue has started to sell the reasons for their need in the enterprise.

3   Defined Process

The need to act with respect to IT Governance is understood and accepted. A baseline set of IT Governance indicators is developed, where linkages between outcome measures and performance drivers are defined, documented and integrated into strategic and operational planning and monitoring processes. Procedures have been standardised, documented and implemented. Management has communicated standardised procedures and informal training is established. Performance indicators over all IT Governance activities are being recorded and tracked, leading to enterprise-wide improvements. Although measurable, procedures are not sophisticated, but are the formalisation of existing practices. Tools are standardised, using currently available techniques. IT balanced business scorecard ideas are being adopted by the organisation. It is, however, left to the individual to get training, to follow the standards and to apply them. Root cause analysis is only occasionally applied. Most processes are monitored against some (baseline) metrics, but any deviation, while mostly being acted upon by individual initiative, would unlikely be detected by management. Nevertheless, overall accountability of key process performance is clear and management is rewarded based on key performance measures.

The IT steering committee is formalised and operational, with defined participation and responsibilities agreed to by all stakeholders. The governance charter and policy is also formalised and documented. The governance organisation beyond the IT steering committee is established and staffed.

4   Managed and Measurable

There is full understanding of IT Governance issues at all levels, supported by formal training. There is a clear understanding of who the customer is and responsibilities are defined and monitored through service level agreements. Responsibilities are clear and process ownership is established. IT processes are aligned with the enterprise and with the IT strategy. Improvement in IT processes is based primarily upon a quantitative understanding and it is possible to monitor and measure compliance with procedures and process metrics. All process stakeholders are aware of risks, the importance of IT and the opportunities it can offer. Management has defined tolerances under which processes must operate. Action is taken in many, but not all cases where processes appear not to be working effectively or efficiently. Processes are occasionally improved and best internal practices are enforced. Root cause analysis is being standardised. Continuous improvement is beginning to be addressed. There is limited, primarily tactical, use of technology, based on mature techniques and enforced standard tools. There is involvement of all required internal domain experts. IT Governance evolves into an enterprise-wide process. IT Governance activities are becoming integrated with the enterprise governance process.

Table 13. IT Governance Maturity Model

There is a fully operational governance structure that addresses a consistent architecture for re-engineering and interoperation of business processes across the enterprise, and ensures competition for enterprise resources and ongoing incremental investments in the IT infrastructure. IT is not solely an IT organisational responsibility but is shared with the business units.

5   Optimised

There is advanced and forward-looking understanding of IT Governance issues and solutions. Training and communication is supported by leading-edge concepts and techniques. Processes have been refined to a level of external best practice, based on results of continuous improvement and maturity modeling with other organisations. The implementation of these policies has led to an organisation, people and processes that are quick to adapt and fully support IT Governance requirements. All problems and deviations are root cause analysed and efficient action is expediently identified and initiated. IT is used in an extensive, integrated and optimised manner to automate the workflow and provide tools to improve quality and effectiveness. The risks and returns of the IT processes are defined, balanced and communicated across the enterprise. External experts are leveraged and benchmarks are used for guidance. Monitoring, self-assessment and communication about governance expectations are pervasive within the organization and there is optimal use of technology to support measurement, analysis, communication and training. Enterprise governance and IT Governance are strategically linked, leveraging technology and human and financial resources to increase the competitive advantage of the enterprise.

The governance concept and structure forms the core of the enterprise IT governing body including provisions for amending the structure for changes in enterprise strategy, organisation or new technologies

ITGI (2001). Board briefing on IT Governance. *Available online: www.itgi.org.*

relational mechanisms. In a complex and turbulent business environment, this framework and the practises will also be influenced by a number of external variables. IT Governance is therefore a very complex and broad concept that can be best approached as a holistic system.

REFERENCES

Broadbent, M., & Weill, P. (1998). *Leveraging the new infrastructure – How market leaders capitalize on Information Technology.* Harvard Business School Press.

Brynjolfsson, E. (1993). The productivity paradox of Information Technology. *Communications of the ACM, 36*(12).

Brynjolfsson, E., & Hitt, L.M. (1998). Beyond the productivity paradox. *Communications of the ACM, 41*(8).

Burn, J.M., & Szeto, C. (2000). A comparison of the views of business and IT management on success factors for strategic alignment. *Information &Management, 37.*

Callahan, J., & Keyes, D. (2003). The evolution of IT Governance @ NB Power. In W. Van Grembergen (Ed.), *Strategies for Information Technology Governance.* Hershey, PA: Idea Group Publishing.

CCTA (1998). *Help desk, The Stationary Office.*

Committee of Sponsoring Organisations of the Treadway Commission (COSO) (1992). Internal Control – Integrated Framework.

Duffy, J. (2002). *IT/Business alignment: Is it an option or is it mandatory?* IDC document #26831.

Duffy, J. (2002). *IT Governance and business value part 1: IT Governance – An issue of critical importance.* IDC document # 27291.

Duffy, J. (2002). *IT Governance and business value part 2: Who's responsible for what?* IDC document #27807.

Earl, J.M. (1993). Experiences in strategic information systems planning. *MIS Quarterly, 17*(1).

Fairchild, A.M. (2003). A view on knowledge management: Utilizing a balanced scorecard methodology for analyzing knowledge metrics. In W. Van Grembergen (Ed.), *Strategies for Information Technology governance.* Hershey, PA: Idea Group Publishing.

Gold, C. (1994). *US measures – A balancing act.* Boston, MA: Ernst & Young Center for Business Innovation.

Gottschalk, P. (2003). Managing IT functions. In W. Van Grembergen (Ed.), *Strategies for Information Technology Governance.* Hershey, PA: Idea Group Publishing.

Guldentops, E. (2002). Knowing the environment: Top five IT issues. *Information Systems Control Journal, 4,* 15-16.

Guldentops, E. (2003). Governing Information Technology through CobiT. In W. Van Grembergen (Ed.), *Strategies for Information Technology Governance.* Hershey, PA: Idea Group Publishing.

Guldentops, E. (2003). *IT Governance: Part and parcel of corporate governance.* CIO Summit, European Financial Management & Marketing (EFMA) Conference, Brussels.

Guldentops, E., Van Grembergen, W., & De Haes, S. (2002). Control and Governance Maturity survey: Establishing a reference benchmark and a self-assessment tool. *Information Systems Control Journal, 6.*

Hammer, M., & Champy, J. (1993). *Reengineering the corporation. A manifesto for business revolution.* New York: Harper Business.

Henserson, J.C., & Venkatraman, N. (1993). Strategic alignment: Leveraging Information Technology for transforming organizations. *IBM Systems Journal, 32*(1).

Henserson, J.C., Venkatraman, N., & Oldach, S. (1993). Continuous strategic alignment. Exploiting Information Technology Capabilities for Competitive Success. *European Management Journal, 11*(2), *Business Quarterly, 55*(3).

Hiles, A. 2000. *The complete guide to IT service level agreements.* Brookfield, CT: Rothstein Associates.

ITGI (2000). CobiT: Governance, Control and Audit for Information and Related Technology. Available online: www.itgi.org.

ITGI (2001). *Board briefing on IT Governance.* Available online: www.itgi.org.

ITGI (2002). *IT Governance executive summary.* Available online: www.itgi.org.

ITGI (2002). *IT Strategy committee.* Available online: www.itgi.org.

Kakabadse, N. K., & Kakabadse, A. (2001). IS/IT Governance: Need for an integrated model. *Corporate Governance, 1*(9), 9-11.

Kaplan, R., & Norton, D. (1992). The balanced scorecard – measures that drive performance. *Harvard Business Review,* (January/February), 71-79.

Kaplan, R., & Norton, D. (1993). Putting the balanced scorecard to work. *Harvard Business Review,* (September/October), 134-142.

Kaplan, R., & Norton, D. (1996). *The balanced scorecard: Translation vision into action.* Harvard Business School Press.

Kaplan, R., & Norton, D. (1996). Using the balanced scorecard as a strategic management system. *Harvard Business Review,* (January/February), 75-85.

Lie, C. L. (2001). Modelling the business value of Information Technology. *Information and Management, 39*(2), 191-210.

Luftman, J. (2000). Assessing Business-IT alignment maturity. *Communications of AIS, 4.*

Luftman, J., & Brier, T. (1999). Achieving and sustaining business-IT alignment. *California Management Review, 42*(1), 109-122.

Ministry Of International Trade And Industry (1999). *Corporate approaches to IT Governance.* Available online: http://www.jipdec.or.jp/chosa/MITIBE/sld001.htm.

OECD. (1999). *OECD principles of corporate governance.* Available online: http://www.oecd.org.

Parker, M. (1996). *Strategic transformation and information technology.* Upper Saddle River, NJ: Prentice Hall.

Patel, N.V. (2003). An emerging strategy for e-business IT Governance. In W. Van Grembergen (Ed.), *Strategies for Information Technology Governance.* Hershey, PA: Idea Group Publishing.

Peterson, R. R. (2003). Information strategies and tactics for Information Technology governance. In W. Van Grembergen (Ed.), *Strategies for Information Technology Governance.* Hershey, PA: Idea Group Publishing.

Porter, M. (1980). *Competitive strategy.* New York: The Free Press.

Porter, M. (1985). *Competitive advantage.* New York: The Free Press.

Porter, M. (2001). *Strategy and the Internet.* Harvard Business Review.

Ribbers, P. M. A., Peterson, R. R., & Parker, M. M. (2002). Designing Information Technology governance processes: Diagnosing contemporary practises and competing theories. *Proceedings of the 35th Hawaii International Conference on System Sciences (HICCS), Maui.* CD-ROM.

Rockart, J. (1979). Chief executives define their own data needs. *Harvard Business Review, 57*(2).

Rockart, J. (1982). The changing role of the Information Systems Executive: A critical success factors perspective. *Sloan Management Review, 245*(1).

Sambamurthy, V., & Zmud, R.W. (1999). Arrangements for Information Technology governance: A theory of multiple contingencies. *MIS Quarterly, 23*(2), 261-290.

Shleifer, A., & Vishny, W. (1997). A survey on Corporate Governance. *The Journal of Finance, 52*(2).

Smaczny, T. (2001). Is an alignment between business and Information Technology the appropriate paradigm to manage IT in today's organizations? *Management Decisions, 39*(10).

Strassman, P. (1990). *The business value of computers.* London: Business Intelligence.

Suomi, R., & Tähkäpää, J. (2003). Governance structures for IT in the health care industry. In W. Van Grembergen (Ed.), *Strategies for Information Technology Governance.* Hershey, PA: Idea Group Publishing.

Van Der Zee, J.T.M., & De Jong, B. (1999). Alignment is not enough: Integrating business and Information Technology management with the balanced business scorecard. *Journal of Management Information Systems, 16*(2).

Van Grembergen, W. (2002). Introduction to the Minitrack: IT governance and its mechanisms. *Proceedings of the 35th Hawaii International Conference on System Sciences (HICCS),* IEEE.

Van Grembergen, W., & Saull, R. (2001). Aligning business and Information Technology through the balanced scorecard at a major Canadian financial group: Its status measured with an IT BSC Maturity Model. *Proceedings of the 34th Hawaii International Conference on System Sciences (HICCS), Maui.* CD-ROM.

Van Grembergen, W., & Van Bruggen, R. (1997). Measuring and improving corporate Information Technology through the balanced scorecard technique. *Proceedings of the European Conference on the Evaluation of Information Technology, Delft, The Netherlands.*

Van Grembergen, W., Kritis, V., & Van Belle, J. L. (1997). *Bedrijfsveranderingen met informatietechnologie (Business transformations through information technology).* Kluwer, Deventer (NL).

Van Grembergen, W., Saull, R., & De Haes, S. (2003). Linking the IT balanced scorecard to the business objectives at a major Canadian financial group. In W. Van Grembergen (Ed.), *Strategies for Information Technology Governance.* Hershey, PA: Idea Group Publishing.

Venkatraman, N. (1999). *Valuing the IS contribution to the business.* Computer Sciences Corporation.

Weill, P. (2002). Research Briefing. *MITSloan, 2,* nr. 2C.

Willcocks, L. (1995). *The evaluation of information systems, investments, Information management.* London: Chapman & Hall.

APPENDIX

Executive Responsibilities of the Board, the CIO and the CEO

Duffy, J. (2002). IT Governance and Business Value Part 2: Who's responsible for what? *IDC document # 27807.*

	Board responsibility	CEO responsibility	CIO responsibility
Executive responsibility for IT/Business partnership	At a time when business and technology are entirely interdependent, the Board has responsibility for confirming that the IT leaders and the IT department are delivering maximum value as defined in the organisation's strategic plan. It is also in the Board's purview to ensure that policy requires the plan to be validated on a regular basis and allows for it to be updated as required.	It is the CEO's responsibility to ensure that business and IT strategies are fully harmonized and that the CIO is provided with a credible management context in which to execute against the plan. It is the CEO's responsibility to ensure that the CIO is a key business player and a full partner in the executive decision-making process. The CEO defines the CIO's roles and responsibilities and supports him or her in responding to the Board's requirements.	It is the CIO's responsibility to interpret the business strategy in terms of IT requirements, to proactively seek ways in which the IT value contribution can be increased, and to develop the vertical and horizontal relationships needed in order to successfully execute against a fully harmonized IT/business strategy.
Executive responsibility for HR organization and management	The role of the Board is value creation, and in that context, the members have the responsibility to ensure that the people appointed to key positions have the appropriate skills and competencies and that performance measures and compensation plans are in the long-term interests of the company and its shareholders. The Board also has the responsibility to ensure that the overall organisational structure (including IT) complements the business model and direction.	The CEO is responsible for ensuring a match between the skills needed by the business and the types of individuals hired. The CEO is also responsible for ensuring that the CIO is given the support needed to hire and retain people with the best IT skills available.	The IT executive has responsibility for maintaining the credibility of the IT organisation, ensuring that the positions and roles critical to driving maximum business value from technology have been clearly defined and staffed with the appropriate people.
Executive responsibility for IT/Business architectures	As the steward responsible for shareholder assets, the Board must review the IT/business architecture and the standards and processes it encompasses to ensure that it mitigates risks associated with legislative and regulatory compliance, ethical use of information, and business continuity. The Board also has responsibility for confirming that the IT/business architecture is designed to drive maximum business value and return.	The CEO is responsible for promoting the IT/business architecture and enlisting the support of other executives. It is also the responsibility of the CEO to give the CIO the authority to effectively develop and manage the IT architecture to ensure full alignment with the business. The CEO ensures that the IT/business architecture reflects the need for legislative and regulatory compliance and the ethical use of information and satisfies the requirement for business continuity.	The CIO has responsibility for planning IT, setting standards, establishing IT policy, and designing and managing architectures that ensure integrated information and technology management across the organisation and throughout the technology life cycle. The CIO is responsible for implementing standards and processes that ensure legislative and regulatory compliance and the ethical use of information and that satisfy the requirement for business continuity.

APPENDIX
Executive Responsibilities of the Board, the CIO and the CEO (continued)

	Board responsibility	CEO responsibility	CIO responsibility
Executive responsibility for operational excellence	Ultimate responsibility for risk management rests with the Board. The Board is responsible for overseeing the management of any arrangements with third parties, confirming that potential risks have been mitigated. It is the Board's responsibility to guide the definition of operational excellence and to monitor the organisation's progress in achieving the goals that have been established and mutually agreed upon, recommending corrective action as needed.	The CEO is responsible for the organisation's system of internal control and ensuring that clear accountability for risk management is embedded in the operations of the organisation. The CEO is responsible for ensuring that arrangements and agreements with third parties do not put the organisation at risk. The CEO is responsible for implementing the policies and processes that underpin operational excellence and ensuring that the appropriate resources are in place to facilitate execution.	It is the CIO's responsibility to ensure that measurable value is delivered on time and on budget. The CIO is responsible for the day-to-day management and verification of IT processes and controls. The CIO is also responsible for ensuring appropriate governance at the individual project or initiative level. It is the CIO's place to inform the CEO and the Board of identified risks. The CIO is responsible for providing liaison with any third parties, minimizing the risk of duplicate effort and redundancy.
Executive responsibility for innovation and renewal	It is the Board's responsibility to ensure that the organisation is sufficiently adaptive to respond to changing demands. The Board is also responsible for ensuring that investment in the future is not sacrificed in order to maintain the status quo.	It is the CEO's responsibility to ensure that the organisation is flexible and adaptive and that it is in the best position to capitalise on its information and knowledge to sense what is happening in the market.	The CIO is responsible for ensuring that IT and IT-related processes are focused on improving business value currently and in the future. The CIO is responsible for monitoring emerging technologies and identifying when and how they would be of benefit to the organisation.
Executive responsibility for ROI strategy and management	The Board is responsible for ensuring that IT delivers on the promise of related strategies through clear expectations and measurement. The Board must work with the CEO to define and monitor performance measures. It is also the Board's responsibility to ensure that IT investments represent a balance of risk and benefit and that budgets are acceptable and reflect the overall organisation's financial direction.	The CEO is responsible for ensuring strong links between business objectives and performance measures. It is the CEO's responsibility to develop an appropriate incentive scheme to drive adherence to the performance measures. The CEO is responsible for integrating the IT budget and investment plan into the overall financial plan, ensuring that it is realistic, balanced, and achievable. The CEO is then responsible for reporting progress to the Board on a regular basis.	The CIO is responsible for developing and managing the IT budget, including short-term and long-term investment strategies. The CIO is responsible for developing a realistic IT performance measurement plan, along with appropriate metrics. In conjunction with the CEO, it is the CIO's responsibility to implement and manage a performance measurement scheme. The metrics used by the CIO should be linked directly to achievement of business goals and, wherever possible, be assigned a financial value.

Chapter II

Integration Strategies and Tactics for Information Technology Governance

Ryan R. Peterson
Information Management Research Center, Instituto de Empresa, Spain

ABSTRACT

Amidst the challenges and changes of the 21ˢᵗ century, involving hyper-competitive market spaces, electronically-enabled global network businesses, and corporate governance reform, IT Governance has become a fundamental business imperative. IT Governance is a top management priority, and rightfully so, because it is the single most important determinant of IT value realization. IT Governance is the system by which an organization's IT portfolio is directed and controlled. IT Governance describes (a) the distribution of IT decision-making rights and responsibilities among different stakeholders in the organization, and (b) the rules and procedures for making and monitoring decisions on strategic IT concerns. The objective of this chapter is threefold. First of all, to describe past developments and current challenges complex organizations are facing in governing the IT portfolio of IT applications, IT development, IT operations and IT platforms. Based upon the lessons we've learned from the past, one of the key objectives is to move beyond 'descriptives', and discuss how organizations can diagnose and design IT Governance architectures for future performance improvement and sustained business growth. The final objective of this chapter is to provide a thorough understanding and holistic picture of effective IT Governance practices, and present a new organizing logic for IT Governance.

INTRODUCTION

One morning in 1997, Ralph Larsen, CEO of Johnson & Johnson, called his controller, JoAnn Heisen, to his office for a meeting. Ralph had just launched a corporate-wide cost-cutting campaign to help finance a drive into highly competitive and costly new drug markets. That morning, Ralph wanted Johnson & Johnson's IT organization to be a bigger part of all of that — and to get smarter about how the company was using IT.

Johnson & Johnson was spending millions annually on IT, yet business executives and customers weren't getting the business information they needed, and the business value they wanted. Hospitals, for example, were asking Johnson & Johnson to help them cut their stashes of supplies, but Johnson & Johnson didn't have the Web-based tracking systems needed to deliver on that request. The electronic networks that did exist suffered frequent breakdowns. JoAnn recalls, "Nobody was talking to each other. And why should they? Nobody asked the business units to talk with each other before, and no one had asked IT how much we were spending on the business." Ralph told JoAnn he wanted to cut IT costs dramatically, but he also wanted to oversight reform.

That morning JoAnn left Ralph's office with a new job — as CIO — and a mission to standardize systems, cut IT costs, and align the IT organization with business strategies, while simultaneously acknowledging the decentralized culture of Johnson & Johnson's numerous business divisions across different countries.

Does the Johnson & Johnson chronicle (Alter, 2001; Scheier, 2001) seem familiar to you? This real-life case illustrates many of the problems and challenges large complex organizations have been facing for over a decade. It's almost cliché that chief executives across the board have experienced many failures and disappointments with IT-enabled business transformations. Expecting strategic value innovation, executives have faced project cancellations, business disruptions, rising customer churn, decreasing share-holder value, and many other disappointments, including losing their jobs.

In fact, executives today are less concerned about getting 'Amazon-ed' than about getting 'Enron-ed'. Corporate responsibility, business sustainability and governance reform are currently high on the strategic agenda in many companies. The growing scrutiny over shareholder interests, lingering economic growth and corporate performance have now also prompted renewed soul-searching and interest into the governance of information and Internet-based technologies. Amidst all these changes and challenging responsibilities, governing IT for sustaining business value has become a fundamental business imperative for thriving in the 'old new' economy.

Re: Information Technology Governance

Boards and business executives have come to recognize that whereas traditionally they could delegate, avoid, or ignore IT decisions, today they cannot conduct marketing, R&D or HR without depending on IT at some point in time. Metaphorically, a 'Speak-See-Hear No Evil' attitude towards IT Governance is no longer viable in today's business landscape (Figure 1).

With the dawning of the 21st century, organizations are experiencing a global digital revolution with profound impacts on their business models and electronic business processes, wherein the interdependency between business and IT is intensely recipro-

Figure 1. IT Governance: No More 'Monkey Business'

cal. This e-business genesis coincides with a business landscape, in which the intensity, unpredictability and diversity of change has accelerated to create a condition of hyper-competition (D'Aveni, 1994), in which there is no stable competitive position, bureaucratic hierarchies become a competitive liability, organizational boundaries are being redefined, core competencies develop into core rigidities, and strategic fit is fleeting, all against a backdrop of global economic turmoil and market volatility.

The growing infusion of e-business technologies in and between organizations, and the 'e-wakening' from the dot.com frenzy, has made both business and IT executives recognize that getting IT right this time will not be about technology, but about governing IT. GartnerEXP's 2002 survey of its 1,500-member CIO community indeed confirms that the top management priorities are all within the domain of IT Governance (Figure 2). As business models and IT become virtually inseparable, managing their integration and co-evolution involves putting the right people in the right place to understand and take direct responsibility for making sure the organization meets its strategic goals, and that all efforts — including IT — are directed toward that end.

Nevertheless, how to govern IT for sustained value innovation remains an enduring and challenging question. In the case of Johnson & Johnson, questions the new CIO needed to answer were:

How can IT best support a complex organization composed of diverse, globally operating business units?

What and how much should we standardize? And who should be involved in making these strategic decisions?

Figure 2. Top Management Priorities in 2002 (GartnerEXP, 2002)

Rank	Priority
1	Strategizing for business-IT alignment
2	Providing leadership and guidance for the Board of Directors and senior executive
3	Demonstrating business value of IT
4	Developing IT senior management team
5	Reducing total IT costs
6	Strengthening program management, and project prioritization and selection
7	Tightening security and privacy safeguards
8	Developing electronic business architectures
9	Attracting, nurturing and sustaining IT resources
10	Reducing IT complexity

How does a company choose the best arrangement for IT Governance? And what is the 'best' arrangement for our company?

At the heart of Johnson & Johnson's journey, as in many other organizations, has been a need to find answers to tough, almost timeless, questions of governance: how to organize for diversity and differentiation, while preserving integration and unity of direction? How to promote local innovation, yet reap the benefits of scale and scope in a global economy? In terms of IT Governance, the general question is how should we be governing our IT activities in order to manage the imperatives of the global, hyper-competitive, digital economy? More specifically, how to direct, organize and control IT in order to realize value for the enterprise and its key constituencies? In this chapter, these and other questions and their answers are explored and discussed.

Objectives and Organization of this Chapter

The purpose of this chapter is, first of all, to describe past developments and current challenges complex organizations are facing in IT Governance. Based upon the lessons we've learned from the past, one of the key objectives is to move beyond 'descriptives', and discuss how organizations can diagnose and design IT Governance solutions for future improvement and growth. The final aim of this chapter is to provide a thorough understanding and holistic picture of effective IT Governance practices, and present a new organizing logic for IT Governance.

This chapter is organized as follows. First, a review of previous studies and developments is presented, and contrasted with current changes and challenges in contemporary businesses. The fundamental problem underlying IT Governance is then described, along with a discussion of design strategies and integrative solutions. Next, the results of a multi-client study are summarized, and a diagnostic model for IT Governance is outlined. Several lessons for future architecting of IT Governance are presented, in which a 'new' logic for IT Governance is described. This chapter concludes with a discussion of the major implications and directions for practice and research on IT Governance.

INFORMATION TECHNOLOGY GRASPING AT GOVERNANCE

In this section, the concept of IT Governance is discussed, and a review of previous studies is presented and contrasted with current business environments. Specifically, this section addresses the body of knowledge on IT Governance. A review of previous studies is presented, and compared to state-of-the-art developments in contemporary organizations. This section concludes with a discussion of existing IT Governance principles, and questions the suitability of the current paradigm.

Information Technology Governance: Definitions and Myths

IT Governance has been the subject of much debate and speculation over the past decades, yet it remains an ephemeral and 'messy' phenomenon, emerging in ever-new forms with growing complexity. Despite more than 30 years of empirical research, and management theory and practices, there are still many breaches to be bridged.

Although questions and concerns regarding IT Governance have been around since the introduction of IT in companies, currently there is no consistent, well-established body of knowledge and skills regarding IT Governance. This is partly attributed to the simultaneous enduring and evolving nature of IT Governance, but partly also due to the specialization and disconnectedness between globally-dispersed IT Governance interest communities. For instance, the differentiation and dissonance among these professional communities has led to the use of various definitions and models on IT Governance (Figure 3), which consequently has not been beneficial to the development of a cumulative body of knowledge and skills on IT Governance[1].

Similar to corporate governance, IT Governance is a topic that has recently been rediscovered, and as yet, is ill-defined and consequently blurred at the edges. To cite Burt (1980), the rich vocabulary emerging from the literature is like a terminological jungle in which any newcomer may plant a tree. Considering the foregoing definitions, and consistent with Cadbury and in line with the OECD's definition of corporate governance, the following definition of IT Governance is used in this chapter (Peterson, 2001):

IT Governance is the system by which an organization's IT portfolio is directed and controlled. IT Governance describes (a) the distribution of IT decision-making rights and responsibilities among different stakeholders in the organization, and (b) the rules and procedures for making and monitoring decisions on strategic IT concerns.

Figure 3. IT Governance Definitions

What is Information Technology Governance?

IT Governance is the organizational capacity exercised by the board, executive management and IT management to control the formulation and implementation of IT strategy and in this way ensure the fusion of business and IT.
- Van Grembergen, 2002.

IT Governance describes a firm's overall process for sharing decision rights about IT and monitoring the performance of IT investments.
- Weill & Vitale, 2002

IT Governance is the responsibility of the board of directors and executive management. It is an integral part of Enterprise Governance and consists of the leadership and organizational structures and processes that ensure that the organization's IT sustains and extends the organization's strategies and objectives.
- IT Governance Institute, 2001

IT Governance defines the locus of enterprise decision-making authority for core IT activities.
- Sambamurthy & Zmud, 2000

IT Governance refers to the patterns of authority for key IT activities.
- Sambamurthy & Zmud, 1999

IT Governance is the degree to which the authority for making IT decisions is defined and shared among management, and the processes managers in both IT and business organizations apply in setting IT priorities and the allocation of IT resources.
- Luftman, 1996

IT Governance describes the locus of responsibility for IT functions.
- Brown & Magill, 1994

IT Governance thus specifies the structure and processes through which the organization's IT objectives are set, and the means of attaining those objectives and monitoring performance. IT Governance is a second order phenomenon, i.e., governance is the set of decisions about *who* and *how* decisions on strategic IT concerns are made. The abovementioned definition also (implicitly) addresses several 'IT Governance myths' that have existed for a long time and still persist, yet, which need to be debunked (explicitly) if business and academic communities are to move forward in their IT Governance thinking and IT Governance practices.

IT Governance Myth: IT Governance is the Responsibility of the CIO

While IT Governance is certainly an essential element of a CIO's portfolio, the CIO is *not* the primary stakeholder. Still, too often, corporate executives and business managers assume that the CIO is taking care of IT Governance. Moreover, abdication of responsibility and accountability by the business, and 'pointing the finger' at IT will not resolve the 'IT productivity paradox', nor the many misalignments between business and IT.

More then a decade ago, Rockart (1988) argued that business management needs to take charge of IT. Remenyi (1997) also indicates that line leadership is an absolute necessity, and placing business management as the principle stakeholder repositions the locus of responsibility for realizing IT value squarely where it should be, i.e., with the business. IT Governance effectiveness is only partially dependent on the CIO and other IT executives, and should be viewed as a shared responsibility and enterprise-wide commitment towards sustaining and maximizing IT business value (Figure 4).

Developing business-IT leadership and IT competencies in business management is pivotal for IT Governance effectiveness and realizing business value with IT (Peterson, 2001; Weill & Broadbent, 1998). In essence, IT Governance relies on the capability of business executives at all levels to set the strategic business — including IT — agenda, understand the business capabilities — not the technicalities — of IT, and monitor business value appropriation from IT.

Figure 4. Primary and Secondary Stakeholders in IT Governance (Peterson, 1998)

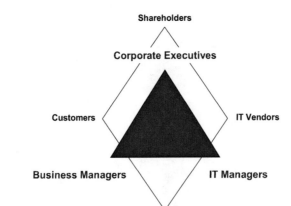

IT Governance Myth: IT Governance is Concerned with Organizing the IT Function

Traditionally, the IT function has been regarded as a single homogenous function. However, given the widespread proliferation and infusion of IT in organizations, involving, e.g., technical platforms, shared IT services centers, and local business-embedded applications, *the notion of a single homogenous IT function is obsolete.*

Weill and Broadbent (1998) indicate that contemporary organizations consist of a portfolio of different interdependent business functions and technical capabilities, some of which are allocated to different levels in the enterprise, and/or to third-party vendors. In discussing 'the new infrastructure', Weill and Broadbent (1998) graphically summarize the different IT functions in a *portfolio* of IT capabilities (Figure 5). IT capabilities are the organizational skills and knowledge that are essential to IT-enabled business transformation, and include IT infrastructure, and the competency to partner with business clients, executive management, dispersed and/or external IT specialists and vendors (Peterson, 2001).

The IT infrastructure is the base foundation of IT capabilities, delivered as reliable shared services throughout the organization, and centrally directed, usually by corporate IT management. The purpose of the IT infrastructure is to enable organization-wide data sharing and cross-business integration. The infrastructure capability describes the degree to which its resources are shareable and reusable. In contrast, local business-embedded applications are concerned with product/service specific needs in order to meet the changing demands of the business and its customers, usually directed by local business management. These applications utilize the infrastructure services and are built on the shared technical platforms.

Figure 5. IT Portfolio (Adapted from Weill & Broadbent, 1998)

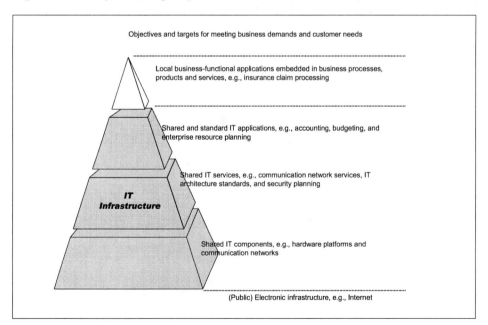

IT Governance Myth: IT Governance is a New Form of 'Old-School' IT Management

Due to the enduring nature of IT Governance, and the perennial, often intractable problems associated with IT value delivery, some may draw the conclusion that IT Governance is simply a new form of 'old-school' IT management. However, although there may be a mere thin dotted line separating IT Governance from IT management, there is a fundamental difference between IT Governance and IT management that goes well beyond theory, which has profound implications for the design and effectiveness of IT Governance in practice.

Whereas the domain of IT management focuses on the efficient and effective supply of IT services and products, and the management of IT operations, IT Governance faces the dual demand of (1) contributing to present business operations and performance, and (2) transforming and positioning IT for meeting future business challenges (Figure 6). This does not undermine the importance or complexity of IT management, but goes to indicate that IT Governance is both internally and externally oriented, spanning both present and future time frames. One of the key challenges in IT Governance is therefore how to simultaneously perform and transform IT in order to meet the present and future demands of the business and the business' customers in a satisfying manner.

The difference between IT Governance and IT management is akin to Parker and Benson's (1988) discussion of enterprise-wide information management, i.e., a shift in emphasis away from information *technology*, to the *business/information* relevance of IT. Whereas elements of IT management and the supply of (commodity) IT services and products can be commissioned to an external IT provider, IT Governance is organization-specific, and direction and control over IT cannot be relegated to the market.

Figure 6. IT Governance and IT Management (Peterson, 2001)

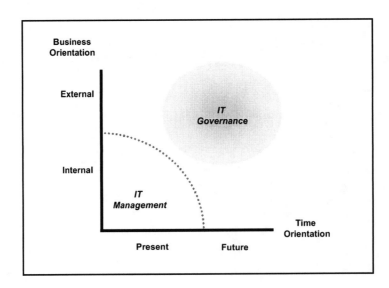

IT Governance Myth: IT Governance Focuses on the (De)Centralization of IT

Acknowledging the rebuttal of the previous myth often leads to a fourth myth, i.e., IT Governance focuses on the locus of IT control, or where IT decision-making authority is allocated in the organization. The discussion on the formal allocation of IT decision-making as vested in organizational positions has led to much rhetoric and speculation on the 'best way' to organize IT Governance, and in the process has rekindled the classical 'centralization versus decentralization' debate. Yet, scholars have recently questioned whether the concept of IT Governance is simply about centralization or decentralization (Peterson et al., 2000; Sambamurthy & Zmud, 2000; Vitale, 2001; Whetherbe, 2001).

The keen reader will recognize that by building forth on the discussion of the previous three myths, the terms centralization and decentralization provide a dichotomy that is meaningless when employed as a generality to IT Governance. The centralization or decentralization can be applied to each of the main IT capabilities in the IT portfolio, yielding eight distinct patterns in IT Governance[2]. In the first pattern, the corporate executive is responsible for IT infrastructure and IT development decisions, whereas (decentral) IT management is responsible for IT application decisions (Figure 7). In the second pattern, however, (decentral) business management is responsible for IT development and IT application decisions.

The discussion of whether to centralize or decentralize IT Governance is based on a rational perspective of the organization, in which choices are reduced to one of internal efficiency and effectiveness (March & Simon, 1958). This view assumes a system of goal consonance and agreement on the means for achieving goals, i.e., rational and logical trade-off between (a) efficiency and standardization under centralization, versus (b) effectiveness and flexibility under decentralization (Figure 8). In general, it is assumed that centralization leads to greater specialization, consistency, and standardized controls, while decentralization provides local control, ownership and greater responsiveness and flexibility to business needs (Brown & Magill, 1998; King, 1983; Rockart et al., 1996). However, flexibility under decentralization may lead to variable standards, which ultimately result in lower flexibility, and specialization under centralization incurs risks due to bounded rationality and information overload (Mintzberg, 1979; Simon, 1961).

Figure 7. Example of Federal Patterns in IT Governance (Peterson et al., 2000)

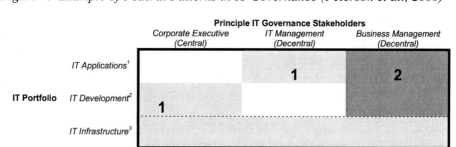

Figure 8. Tradeoffs and the Best of Both

Design / Drivers	Centralized IT Governance	Decentralized IT Governance	Federal IT Governance
Synergy	+	-	+
Standardization	+	-	+
Specialization	+	-	+
Customer responsiveness	-	+	+
Business ownership	-	+	+
Flexibility	-	+	+

A *socio-political view* of IT Governance suggests that the debate concerning centralization versus decentralization is used to further the goals of specific organizational actors in a *satisficing* manner (Simon, 1961), in ways that might not help to meet organizational goals (Cyert & March, 1963). There are important differences among factions within organizations, leading to the presence of conflict and disagreement over organization goals and the means for achieving them. Consider, for instance, the notorious strategic conflicts between business and IT stakeholders. These stakeholders represent groups or individuals that influence, and are influenced by, strategic decisions regarding IT, and may thereby have different, often competing stakes in IT. Power struggles and cultural clashes are endemic to IT Governance, and the question is often '*whose way is it going to be?*', rather than, '*which way is the best?*'

The potential risk in contemporary business environments is that either centralization or decentralization fit the organization into a fixed structure. The challenge is therefore to balance the benefits of decentralized decision-making and business innovation and the benefits of central control and IT standardization. As Mintzberg (1979) points out, centralization and decentralization should not be treated as absolutes, but as two ends of a continuum.

Over the past decade, organizations have set out to achieve the 'best of both worlds' by adopting a federal IT Governance structure. In a federal IT Governance model, IT infrastructure decisions are centralized, and IT application decisions are decentralized (Brown & Magill, 1998; Rockart et al., 1996; Sambamurthy & Zmud, 1999; Weill & Broadbent, 1998). The federal IT Governance model thus represents a hybrid model of both centralization and decentralization. Across the Atlantic, a number of prominent companies in different industries have been actively experimenting with this relatively new model for IT Governance (Figure 9).

Recall, for instance, the CIO's mission at Johnson & Johnson: to standardize systems, cut IT costs, and align the IT organization with business strategies, while simultaneously acknowledging the decentralized culture of the business divisions across different countries. The competing drives towards improving cost-efficiencies, but also business responsiveness, and IT standardization, yet also IT innovation, led the CIO at Johnson & Johnson to adoption and use of a federal IT Governance model (Alter, 2001; Scheier, 2001).

While the implementation of the federal IT Governance model has paid off for Johnson & Johnson (e.g., significant cost reduction, cheaper maintenance costs, eliminated duplicate IT developments, enhanced pharmaceutical R&D, improved time-to-market for new products, profit growth), and they have been able to develop

Figure 9. Examples of Leading Companies Adopting Federal IT Governance

Companies Adopting Federal IT Governance	
Verizon Communications Inc.	Cisco Systems Inc.
United Technologies Corp.	Royal Dutch Shell Company
Chevron Texaco Corp.	British Petroleum PLC
General Motors	ABN AMRO
Avnet Inc.	Royal Dutch Airlines KLM
Siemens AG	HCA Inc.

unprecedented levels of cooperation between traditionally independent business units, it was a perilous and painstaking transformation.

Earlier attempts to (re-) centralize IT failed due to cultural barriers and business' resistance to change and relinquish of IT control. The CIO indicates that the federal IT Governance model is tricky to manage, and after commencing new IT Governance regime the CIO states (Alter, 2001; Scheier, 2001):

"It's hard getting all of Johnson &Johnson's businesses to go along with some of even the most benign changes in policy. Originally, we hoped to create a single, centralized strategy, but soon realized that only a federalized approach would work. We are too complex and independent from one business unit to the next to devise one strategy."

"I get '190 land mines' in any given day. Some business units, for example, try to convince me they can't adopt some corporate technology standards, or kick in their share of the cost for upgrades in infrastructure."

In sum, the federal IT Governance model is often easier contemplated than implemented. A federal approach towards IT Governance challenges managers in local business units to surrender control over certain business-specific IT domains for the well-being of the enterprise, and to develop business-to-corporate and business-to-IT partnerships. As the Johnson & Johnson case illustrates, this is a feat that many companies struggle with, especially considering the 'IT Governance legacy' large and complex organizations face.

Determinants of IT Governance: Reviewing the Empirical Evidence

While the previous section discussed some of the myths regarding IT Governance and presented anecdotal cases concerning contemporary IT Governance practices, this section takes a more scientific view of IT Governance by reviewing the empirical data to date. Previous studies have sought an answer to the 'best way' of designing IT Governance, recognizing that this 'best way' is contingent upon internal and external factors (Brown & Magill, 1994; Sambamurthy & Zmud, 1999).

These studies examined the influence of various determinants, including *organization size* (Ahituv et al., 1989; Brown & Magill, 1994; Clark, 1992; Ein-Dor & Segev, 1982; Sambamurthy & Zmud, 1999; Tavakolian, 1989); *business strategy* (Brown & Magill, 1994; Peterson, 2001; Sambamurthy & Zmud, 1999; Tavakolian, 1989); and *business governance structure* (Ahituv et al., 1982; Brown & Magill, 1994; Peterson, 2001; Sambamurthy & Zmud, 1999; Tavakolian, 1989).

The cumulative of these empirical findings indicates that (Figure 10):

1. Central IT Governance is associated with small-sized organizations following a cost-focused business (competitive) strategy, and characterized by a centralized business governance structure, environmental stability, low information-intensive business products/services, and low business experience and competency in managing IT.

2. Decentral IT Governance is associated with large, complex organizations following an innovation-focused business (competitive) strategy, and characterized by a decentralized business governance structure, environmental volatility, high information-intensive business products/services and processes, and high business experience and competency in managing IT.

But what type of IT Governance arrangement should a company adopt in the following (though hypothetical, certainly not unrealistic) scenarios?

Company A is a large bank focused on operational excellence and new product and service development, and whose business management, however, lacks the professional skills and knowledge to manage IT.

Company B is a small pharmaceutical company focused on product excellence, rapid innovation and quick commercialization, and whose business management is leading in IT innovation.

Figure 10. Determinants of IT Governance (Peterson, 1998)

DETERMINANTS	Centralized Model		Federal Model		Decentralized Model
Business strategy	Cost-focus	←		→	Innovation-focus
Business governance	Centralized	←		→	Decentralized
Firm size	Small	←		→	Large
Information-intensity	Low	←		→	High
Environment stability	High	←		→	Low
Business competency	Low	←		→	High

Company C is a large insurance company focused on achieving cost-efficiencies and providing customer value through customized products, and whose business management has developed the required competencies to manage IT projects.

Each of these scenarios describes a situation in which multiple contingency factors are active in determining IT Governance, yet provide conflicting solutions. Should Company A adopt decentral IT Governance because of its size and strategy? Yet, how does it reconcile this with the lack of business competency? Should Company B adopt central IT Governance because of its size? Yet, how does it combine this with its focus on innovation and business leadership? Should Company C decentralize due to size and customization, or should it centralize due to a focus on cost-efficiencies?

Although previous studies have increased our understanding of IT Governance determinants, they have failed to assess the multiplicity of these determinants. In practice, IT Governance models are influenced by many factors simultaneously, and determining the right structure is a complex endeavor. While traditionally IT Governance focused on either efficiency or flexibility, often in a sequential manner — recall the continuous pendulum swing of centralization and decentralization — currently, IT Governance faces the dual demands for (a) flexibility and speed, and (b) efficiency and reliability. The latter concern is of long standing, in which IT Governance was concerned with efficiency and cost reduction, often directed at the operational level. Subsequently, IT Governance focused on managing IT as a strategic resource, in which the primary aim was to align IT with the business strategy in order to gain competitive advantage. Having emerged from both practices with ambiguous results and experiences, executives are recognizing the need to meet the demands for both (a) delivering customized, high quality IT products and services, and (b) compressing costs, risks and time in order to meet business needs in an efficient, reliable and effective manner.

Consider, for instance, the case of Johnson & Johnson. The competing drives towards improving cost-efficiencies but also business responsiveness, and IT standardization, yet also IT innovation, led Johnson & Johnson to adopt a federal IT Governance model. The complexity (size and span) of this organization would dictate a decentralized approach, which was the traditional IT Governance approach. Yet, the need to cut costs, standardize IT, and improve IT performance led Johnson & Johnson to centralize IT infrastructure decisions. Thus, the case of Johnson & Johnson illustrates how IT Governance is subject to the pulls and pressures of multiple, rather than singular, contingency forces.

Recent empirical evidence indicates that organizations adopt federal IT Governance when pursuing multiple competing objectives (Peterson, 2001; Sambamurthy & Zmud, 1999). Due to the relentless pace and unpredictable direction of change in contemporary business environments, it should be no surprise that many organizations need to focus on both standardization and innovation, and in the process have adopted a federal IT Governance model.

Strategic Flexibility: The New Enterprise Logic

Business has embarked on a new era of competition that is faster, more turbulent, and increasingly global and digital, simultaneously requiring relentless cost-efficiencies as well as the flexibility to find new ways to innovate and create value. It is no secret that the contemporary business landscape is characterized by (D'Aveni, 1999; El Sawy et al., 1999):

- Time and cost compression in product-life and design cycles;
- Accelerating technological advancements;
- Fickle customer loyalty;
- Tailored, knowledge-intensive products and services;
- Unexpected entry by new competitors, and repositioning of incumbents;
- Redefinition of industry and organizational boundaries; and
- Global market volatility.

As the mosaic of these developments transpires simultaneously in unpredictable patterns, organizations face significant uncertainty and ambiguity in determining their strategic direction. Organizations operating in turbulent fields experience competing goals and performance demands, including, e.g., pressures to innovate and customize products and services, improve levels of responsiveness and speed, and increase productivity and efficiency (Daft, 1998; Mintzberg, 1979). These competing demands cause conflicting contingencies, which are endemic to complex open social systems and require both exploitation and exploration strategies (March, 1991).

Exploitation strategies involve taking advantage of what is already known, i.e., cashing in on the investments made in existing capabilities. *Exploration* strategies, on the other hand, involve the creation of new knowledge and capabilities. Whereas exploitation builds forth on the efficient supply of extant products and services, exploration is geared at developing new products and services in order to meet changing and ambiguous environmental demands (March, 1991). Organizations must determine the proper balance between these competing strategies for developing capabilities and preserving organizational conditions. Thus, turbulent environments require the ability to explore new opportunities effectively, and exploit existing opportunities efficiently.

Contemporary organizations do not have single goals, and face multiple, often conflicting, contingencies. The 'low-cost versus differentiation' dichotomy (Porter, 1980) is currently fallacious, as organizations are effectively pursuing both strategies simultaneously in order to meet the competing demands of volatile global electronic markets. D'Aveni (1999) concludes that under these conditions, there is no sustainable competitive advantage based on either a low-cost strategy or a differentiation strategy. Companies can only build temporary advantages in order to sustain strategic momentum through a series of initiatives, rather than achieve 'fit' with the external environment.

Instead, companies need to adopt *simultaneous strategic thrusts*, in rapid and surprising manners, in order to offset competitors and satisfy customer needs (D'Aveni, 1999). Adopting simultaneous strategic thrusts requires an organization to focus on multiple value-creating drivers, involving excellence in business operations, product development, *and* customer service delivery (Figure 11). Market leaders focus on *at least* one value driver, and meet a minimum threshold of competence in the other two (Treacy & Wiersema, 1995).

Avnet Inc. is a perfect example of how market leaders are transforming towards meeting multiple value drivers (El Sawy et al., 1999). In 1991, Avnet Inc., then Marshall Industries, realized that customers wanted products and services at the lowest possible

Figure 11. Complementary Value-Creating Drivers (Adapted from Treacy & Wiersema, 1995)

Value Driver	Operational Excellence	Product Excellence	Service Excellence
Description	Providing reliable products or services at competitive prices, and delivered with minimal inconvenience	Offering customers leading edge products and services that consistently enhance the customer's use of the product	Segmenting and targeting markets precisely and tailoring offerings to match exactly the demands of those customers
Focus	Competitive prices, minimize overhead costs, reduce transaction costs, optimize business processes across functional boundaries, improve reliability, streamline enterprise workflows	Innovation, speed, state-of-the-art technologies, commercialization, short cycle times, time-to-market, selective controlled experimentation, directed 'grass root' innovation	Customization of products and services, build customer loyalty and life-time value to the company, responsiveness, cultivate relationships and understanding, satisfy unique customer needs
Threshold	Operational Efficiency	Product Enhancement	Customer Responsiveness

cost, highest possible quality, greatest possible customization, and fastest possible delivery time. Avnet Inc. recognized that they needed to focus on all three value drivers. They realized that it was no longer adequate to think of operational excellence in terms of cost of individual transactions, and expanded that to include the total cost of value-added services (El Sawy et al., 1999). Avnet Inc. also focused on product excellence by enhancing product features, customization, and the anticipation of future needs. Service excellence was also improved by 24x7 services, reduced delivery time, and reduced time-to-market for customers' products (El Sawy et al., 1999).

Empirical studies indicate that value-creating organizations have learned to master more than one value driver (Boynton, 1993; Buenger et al., 1996; Khandwalla, 1976; Miller & Friesen, 1978; Peterson, 2001; Quinn & Rohrbaugh, 1983; Quinn et al., 2000; Treacy & Wiersema, 1995). Organizations learn to manage the conflicting pulls by becoming 'bi-focal', and recognize the requisite *complementary* — not competing — nature of value creation.

In order to satisfy complementary value-creating drivers, an organization should have a variety of capabilities at least as great as the demands and disturbances in the environment (Ashby, 1956; Hitt et al., 1998). Consequently, organizations develop a repertoire of competencies to respond to and influence their external environment. Whereas in the 70s and 80s organizations would integrate vertically, today organizations have recognized the need to focus on their core competencies. Yet, continuous differentiation leads to fragmentation, unless a corresponding process of integration complements it. The uncertainty and ambiguity associated with the complex of external demands and differentiated capabilities creates the need for integration to achieve clarity of direction and unity of purpose in responding decisively and swiftly (Lawrence & Lorsch, 1969; Hitt et al., 1998; Venkatraman, 2000).

The concentration on core competencies in the early 1990s, and the recent focus on (inter- and intra-) organizational collaboration are exemplary of the need to both differentiate and integrate in turbulent environments. Focusing on core competencies alone leads to a situation of professional disconnectedness, whereas collaboration without core competencies only provides for administrative coupling (Peterson, 2001).

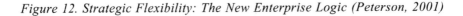

Figure 12. Strategic Flexibility: The New Enterprise Logic (Peterson, 2001)

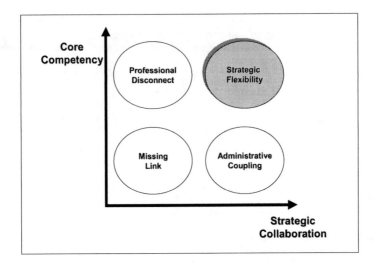

Value-creating organizations thus focus on both their core competencies and strategic collaboration (Figure 12). The degree to which organizations can achieve this is a measure of an organization's *strategic flexibility*, i.e., developing differentiated capabilities to pro-act in an integrated manner to unanticipated changes (Hitt et al., 1998). Organizations improve their chances of success and survival in a turbulent environment by creating strategic flexibility that gives them the ability to pursue alternative courses of actions in response to unexpected environmental conditions.

Strategic flexibility in a turbulent and competitive business environment requires organizations to be dynamically stable (Marchand, 2000). Organizations need to simultaneously develop a variety of differentiated capabilities in order to serve a range of changing customer and market demands, and integrate these for developing joint expertise and providing direction (Hitt et al., 1998; Nadler & Tushman, 1998). The underlying enterprise architecture is an interconnected network of differentiated strategic capabilities in which IT is an integral part, wherein the interdependency between IT and business is intensely reciprocal. Consequently, the efficiency *and* flexibility with which IT capabilities are infused in the enterprise architecture become business critical. The business imperatives of strategic flexibility and dynamic stability thus require the business of IT to:

1. Develop and deliver applications that facilitate business responsiveness to customer demands in a rapid and efficient manner;
2. Provide cost-effective, scalable infrastructures and operations that enable cycle time improvement and streamlined, enterprise-wide business processes; and
3. Add value to the enterprise by focusing on operational excellence, product excellence, and service excellence.

Similar to the business enterprise, IT needs to develop distinct strategic capabilities in order to satisfy multiple value drivers (Figure 13). The primary value drivers for IT focus

Figure 13. Primary IT Value Drivers (Peterson et al., 2000)

Value Driver	Service Delivery	Solution Integration	Strategic Innovation
Description	Providing reliable IT operations and services, delivered with maximum reliability and availability	Offering business leading-edge products and services that consistently enhance the business' use of products and services	Targeting business value drivers, and tailoring offerings that supersede the demands of the business
Focus	Provisioning of utilities, provide reliable, cost-effective, and secure IT services, manage cross-unit synergies across the corporation, manage IT infrastructure standards, establish IT infrastructure flexibility and scalability.	Strategic analysis of business needs for IT, deciding on the necessary applications, and delivering them either through internal development, external contracting or packaged software. Focus on ensuring timely and cost-effective delivery of IT applications.	Focus on ways in which IT can be used to strengthen business competencies, customer relationships and business partner networks. Ensure IT applications have a value focus (operational excellence, product excellence, service excellence).
Threshold	Infrastructure	Integration	Impact

on service delivery, solution integration and strategic innovation (Feeny & Willcocks, 1998; Sambamurthy & Zmud, 2000; Peterson et al., 2000).

Value-adding IT organizations focus and channel energy at excelling in a specific dimension, yet maintain threshold standards on the other performance dimensions. The strategic importance of measuring different performance criteria for IT has also been recognized in the development of a 'Balanced Scorecard' for IT (Van Grembergen & Bruggen, 1997). IT organizations today cannot afford to focus on service delivery at the expense of solution integration, or vice versa. Furthermore, strategic innovation is difficult, if not impossible, to achieve without some base-line performance in service delivery and solution integration.

In terms of cause-effect relationships, strategic innovation is the value link between IT performance and business performance. Whereas service delivery and solution integration provide the foundation for strategic innovation, the latter provides the strategic IT capability for operational, product and/or service excellence. Moreover, it is pivotal that key stakeholders involved in IT Governance recognize the multiplexity of value drivers, and develop a shared understanding of the value of IT (Figure 14).

Similar to the business, the new enterprise logic for the IT business is based on developing the agility and flexibility to sense and respond to existing and emerging demands in an integrated and proactive manner (Peterson, 2001). The challenge for IT Governance is then, as discussed earlier, how to simultaneously perform and transform IT in order to meet the present and future demands of the business in a satisfying manner. Thus, the IT business also needs to manage the balance between exploitation and exploration.

Consequently, the IT business also needs to transform. Analogous to the business, IT requires a dynamically stable organizational architecture to deliver value to the business, i.e., stable enough to exploit IT capabilities for service delivery and solution integration, and sufficiently dynamic to explore new value innovation opportunities. Consistent with the 'law of requisite variety' (Ashby, 1956), the IT business should have a variety of capabilities at least as great as the demands and disturbances in the business enterprise environment.

Figure 14. IT Value Drivers: An Integrated Perspective (Peterson et al., 2000)

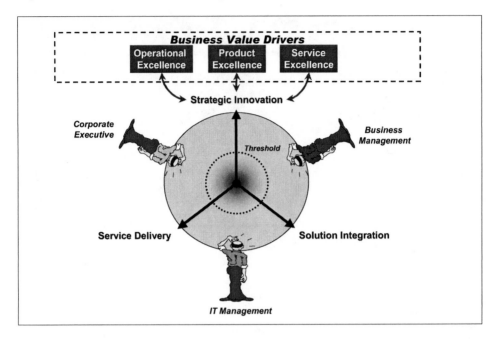

The business of IT has indeed experienced a fundamental transformation over the past few decades. In the 1950s, accounting and controller departments mainly used electronic computers. Some of these departments were centralized, while others were decentralized, and decision-making defaulted to those departments that made use of the technology. The benefits gained from automating the business's processes and functions led most organizations to integrate and centralize their technologies in the 1960s and 1970s. Specialists skilled in a number of hardware and software fields were required, and often reported to a central IT group. This central or corporate IT group primarily served a manufacturing role in a function-oriented organization with different functional IT departments, e.g., development, operations and technology (Figure 15).

By the mid 1980s, both the business and IT environments had changed significantly. Business markets became more complex and competitive. Businesses adopted divisional structures, each with their own products and market services, and concurring responsibilities and accountabilities. The proliferation of IT also became more complex with the dispersion of IT to the business units. Local IT managers were given authority over IT in order to respond to the local needs of the business in the competitive environment.

In the early 1990s, as companies experienced the demise of traditional geographic and business boundaries, and the emergence of 'the new infrastructure' (Weill & Broadbent, 1998), IT was again resorted to a central IT group. The title and function of CIO emerged on organizational charts, and many organizations were characterized by both a corporate IT department, often led by a CIO or IT director, and several local IT departments, often still functionally organized. During this period, IT organizations also began to see the benefits of employing a process-based IT organization (Figure 15), and

Figure 15. Transformation of the IT Organization (Peterson, 1998)

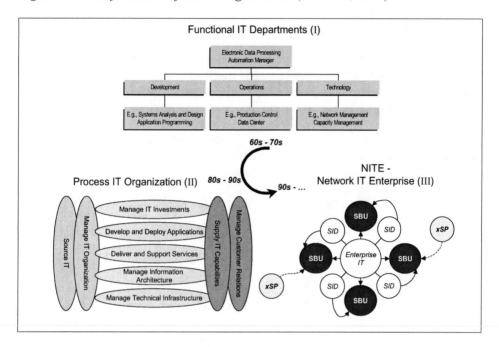

reengineering IT around value added process, both upstream (e.g., source IT, manage IT investment) and downstream (deploy IT applications, deliver IT services, manage customer relations).

Consider, for instance, the transformation of British Petroleum IT's function. In 1989, the IT function set out to transform itself, in line with its mission to "...become the best in class in its support of BP's goal to be the best upstream company in the world" (Cross et al., 1997). They supported this mission by changing their direction, with the explicit purpose of moving from "system provider" to "infrastructure planner". By 1994, the IT function was truly an "infrastructure planner", overseeing technical integrity and value creation through information sharing. Although some IT resources remain in the local businesses, the top IT management team provides global vision for infrastructure planning (Cross et al., 1997).

Today, a new transformation is transpiring. IT organizations are transforming toward a 'core-peripheral' organization design, in which the IT infrastructure is directed by Corporate IT management, and local business applications are managed by business or IT management. At the 'core' of the organization, IT infrastructure decisions are centralized and allocated to the corporate unit, whereas IT application decisions are decentralized and allocated to the different operational business units at the organization 'periphery'. This Network IT Enterprise (NITE) is built around solution integration delivery (SID) teams that focus on the needs of the business, and contracts with external service providers (xSPs). The (internal) SID and (external) SP components 'hover' around a center that manages and provides core IT competencies (e.g., IT services and IT skills). NITE resembles the enterprise architecture in contemporary organizations, and embodies the principles of strategic flexibility and dynamic stability.

A good example of NITE is found at Verizon Communications Inc., one of the world's largest investors in global communications markets. Verizon Communications, formed by the merger of Bell Atlantic and GTE, is a leading provider of communications services. Verizon companies are the largest providers of wireline and wireless communications in the US, with nearly 134 million access line equivalents and more than 29 million wireless customers. Verizon is also the world's largest provider of print and online directory information. With more than 247,000 employees and $67 billion in 2001 revenues, Verizon's global presence extends to 45 countries in the Americas, Europe, Asia and the Pacific.

In 1996, Verizon (then Bell Atlantic) adopted a core-peripheral design in molding a network IT enterprise. The NITE at Verizon reflects a 'centers of excellence' (CoE) approach (Figure 16), in which roles and processes are differentiated for conceptualizing IT applications, delivering IT services and building the requisite IT competencies (Clark et al., 1997). The implementation of NITE has enabled Verizon to develop a flexible IT organization focused on IT-based value innovation.

Discussion

In retrospect, comparing the empirical evidence on IT Governance with the new business imperatives of strategic flexibility and dynamic stability, it is clear how and why the federal IT Governance model has emerged as the dominant design in contemporary organizations. Already forecasted by Zmud et al. (1986), different studies indicate that this federal model is the dominant IT Governance practice in contemporary organizations (Feeny et al., 1989; Hodgkinson, 1996; Sambamurthy & Zmud, 1999). The federal model is also propagated as the best model, 'capturing the best of both — centralized and decentralized — worlds' (Von Simson, 1990). Rockart et al. (1996) describe the federal

Figure 16. Verizon's NITE (Adapted from Clark et al., 1997)

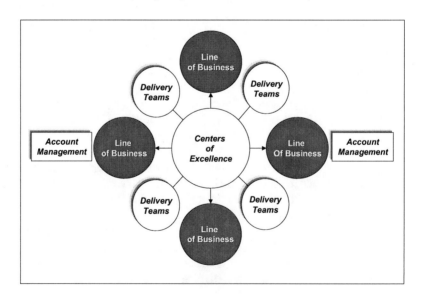

model as one of the fundamental imperatives of IT in the late 1990s, and urge organizations to adopt the federal model, regardless of organizational contingencies. Recently, Earl (2000) argues that every company needs to build a degree of IT federalism.

From a strategic perspective, both business and IT are facing multiple, competing objectives to reduce costs, standardize, innovate and provide customer value. From an organizational perspective, the enterprise architecture is characterized by a dynamic network of integrated business-IT capabilities (Figure 17). The federal IT Governance model, by dividing direction and control over IT between central and local offices, across different business and IT constituencies, creates a structure that is consistent with the enterprise architecture, i.e., both stable and dynamic, and enables IT-based strategic differentiation of the business.

However, some researchers have recently questioned whether 'the truth' regarding IT Governance is still out there. Is the federal IT Governance model really where the buck stops? Is dividing the locus of IT decision-making authority the answer in our quest for IT value?

Peppard and Ward (1999) state that while appealing at present, the federal IT Governance model is more of a theoretical construction than a direct practical solution. The problem, according to Sauer and Yetton (1997), is that on the surface, the federal model seems to provide a solution to the competing demands of corporate efficiency and business flexibility. However, the structure it creates depends on IT managers ensuring that activities of the different business units, which have competing interests, are

Figure 17. Evolution of Value Drivers, Enterprise Architectures and the Governance of IT (Peterson, 2001)

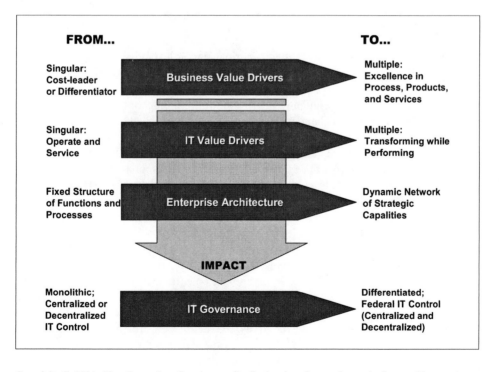

integrated across the organization. The reality is that local IT units tend to be 'captured' by the goals of the business unit, while the central IT unit tends to become divorced from the business (Sauer & Yetton, 1997).

Sambamurthy and Zmud (2000) argue that there are some signs that the accumulated wisdom on IT Governance might still be inadequate. Further, Peterson (1998; 2001) states that our understanding of IT Governance is not only inadequate, it is incomplete. To cite Simon (1960), the current federal IT Governance model is like designing an unreliable ship with risky steersmanship:

"...there is no use in one group of experts producing the hull, another the design for the power plant, and a third the plans for the passenger quarters, unless great pains are taken at each step to see that all these parts fit a seaworthy ship."

DESIGNING EFFECTIVE IT GOVERNANCE ARCHITECTURES

In this section, the fundamental IT Governance problématique is addressed, and architectural solutions are discussed. Specifically, this section describes how and why our current framing of IT Governance is incomplete, and needs a fundamentally different approach, and how strategic risks can be mitigated through effective IT Governance architectures, and integration strategies and tactics. This section concludes by discussing the need to align IT Governance architectures with value drivers and enterprise architectures.

Federal IT Governance: What is Wrong with this Picture?

In *The Blind Men and the Elephant*, American poet John Godfrey Saxe (1816-1887) retells an ancient Indian fable of six blind men who visit the palace of the Rajah and encounter an elephant for the first time (Figure 18). The first blind man put out his hand and touched the side of the elephant: "How smooth! An elephant is like a wall." The second blind man put out his hand and touched the trunk of the elephant: "How round! An elephant is like a snake." The third blind man put out his hand and touched the tusk of the elephant: "How sharp! An elephant is like a spear." The fourth blind man put out his hand and touched the leg of the elephant: "How tall! An elephant is like a tree." The fifth blind man reached out his hand and touched the ear of the elephant: "How wide! An elephant is like a fan." The sixth blind man put out his hand and touched the tail of the elephant: "How thin! An elephant is like a rope."

Although a well-known, simple story, the morale of this ancient Indian fable applies equally well to IT Governance, and in particular, the federal IT Governance model. Much like an elephant, or any other large living organism, the federal IT Governance model is a complex system, involving different stakeholder constituencies with specific perceptions, views and motivations. Similar to the blind men, these stakeholders have specific stakes and responsibilities in governing IT. This is especially the case with federal IT Governance, in which corporate executives are focused on creating enterprise-wide

Figure 18. Framing Federal IT Governance

synergies, standardizing IT, and controlling the IT infrastructure, whereas local business and IT managers are concerned with improving flexibility and responsiveness to business-specific problems and opportunities. Though each constituency may be correct in pursuing their own strategic objectives, their 'blinded' focus impedes effective governance of IT.

While research insists on the dominance and importance of the federal IT Governance model, previous studies have failed to specify how and which of the eight distinct patterns of federal IT Governance are adopted, and under what circumstances these distinct patterns are effective. The federal model is not a single design, but consists of different patterns in dividing IT decision-making authority (Figure 18). Furthermore, the federal IT Governance model introduces a 'new division', in which the decision-making actions of individual units are divided, thus becoming interdependent, thereby requiring coordination, especially considering the dynamic task environments. Previous studies, however, assume that once IT decision-making is allocated to corporate executives, business management, and IT management, coordination will follow automatically through hierarchical lines of reporting. Thus, while previous studies focus on the differentiation of decision-making for IT, they do not address the integration of decision-making for IT.

In terms of Mintzberg (1979), previous studies address the division of responsibilities, but fail to take into account coordination to accomplish activities. Lorsch and Lawrence (1970) state that the real difference between centralization and decentralization is much more complex than patterns of decision-making:

"Another shortcoming in the traditional views about centralization and decentralization is a failure to recognize that the issue is really one of a vertical division of labor and coordination. Therefore, it is not just a question of dividing responsibility up and down the hierarchy, but it is also a question of organizing the flow of information and coordinating devices. If these labels are to capture the realities of how complex organizations operate, they must refer to systems of organizational variables which include division of work and differentiation; the integration among

Figure 19. Symptoms and Strategic Risks Associated with the Lack of Integration (Peterson, 2001)

Symptoms and Strategic Risks	
• Lack of IT prioritization	• Business management takes no responsibility for IT-enabled business change
• Wasted IT investments	
• IT organization fails to meet commitments	• IT organization fails to realize business value
• IT management does not understand the business	• Increasing customer dissatisfaction
• Unresolved conflicts between business and IT management	• Less revenues
	• Loss of market share
• Executives do not support IT	• Weakened competitive position

divisions and the headquarters; the types of integrative structural devices used, as well as the information flows and decision-making processes operating within the organization."

Although this discussion may seem utterly theoretical and outdated, consider the symptoms and strategic risks associated with the lack of integration in companies today (Figure 19), involving not only internal IT troubles, but moreover, major business tribulations.

To create value for the business and mitigate strategic risks, IT Governance should have a variety of capabilities at least as great as the demands posed by the multiplicity of value drivers. Consequently, IT Governance requires a fusion of different business and IT competencies, involving both corporate executives and business and IT management. However, the differentiated fusion of business and IT competencies leads to fragmentation, unless there is a corresponding balanced process of integration. The complexity associated with the multiplicity of value drivers and stakeholder constituencies creates the need for integration to achieve clarity of direction and unity of purpose in governing IT effectively (Peterson, 2001).

The organizing logic is similar to that of the enterprise, albeit at a different level of governance. Whereas the enterprise logic focuses on core competencies and strategic collaboration at the inter-organizational level, the organizing logic for IT Governance focuses on the differentiation and integration of IT decision-making and monitoring at the intra-organizational level (Figure 20).

In the past, however, there has been a strong bias toward defining and designing IT Governance in a disjointed manner, i.e., reductionism. This predisposition is often instilled upon us — both executives and researchers — through years of professional education and indoctrination. We have been taught, especially in Western cultures, when faced with a complex problem, to divide the problem into smaller pieces, and solve each piece separately. This 'divide-and-conquer' mentality prevails in much problem-solving in both business and information systems communities (e.g., IT business investment processes, IT program management, algorithm design). However, solving each piece of the problem does not address the complete problem, i.e., addressing subsystems sequentially or in parallel is no guarantee that the supra-system will function effectively (Von Bertalanffy, 1968). The classic Indian parable thus illustrates the need for a holistic systems view of IT Governance.

Figure 20. Design Logic at Different Levels of Governance

Level of Governance	Design Logic	
Inter-Organizational	Core Competency	Strategic Collaboration
Intra-Organizational	Differentiation	Integration

Designing Effective IT Governance Architectures

A holistic systems[6] view emphasizes the need to view IT Governance as a complex open social system interacting with its environment, and consisting of a set of interdependent subsystems that produce a purposeful whole (Peterson, 2001). Complex systems are characterized by reciprocal interdependence (Thompson, 1967), in which decisions made by subunits are mutually dependent and influential, thereby increasing the need to exchange information. In complex governance systems, each decision-making unit presents direct decision contingencies for every other unit (Lorsch & Lawrence, 1970). Interacting subsystems in a social system imply that stakeholders are interdependent and need to work together in a coordinated fashion to achieve objectives. A systems thinking approach towards IT Governance acknowledges its complex and dynamic nature, and underscores the importance of personal mastery and mental models, and team learning and shared vision (Senge, 1990).

The systems logic underlying governance is the division and coordination of decision-making in order to direct the operational system towards the realization of the goals of the organizational system (Simon, 1961). This systems logic is the foundation for Organization Design Economics[7]. The design logic of organizing revolves around the processes of (a) *division* of labor into various tasks to be performed, and (b) *coordination* of these tasks to accomplish the activity and achieve the organization's goals. Thompson (1967) describes the basic logic of organizing in the following manner:

"By delimiting tasks, responsibilities and other matters, organizations provide their participating members with boundaries within which efficiency may be achieved. But if structure affords numerous spheres of bounded rationality, it must also facilitate the coordinated action of those interdependent elements."

In terms of IT Governance, and specifically the 'new division' of federal IT Governance, this involves centralizing and decentralizing specific aspects of the IT portfolio. In order to achieve the organization's objectives, IT Governance also needs to integrate these different aspects of the IT portfolio across the principle stakeholders, i.e., corporate executives, IT management and business management.

The manner in which responsibilities and accountabilities for the IT portfolio are organized and integrated is defined as an ***IT Governance architecture***. An IT Governance architecture describes the *differentiation and integration of strategic decision-making for IT* (Peterson, 2001). The IT Governance architecture specifies the strategic policies and business rules that provide direction to strategic IT decision-making, and

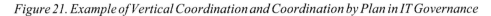

Figure 21. Example of Vertical Coordination and Coordination by Plan in IT Governance

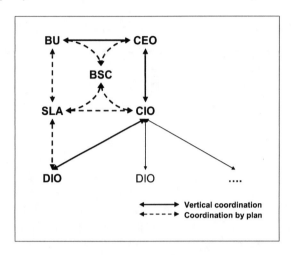

plots a path for achieving business objectives (Weill & Broadbent, 1998). Designing an effective IT Governance architecture is thus dependent on both the differentiation and integration of strategic decision-making for IT (Peterson et al., 2000). However, whereas executives and scientists have been keen to adopt and propagate a federal model of IT Governance, i.e., differentiate IT decision-making, following through on integration has been somewhat undermined and still remains a challenge.

Traditionally, organizations and IT Governance relied on the hierarchy and standardization for coordination. *Hierarchy* or *vertical coordination* describes the hierarchical referral of infrequent situations for which standardized programs have no solution (Galbraith, 1973). The hierarchy achieves coordination by having one person (e.g., CIO-Chief Information Officer) take responsibility for the work of others, issuing instructions and monitoring actions (Figure 21). If the hierarchy gets overloaded, additional levels or positions can be added to the hierarchy (e.g., DIOs-Division Information Officers). *Standardization* or *coordination by plan*, on the other hand, describes the use of standard programs, formal rules and procedures, and the specification of outputs, goals and targets (Galbraith, 1973; March & Simon, 1958). The adoption and use of Balanced Scorecard methodologies (BSC) and service level agreements (SLAs) are typical examples of how contemporary organizations coordinate by plan (Peterson & Ribbers, 2002).

Vertical coordination and standardization, however, only provide limited coordination capability in complex and uncertain environments (Daft, 1998; Galbraith, 1973, 1994; Mintzberg, 1979). With the profusion of electronically-enabled, globally-operating organizations characterized by a multiplicity of value drivers in a dynamic network of strategic capabilities, the best CIOs, DIOs, BSCs and/or SLAs will not suffice in designing effective IT Governance architectures. Instead, IT Governance needs to focus on *lateral coordination capabilities*.

Traditionally described as 'informal mutual adjustment' (March & Simon, 1958; Thompson, 1967), lateral coordination and horizontal relationships represent the most significant contemporary development in practice and theory. Lateral coordination has recently been rediscovered as a strategic organizational capability for competing in contemporary hyper-competitive environments (Galbraith, 1994; Hitt et al., 1998). In turbulent environments, performance is driven by an organization's resources that are valuable and unique (Collis & Montgomery, 1995). From a resource-based perspective, lateral coordination is a resource that is hard to imitate, cannot be purchased, and is time-dependent and socially complex (Hitt et al., 1998; Powell, 1992).

How, then, can organizations develop this requisite lateral coordination capability, and what (horizontal) integration mechanisms should be used for governing IT effectively?

Integration Strategies and Tactics for IT Governance

Integration strategies for IT Governance can be classified according to two dimensions. Vertically, integration mechanisms can focus either on integration structures or integration processes, whereas horizontally, a division is made between formal positions and processes, and relational networks and capabilities (Peterson, 2001). Collectively, this provides four types of integration strategies for IT Governance (Figure 22):

1. Formal integration structures;
2. Formal integration processes;
3. Relational integration structures;
4. Relational integration processes.

Formal integration structures involve appointing IT executives and accounts, and institutionalizing special and standing IT committees and councils. The use of account and/or relationship managers aid IT managers to develop an improved understanding of business needs, and aid in proactive — versus reactive — behavior by IT managers (Peterson, 2001). Committees and/or executive teams can take the form of temporary task forces — e.g., project steering committees — or can alternatively be institutionalized as an overlay structure in the organization in the form of executive or IT management councils (Figure 22). Committees vary in the degree to which they have an advisory function or have formal decision-making authority. Advisory steering committees are also referred to as advisory, review or guidance committees. Contrary to specialized task forces, steering committees and advisory boards bring together different stakeholders on a relatively permanent basis for resolving Business/IT decision-making questions and problems.

Formal integration processes describe the formalization and institutionalization of strategic IT decision-making/monitoring procedures and performance. Formal integration processes vary with levels of (Peterson, 2001):

1. *Comprehensiveness:* the degree to which IT decision-making/monitoring activities are systematically and exhaustively addressed, i.e., (1) the identification and formulation of the business case and/or 'business rationale' for IT decisions, (2) the prioritization and selection of business-IT course-of-action, and (3) monitoring and evaluating of IT decision-implementation and IT performance (Parker & Benson, 1988; Luftman & Brier, 1999; Willcocks, 1996).
2. *Formalization:* the degree to which IT decision-making/monitoring follows specified rules and standard procedures, often embedded in formalized decision-making

Figure 22. Integration Strategies and Tactics for IT Governance (Peterson, 2001)

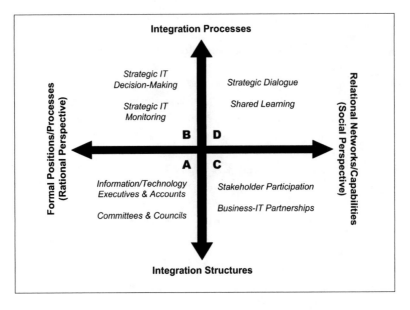

methodologies and management frameworks (e.g., *BITS*-Balanced IT Scorecard, *SAM*-Strategic Alignment Model, *CSF*-Critical Success Factors, and/or *IE*-Information Economics, etc.).

3.   *Integration:* the degree to which business and IT decisions are integrated, i.e., (1) *administrative integration*, in which budgets and schedules are pooled between business and IT, (2) *sequential integration*, in which business decisions provide directions for IT decision-making, (3) *reciprocal integration*, in which business and IT decisions are mutually influential, and (4) *full integration*, in which business and IT decisions are made concurrently in the same process (Teo & King, 1999).

Whereas the foregoing formal integration mechanisms tend to be mandatory, tangible, and often implemented in a 'top-down' manner, relational integration mechanisms are 'voluntary actions', which cannot be programmed and/or formalized, and which are often intangible and tacitly present in the organization (Peterson, 2001). Kahn and McDonough (1997) refer to integration as a composite of interaction and collaboration. Interaction describes the formal structures and processes used for information-exchange and communication, whereas collaboration is described as the affective, participative and shared element of integration, corresponding to a willingness to work together. Galbraith (1994) and Malone and Crowston (1994) describe these levels as a layered system of successively deeper levels of coordination, i.e., higher levels of lateral coordination capability (Figure 23). Recent evidence indicates that while formal integration mechanisms are necessary, they are insufficient for designing effective IT Governance architectures in competitive environments (Peterson et al., 2000).

Figure 23. Operationalization of Integration Strategies and Tactics for IT Governance (Peterson et al., 2000)

Integration Strategy	(a) Formal Integration Structures	(b) Formal Integration Processes	(c) Relational Integration Structures	(d) Relational Integration Processes
Tactics	IT Executives & Accounts Committees & Councils	Strategic IT Decision-Making Strategic IT Monitoring	Stakeholder Participation Business-IT Partnerships	Strategic Dialogue Shared Learning
Mechanisms	- CIO on Board - IT program managers - IT relationship managers - IT executive councils - eBusiness advisory board - eBusiness task force - IT standing teams	- Balanced (IT) Scorecard - Critical Success Factors - Scenario analysis - SWOT analysis - Strategic Alignment - Information Economics - Service Level Agreements - IT benefits management	- Active participation by principle stakeholders - Collaboration between principle stakeholders - Partnership rewards and incentives - Business/IT co-location - Business/IT 'virtual connection'	- Shared understanding of business/IT objectives - Active conflict resolution ('non-avoidance) - Cross-functional business/IT training - Cross-functional business/IT job rotation
Low		← Lateral Coordination Capability →		High

Relational integration structures involve the active participation of and collaborative relationships between corporate executives, IT management, and business management. Central to relational integration is the participative behavior of different stakeholders to clarify differences and solve problems in order to find integrative solutions. The ability to integrate relationally allows an organization to find broader solutions, and unleashes the creativity involved in joint exploration of solutions that transcend functional boundaries and define future possibilities (Peterson, 2001).

Relational integration structures are characterized by their participative and shared nature. Participation is a process in which influence is exercised and shared among stakeholders, regardless of their formal position or hierarchical level in the organization. Active stakeholder participation balances the involvement of business and IT communities in information processing, decision-making and problem-structuring/solving (Peterson, 2001). Collaboration integration refers to a close, functionally interdependent relationship in which organizational units strive to create mutually beneficial outcomes. Henderson (1990) describes this as a strategic partnership that reflects a working relationship of long-term commitment, a sense of mutual collaboration, and shared risks and benefits. Mechanisms that facilitate relational integration include joint performance incentives and rewards, co-location of business and IT managers, and the creation of 'virtual meeting-points' for business and IT managers (Peterson et al., 2000; Ross et al., 1996; Weill & Broadbent, 1998).

Relational integration processes describe strategic dialogue and shared learning between principle business and IT stakeholders. Strategic dialogue involves exploring and debating ideas and issues in depth prior to decision-making or outside the pressure of immediate IT decision-making (Van der Heijden, 1996). A strategic IT dialogue incorporates a wide range of initially unstructured business perspectives and IT views, and involves rich conversation and communication to resolve diverging perspectives and stakeholder conflicts. Conflicts are resolved through the use of active and passive resolution strategies. Active conflict resolution involves confrontation and competition strategies, whereas passive conflict resolution involves avoidance and smoothing-over strategies, i.e., conflicts remain and are not explicitly resolved (Lawrence & Lorsch, 1969; Robbins, 1994).

Shared learning is defined as the co-creation of mutual understanding by members of organizational subunits of each other's goals and objectives (Peterson, 2001). The essence of organizational coordination capability is the integration of domain-specific expertise and tacit knowledge. Shared learning is developed when people in close collaboration enact a single memory, *with differentiated competencies* and *responsibilities* (Weick & Roberts, 1993). Shared learning resides in specialized relationships among stakeholders, and in particular, the information flows and decision-making processes that shape their dealings with each other (Lorsch & Lawrence, 1970). Parker et al. (1997) argue that identifying acceptable solutions to ambiguous problems in complex and dynamic environments requires the collaboration of different stakeholders, working with different paradigms, and offering different insights. Shared learning is inherently dynamic, and results in coordinated decision-making and collaborative relationships, which are particularly relevant and beneficial when the need for reliability is high, and decision-making is non-routine, involving interactive complexity, i.e., the combination of complex interpersonal interactions with a high degree of interdependence (Weick & Roberts, 1993).

Research indicates that when business and IT managers understand each other's perspectives in IT decision-making, they can accurately interpret and anticipate actions, and coordinate adaptively (Peterson, 2001). Within the context of IT Governance, shared learning describes the mutual understanding of business and IT objectives and plans by business and IT executives (Reich & Benbasat, 1996; Weill & Broadbent, 1998). Mechanisms that support shared learning include strategic dialogues between business and IT executives, cross-functional business-IT training, and cross-functional business-IT job-rotation/transfers (Luftman & Brier, 1999; Peterson et al., 2000; Ross et al., 1996; Weill & Broadbent, 1998).

Discussion

Designing an effective IT Governance architecture is dependent on both the differentiation and integration of strategic decision-making for IT. Whereas differentiation focuses on the distribution of IT decision-making rights and responsibilities among different stakeholders in the organization, integration focuses the coordination of IT decision-making/-monitoring processes and structures across stakeholder constituencies. The notion of an IT Governance *architecture* emphasizes the need to define and control the interfaces between the separate components of the IT Governance system. Designers of IT Governance architectures thus need to consider and implement integration strategies and tactics for governing IT effectively.

Contrary to the IT Governance evolution described earlier in this chapter (e.g., Figure 17), the governance of IT has evolved not only from a monolithic toward a differentiated structure, but moreover, towards an integrated system of differentiated business and IT capabilities. Similar to the enterprise architecture, but then at a 'micro-level', an IT Governance architecture resembles a dynamic network of strategic capabilities for directing and controlling IT in line with the business, and the business' objectives. Embedding the IT Governance architecture in the enterprise architecture, and aligning IT Governance with IT value drivers enables an organization to realize its business value drivers (Figure 24). Effective IT Governance architectures should be consistent with the enterprise architecture and IT value drivers, whereas enterprise architectures and IT value drivers should be in sync with business value drivers (Peterson, 2001). Given the multiplexity of value drivers and the dynamic stability of the enterprise architecture, aligning value drivers, enterprise architectures and IT Governance architectures is inherently dynamic.

The key question which then arises is, of course, how do we — executives and researchers — know whether an IT Governance architecture is effective, and what measures can be taken to redesign and improve the effectiveness of the IT Governance architecture? In the next section, this challenging question is addressed.

DIAGNOSING
IT GOVERNANCE EFFECTIVENESS

In this section, the design of a measurement system for diagnosing IT Governance effectiveness is discussed. In the following sections, the rationale and foundation for an IT Governance diagnostic is described. This section presents the IT Governance diagnostic model, and discusses how it can be used to diagnose and track IT Governance effectiveness.

Figure 24. Aligning Value Drivers, Enterprise Architectures and IT Governance Architectures (Peterson, 2001)

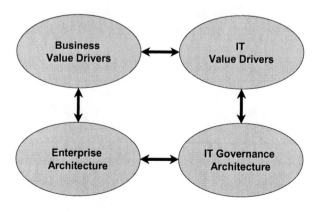

Motivation and Methodology

A critical part in designing effective IT Governance architectures is devising a measurement system to assess and affect IT Governance effectiveness. However, while our descriptive knowledge of IT Governance has increased dramatically over the past three decades, prescriptive actions with regard to designing effective IT Governance architectures are scant. Recently, Weill and Vitale (2002) indicate that executives moving into e-business should examine their IT Governance and determine whether IT Governance encourages the necessary communication and control of e-business initiatives. Diagnosing IT Governance effectiveness and establishing an IT Governance measurement system that links cause and effect is consequently strategically important and highly relevant (Weill & Woodham, 2002).

Drawing upon the results of a longitudinal study on the design and effectiveness of IT Governance (Peterson, 1997, 1998, 2000, 2001), an IT Governance diagnostic model is developed and implemented in several large, multi-business-unit organizations across different industries in Europe. This comprehensive multi-client study was based on several in-depth case studies conducted in different sectors of the economy, including Banking, Insurance, Manufacturing, Health Care, Telecom & IT, and Travel & Leisure. Data was collected and analyzed on (1) the strategic challenges and IT value drivers in business transformations, (2) the locus and distribution of IT decision-making/monitoring responsibilities, (3) the integration strategies and tactics, and coordination capability and maturity of IT Governance, and (4) the impact of IT Governance on IT value realization.

More than 100 interviews were conducted with senior executives, directors and managers in both business and IT organizations and departments. These interviews were complemented by the collection and analysis of strategic company reports (e.g., IT investment analysis, Service Delivery Audits), and industry trends publications. In several of the client-organizations, executive workshops were organized to (1) present the main findings of the study, and (2) discuss the implications for the design and effectiveness of IT Governance in their organization. In the next section, the results of this comprehensive study are described.

The IT Governance Diagnostic Diamond

This section (Peterson, 2001) describes a holistic, high-level assessment model of IT Governance design and effectiveness. The model is theoretically inspired by Organization Design Economics, specifically, the *congruency model of organizations* (Galbraith & Lawler, 1993; Nadler & Tushman, 1998), a *balanced strategic management system* (Kaplan & Norton, 1992, 1996), and *conversion effectiveness* (Weill & Broadbent, 1998). Each of these concepts and underlying models are used extensively in practice, and has been empirically validated.

The congruency model directs our attention towards the contextualistic nature of effective designs, emphasizing the interplay between context, design and effectiveness. The balanced strategic management system reminds us that different (types of) measures should be used to assess performance, and that these measures are embedded in a process linking strategic objectives with adaptive actions (i.e., strategic flexibility). In contrast to the micro-economic 'black-box' model of IT value, conversion effectiveness recognizes that realizing IT value is based on the level of *Value Conversion Process Maturity* — VCPM — which is characterized by (Peterson, 2001):

1.   Senior (business) executive input and involvement (not 'abdication');
2.   Collaborative relationships and shared learning (not 'political clout');
3.   Integrated business-IT decision-making/monitoring (not 'management-by-budget');
4.   Business-IT competence (not 'ignorance').

Note: high VCPM is primarily dependent upon the business organization and its executives, and only secondly on the IT organization. Furthermore, high VCPM is directly related to the level of lateral coordination capability, i.e., the adoption and use of relational integration structures and processes (see Figure 23).

The IT Governance diagnostic diamond is organized into two axes (Figure 25). *Vertically*, the model depicts the IT Governance architecture consisting of the:

1.   *differentiation of IT decision-making*, i.e., who has what authority and responsibility to make decisions regarding the IT portfolio?;
2.   *integration of IT decision-making*, i.e., what integration strategies, tactics and mechanisms are used to coordinate IT Governance?

Answering the first question regarding differentiation provides (1) a specific federal profile of IT Governance (see Figure 8), and (2) improves transparency into the distribution and allocation of IT decision-making authority and responsibility. Answering the second question regarding integration provides (1) a description of the integra-

Figure 25. IT Governance Diagnostic Diamond (Peterson, 2001)

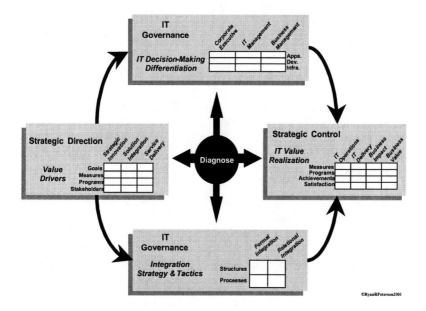

tion strategies and tactics used, and (2) an assessment of the current level of lateral coordination capability (see Figure 23).

Horizontally, the IT Governance diagnostic diamond distinguishes between *strategic direction* and *strategic control*, i.e., what are the value drivers and strategic intents, and to what extent have these value drivers been realized, or is IT contributing to business performance? The latter is the ultimate measure of true IT Governance effectiveness.

In assessing the value drivers, organizations focus on the value propositions for IT (i.e., Strategic Innovation, Solution Integration, and Service Delivery), and the operationalization of these value drivers into operational goals, and specific measures and targets. Moreover, it is vital that executives also indicate *what* initiatives are currently underway, and planned for the future, that address this value driver, and *who* are involved and affected by these programs.

In determining IT value realization, the IT Governance diagnostic diamond recognizes the dilution of IT impacts, and consequently distinguishes four generic measures for assessing IT value realization, i.e.:

1. *IT services,* consistent with 'Service Delivery' value driver (see Figure 14);
2. *IT solutions,* consistent with 'Solution Integration' value driver (see Figure 14);
3. *Business impact,* addressing the business value drivers, i.e., 'Operational Excellence', 'Product Excellence', and 'Service Excellence', and the respective foci and operationalization (see Figure 12);
4. *Business value,* addressing (future) financial performance, e.g., revenue growth, sales growth, return on assets, market share, shareholder value.

Similar to strategic direction, assessing strategic control involves the specification of the measures and targets involved, and to what extent the specific initiatives and programs have achieved their targets and objects, including level of stakeholder satisfaction.

Diagnosing IT Governance effectiveness and applying the IT Governance diagnostic diamond follows a 'left-to-right' logic in a step-wise process. The following four steps are followed for both the '*Ist*' (current 'as-is' position), as well as for the '*Soll*' (desired 'to-be' position) situation.

1. *Describe and assess the current/future business and IT value drivers for the organization:*
    a. What are the main business value drivers?
    b. What are the main IT value drivers?
    c. What specific goals and targets, and initiatives and programs are underway that address the IT value drivers and business value drivers?
2. *Describe and assess the current/future differentiation of IT decision-making authority for the IT portfolio:*
    a. Who is responsible for IT-business applications decisions, IT innovation decisions, and/or IT infrastructure decisions?
    b. How clearly are these responsibilities formulated, and made transparent to the organization, especially other key stakeholders?
    c. Is the differentiation of IT decision-making in line with the value drivers?

3.   *Describe and assess the current/future integration of IT decision-making:*
     a. What integration strategies, tactics and mechanisms are employed by the organization?
     b. What is the level of lateral coordination capability and value conversion process maturity?
     c. Are IT Governance integration strategies, tactics and mechanisms in line with the type value drivers and IT Governance differentiation employed by the organization (now and in the future)?
4.   *Describe and assess the current/future IT value realization:*
     a. What is the contribution of IT to improved business performance?
     b. What are the main business impacts, and how do these relate to the business value drivers, goals and measures?
     c. How is the IT organization performing on service delivery, solution integration, and strategic innovation?

Applying the IT Governance diagnostic diamond and following this step-wise approach for both the current and the desirable situation provides a diagnosis of the suitability of the existing IT Governance architecture, and identifies strategic discrepancies with the future, desirable position, and measures to redesign and improve the IT Governance architecture.

CONCLUSION: TOWARD A NEW IT GOVERNANCE PARADIGM

This section concludes the chapter on Integration Strategies and Tactics for IT Governance. In this final section, the main lessons we have learned are summarized, and based upon the previous sections, a new IT Governance paradigm is presented. In the final section, directions and opportunities for practitioners and researchers of IT Governance are outlined.

What Have We Learned? Implications for Management

Despite the many bridges that still remain to be breached in the (theories and practices) of IT Governance, there are a number of important lessons we have learned over the past decades, which were addressed in this chapter. When contemplating and/or studying IT Governance, both executives and researchers would do well to consider these important lessons learned.

Amidst the challenges and changes of the 21st century, involving hyper-competitive market spaces, electronically-enabled global network businesses, and corporate governance reform, IT Governance has become a fundamental *business* imperative. IT Governance is a top management priority, and rightfully so, because it is the single most important determinant of IT value realization.

IT Governance is the system by which an organization's IT portfolio is directed and controlled. IT Governance describes (1) the distribution of IT decision-making rights and responsibilities among different stakeholders in the organization, and (2) the rules and procedures for making and monitoring decisions on strategic IT concerns. IT Governance is thus corporate governance focused on IT, and it is not (only) the responsibility of the CIO or the organization of the IT function, in which an 'old new' IT management form is introduced to choose between centralization or decentralization.

The terms centralization and decentralization provide a dichotomy that is meaningless when employed as a generality to IT Governance. The centralization or decentralization can be applied to each of the main IT capabilities in the IT portfolio, yielding distinct patterns of federal IT Governance. The federal IT Governance model is currently the dominant model in contemporary organizations, as they seek to combine the benefits of synergy, standardization and specialization with the advantages of autonomy, innovation and flexibility.

The dominance of the federal IT Governance model is a strategic response to the needs of strategic flexibility and dynamic stability in contemporary organizations and markets, in which both business and IT organizations are adopting multiple, complementary value drivers. However, it should be emphasized that redesigning for federalism challenges managers in local business units to surrender control over certain business-specific IT domains for the well-being of the enterprise, and to develop business-to-corporate and business-to-IT partnerships.

An IT Governance architecture recognizes this coordination challenge, and emphasizes the need to take a holistic systems view of IT Governance. An IT Governance architecture is defined as the manner in which responsibilities and accountabilities for the IT portfolio are organized and integrated, and describes the differentiation and integration of strategic decision-making for IT (Peterson, 2001). Four types of integration strategies, each consisting of several integration tactics and integration mechanisms, can be employed to assure the level of coordination capability in the IT Governance architecture. Embedding the IT Governance architecture in the enterprise architecture and aligning IT Governance with IT value drivers enables an organization to realize its business value drivers.

Consequently, diagnosing the IT Governance architecture and developing a balanced strategic measurement system to assess and track IT Governance effectiveness is essential. Applying the IT Governance diagnostic diamond and utilizing the proposed step-wise approach provides a diagnosis of the suitability of the existing IT Governance architecture, and identifies strategic discrepancies with the future, desirable position, and measures to redesign and improve the IT Governance architecture in terms of strategic flexibility and dynamic stability.

Toward a New IT Governance Paradigm

Until recently, efficiency was the keyword in designing IT Governance. This made sense in a world characterized by a stable placid environment, in which neither the core technology nor the markets in which companies were operating changed drastically over time. Organizations could afford to use a 'command-and-control' structure to govern IT. However, with the business imperatives and new enterprise logic of strategic flexibility and dynamic stability, this 'old' IT Governance paradigm no longer seems viable, nor prudent.

Rather than being a system of command-and-control, focusing on the locus of IT decision-making authority, this chapter concludes that effective IT Governance in contemporary organizations is more likely to resemble a network of multiple business-

IT collaborative relationships based on competencies and flexibility (Figure 26). IT is less about who is hierarchically positioned to be in control, and more about the complementary — business and IT — competencies an organization possess, and how it can integrate these in order to develop the required strategic flexibility for realizing and sustaining business value from IT in a complex and dynamic environment. The emerging IT Governance paradigm is based on principles of collaboration, competency and flexibility — not control, authority and efficiency.

These principles of the emerging IT Governance paradigm underscore and reaffirm the importance of flexible management systems in complex and uncertain environments. The organizing logic in the emerging IT Governance paradigm is characterized by a collaborative network structure, where communication is more likely to be lateral, task definitions are more fluid and flexible — related to competencies and skills, rather than being a function of position in the organization — and where influencing of business-IT decisions is based on expertise rather than an individual (or group's) position in the hierarchy.

In collaborative relationships between business and IT stakeholder constituencies, managers work together to understand business and IT competencies, opportunities, risks and benefits. This collaborative relationship demands that both business and IT managers take responsibility for business operations and IT innovation, which is achievable only when stakeholder constituencies share their unique expertise and competencies. This emerging paradigm for IT Governance is based on a 'philosophy' of collaboration where the need for distinct competencies are recognized and developed, and shared adaptively across functional, organizational, cultural and geographic boundaries.

Directions and Opportunities for Future Research

In Spanish there is a saying, *"el camino se hace al caminar"*, which roughly translates into English as the road is paved as you go. While we have come a long way in understanding and implementing strategies for IT Governance, and designing effec-

Figure 26. The Emerging IT Governance Paradigm (Peterson, 2001)

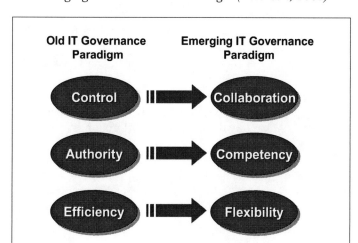

tive IT Governance architectures, the real journey still lays ahead of us. Specifically, there are (at least) five areas of theoretical and empirical study that seem particularly fruitful for future research and practice of IT Governance. These are:

1.   The development and integration of multiple complementary business and IT value drivers, e.g., how do organizations and networks of organizations develop and integrate the multiplicity of business and IT value drivers? How do organizations develop the requisite capabilities to become strategically flexible and dynamically stable?

2.   The development of lateral coordination capability and value conversion process maturity, e.g., what are the interrelations between the different integration strategies and tactics? Is relational integration dependent upon formal integration, or does it enable formal integration? Are certain integration mechanisms 'multi-modal', i.e., do they facilitate more than one integration strategy?

3.   The alignment of the IT Governance architecture with IT value drivers and the enterprise architecture, e.g., how do organizations evolve and maintain such a complex and dynamic alignment? Are there different paths to aligning the IT Governance architecture?

4.   The application of the IT Governance diagnostic diamond, e.g., what are the experiences of (other) organizations applying the IT Governance diagnostic diamond? What actions does an organization take after diagnosing IT Governance effectiveness?

5.   The emerging IT Governance paradigm, e.g., how can the new principles be operationalized for research and practice? Is this emerging paradigm typical for contemporary firms in Europe, or are these principles also emerging in other continents? And what about the 'dot.coms'?

These five directions provide ample opportunities to expand our knowledge on strategies for IT Governance, and will enable us — executives and researchers — to improve IT Governance effectiveness.

REFERENCES

Ahituv, N., Neumann, S., & Zviran, M. (1989). Factors affecting the policy for distributing computing resources. *MIS Quarterly, 13*(4), 389-401.

Alter, A. (2001). Thinking out loud: Interview with JoAnn Heissen. *CIO INSIGHT*, (December).

Ashby, W. R. (1956). *Introduction to cybernetics.* New York: John Wiley & Sons.

The Blind Men and the Elephant (1963). McGraw-Hill Book Company.

Boynton, A. C. (1993). Achieving dynamic stability through information technology. *California Management Review, 35*(2), 58-77.

Brown, C. V., & Magill, S. L. (1994). Alignment of the IS function with the enterprise: Toward a model of antecedents. *MIS Quarterly, 18*(4), 371-403.

Brown, C. V., & Magill, S. L. (1998). Reconceptualizing the context-design issue for the information systems function. *Organization Science, 9*(2), 176-194.

Buenger, V., Daft, R. L., Conlon, E. J., & Austin, J. (1996). Competing values in organizations: Contextual influences and structural consequences. *Organization Science, 7*(5), 557-576.

Burt, R. S. (1980). Models of network structure. *Annual Review of Sociology, 6,* 79-141.

Clark, C., Cavanaugh, N., Brown, C. V., & Sambamurthy, V. (1997). Building change-readiness IT capabilities: Insights from the Bell Atlantic experience. *MIS Quarterly, 21*(4), 425-455.

Clark, T. D. (1992). Corporate systems management: An overview and research perspective. *Communications of the ACM, 35*(2), 61-75.

Collis, D. J., & Montgomery, C. A. (1995). Competing on resources: Strategy in the 1990s. *Harvard Business Review, 73*(4), 118-128.

Cross, J., Earl, M., & Sampler, J. L. (1997). Transformation of the IT function at British Petroleum. *MIS Quarterly, 21*(4), 401-423.

Cyert, R. M., & March, J. G. (1963). *A behavioral theory of the firm.* Englewood Cliffs, NJ: Prentice Hall.

Daft, R. L. (1998). *Organization theory and design.* (6th ed.). Cincinnati, OH: South Western College Publishing.

D'Aveni, R. A. (1994). *Hypercompetition: Managing the dynamics of strategic maneuvering.* New York [etc.]: Free Press.

D'Aveni, R. A. (1999). Strategic supremacy through disruption and dominance. *Sloan Management Review, 40*(3), 127-135.

Earl, M. (2000). Local lessons for global businesses. In D.A. Marchand, T.H. Davenport, & T. Dickson, T. (Eds.), *Mastering Information Management* (pp. 86-91). London: Financial Times/Prentice Hall.

Ein-Dor, P., & Segev, E. (1982). Organizational context and MIS structure: Some empirical evidence. *MIS Quarterly, 6*(3), 55-68.

El Sawy, O., Malhotra, A., Gosain, S., & Young, K. (1999). IT-intensive value innovation in the electronic economy: Insights from Marshall industries. *MIS Quarterly, 23*(3), 305-335.

Feeny, D. F., & Willcocks, L. (1998). Core IS capabilities for exploiting information technology. *Sloan Management Review, 39*(3), 9-21.

Feeny, D. F., Edwards, B. R., & Earl, M. (1989). *Complex organizations and the information systems function.* Oxford Institute for Information Management, Oxford, RDP 87/7.

Galbraith, J. R. (1973). *Designing complex organizations.* MA : Addison-Wesley.

Galbraith, J. R. (1974). Organization design: An information processing view. *Interfaces, 4*(3), 28-39.

Galbraith, J. R. (1987). Organizational design. In J. W. Lorsch (Ed.), *Handbook of Organizational Behavior.* Englewood Cliffs: Prentice Hall.

Galbraith, J. R. (1994). *Competing with flexible lateral organizations.* MA: Addison-Wesley.

Galbraith, J. R., & Lawler, E. E. (1993). *Organizing the future: The new logic of managing complex organizations.* San Francisco, CA: Jossey-Bass.

Henderson, J. C. (1990). Plugging into strategic partnerships: The critical IS connection. *Sloan Management Review, 31*(3), 7-18.

Hitt, M. A., Keats, B. W., & DeMarie, S. M. (1998). Navigating the new competitive landscape: Building strategic flexibility and competitive advantage in the 21st century. *Academy of Management Executive, 12*(4), 22-42.

Hodgkinson, S. L. (1996). The role of the corporate IT function in the federal IT organization. In M. Earl (Ed.), *Information Management: The Organizational Dimension* (pp. 247-268). Oxford: Oxford University Press.

IT Governance Institute. (2001). *Board briefing on IT Governance.* Available online: http://www.itgovernance.org.

Kahn, K. B., & McDonough, F. (1997). An empirical study of the relationships among co-location, integration, performance, and satisfaction. *Journal of Production Innovation Management, 14,* 161-178.

Kaplan, R. S., & Norton, D. P. (1992). The balanced scorecard: Measures that drive performance. *Harvard Business Review, 70*(1), 71-79.

Kaplan, R. S., & Norton, D. P. (1996). Using the balanced scorecard as a strategic management system. *Harvard Business Review, 74*(1), 75-85.

Khandwalla, P. N. (1976). The techno-economic ecology of corporate strategy. *Journal of Management Studies, 13*(1), 62-75.

King, J. L. (1983). Centralized versus decentralized computing: Organizational considerations and management options. *Computing Surveys, 15*(4), 319-349.

Lawrence, P.R., & Lorsch, J.W. (1967). Differentiation and integration in complex organizations. *Administrative Science Quarterly, 12*(1), 1-47.

Lawrence, P.R., & Lorsch, J.W. (1969). *Developing organizations: Diagnosis and action.* Addison-Wesley.

Lorsch, J. W., & Lawrence, P. R. (1970). *Studies in organization design.* Georgetown, Ontario: Irwin-Dorsey Limited.

Luftman, J., & Brier, T. (1999). Achieving and sustaining business-IT alignment. *California Management Review, 42*(1), 109-122.

Malone, T. W., & Crowston, K. (1994). The interdisciplinary study of coordination. *ACM Computing Surveys, 26*(1), 87-118.

March, J. G. (1991). Exploration and exploitation of organizational learning. *Organization Science, 2,* 71-87.

March, J. G., & Simon, H. A. (1958). *Organizations.* (2nd ed.). John Wiley & Sons.

Marchand, D. A. (2000). Hard IM choices for senior managers. In D. A. Marchand, T. H. Davenport, & T. Dickson (Eds.), *Mastering Information Management* (pp. 295-300). London: Financial Times/Prentice Hall.

Marchand, D. A. (2000). How to keep up with the hypercompetition. In D. A. Marchand, T. H. Davenport, & T. Dickson (Eds.), *Mastering Information Management* (pp. 120-126). London: Financial Times/Prentice Hall.

Miller, D. (1988). Relating Porter's business strategies to environment and structure: Analysis and performance implications. *Academy of Management Journal, 31*(2), 280-308.

Miller, D., & Friesen, P. H. (1978). Archetypes of strategy formulation. *Management Science, 24*(9), 921-933.

Mintzberg, H. (1979). *The structuring of organizations.* Englewood Cliffs: Prentice Hall.

Nadler, D., & Tushman, M. L. (1998). *Competing by design: The power of organizational architecture.* Oxford: Oxford University Press.

Parker, M., Benson, R., & Trainor, H. (1988). *Information economics: Linking business performance to information technology.* Englewood Cliffs, NJ: Prentice Hall.

Parker, M., Trainor, H., & Benson, R. (1989). *Information strategy and economics: Linking business performance to information technology.* Englewood Cliffs, NJ: Prentice Hall.

Parker, M. M., & Benson, R.J. (1989). Enterprisewide information management: State-of-the-art strategic planning. *Journal of Information Systems Management,* (Summer), 14-23.

Parker, M. M., Ribbers, P. M., & Parker, S. R. (1997). *Critical IT issues for business transformation.* Handbook BIK, D 1005-1, Tilburg University.

Peppard, J., & Ward, S. (1999). Mind the gap: Diagnosing the relationship between the IT organization and the rest of the business. *Journal of Strategic Information Systems, 8,* 29-60.

Peterson, R., & Ribbers, P. M. A (2002). *Service Level Agreements: Noodzakelijk en toereikend?* Available online: http://www.sbit.nl/ego/archief.php.

Peterson, R .R. (1998). Successful exploitation of Information Technology in financial services. The role of IT Governance. *Proceedings of the International Conference on Information Systems (Doctoral Consortium), Helsinki, Finland* (December 13-16, 1998).

Peterson, R. R. (2000). Emerging capabilities for IT Governance: Exploring stakeholder perspectives in Financial Services. *Conference Proceedings European Conference on Information Systems 2000, Vienna, Austria.*

Peterson, R. R. (2000). Examining ambidextrous designs for information governance: Evidence from a longitudinal comparative case study in Financial Services. *Conference proceedings of the European Conference on IT Evaluation 2000, Dublin, Ireland.*

Peterson, R. R. (2001). Configurations and coordination for global information governance: Complex designs in a transnational European context. *Proceedings of the 34th HICSS Conference, Hawaii.*

Peterson, R. R. (2001). *Information Governance: An empirical investigation into the differentiation and integration of strategic decision-making for IT.* The Netherlands: Tilburg University.

Peterson, R. R., O'Callaghan, R., & Ribbers, P. M. A. (2000). Information Technology governance by design. *Conference proceedings of the International Conference on Information Systems 2000, Brisbane, Australia.*

Peterson, R. R., Smits, M., & Spanjers, R. (2000). Exploring IT-enabled networked organizations in health care: Emerging practices and phases of development. *Conference Proceedings the European Conference on Information Systems 2000, Vienna, Austria.*

Porter, M. E. (1980). *Competitive advantage.* Harvard Business Press.

Powell, T. C. (1992). Organizational alignment as competitive advantage. *Strategic Management Journal, 13,* 119-134.

Quinn, R. E., & Rohrbaugh, J. (1983). A spatial model of effectiveness criteria: Towards a competing values approach to organizational analysis. *Management Science, 29*(3), 363-377.

Quinn, R. E., DeGraff, J., & Thakor, A. (2000). Creating shareholder value — and dispelling some myths. In T. Dickson (Ed.), *Mastering Strategy*. London: Pearson Education Limited.

Reich, B. H., & Benbasat, I. (1996). Measuring the linkage between business and information technology objectives. *MIS Quarterly, 20*(1), 55-81.

Remenyi, D. (2000). The elusive nature of delivering benefits from IT investment. *Electronic Journal of Information Systems Evaluation, 3*(1). Available online: http://www.iteva.rug.nl/ejise/index.html.

Robbins, S. P. (1994). *Essentials of organizational behavior.* London: Prentice Hall International.

Rockart, J. F. (1988). The line takes the leadership: IS management in a wired society. *Sloan Management Review, 29*(4), 57-64.

Rockart, J. F., Earl, M., & Ross, J. W. (1996). Eight imperatives for the new IT organization. *Sloan Management Review, 38*(1), 43-55.

Ross, J. W., Beath, C. M., & Goodhue, D. L. (1996). Develop long-term competitiveness through IT assets. *Sloan Management Review, 38*(1), 31-42.

Sambamurthy, V., & Zmud, R. W. (1999). Arrangements for information technology governance: A theory of multiple contingencies. *MIS Quarterly, 23*(2), 261-290.

Sambamurthy, V., & Zmud, R. W. (2000). Research commentary. The organizing logic for an enterprise's IT activities in the digital era: A prognosis of practice and a call for research. *Information Systems Research, 11*(2), 105-114.

Sauer, C., & Yetton, P. W. (1997). The right stuff: An introduction to new thinking about IT management. In C. Sauer, P. W. Yetton, & Associates (Eds.), *Steps to the Future: Fresh Thinking on the Management of IT-Based Organizational Transformation* (pp. 1-25). San Francisco, CA: Jossey-Bass.

Scheier, R. (2001). Central intelligence: Johnson & Johnson case study. *CIO INSIGHT*, (December).

Senge, P. M. (1990). *The fifth discipline.* New York: DoubleDay.

Simon, H. A., & Barnard, C. I. (1961). *Administrative behavior: A study of decision-making processes in administrative organization.* New York: The Macmillan Company.

Simson, E. M. v. (1990). The centrally decentralized IS organization. *Harvard Business Review, 68*(4), 158-160.

Simson, E. M. v. (1995). The recentralization of IT. *Computerworld, 1*(11), 2-7.

Tavakolian, H. (1989). Linking the information technology structure with organizational competitive strategy: A survey. *MIS Quarterly, 13*(3), 309-317.

Teo, T. S. H., & King, W. R. (1999). An empirical study of the impacts of integrating business planning and information systems planning. *European Journal of Information Systems, 8,* 200-210.

Thompson, J. (1967). *Organizations in action: Social sciences bases of administrative theory.* New York: McGraw-Hill.

Treacy, M., & Wiersema, F. (1993). Customer intimacy and other value disciplines. *Harvard Business Review, 71*(1), 84-93.

Treacy, M., & Wiersema, F. (1995). *The discipline of market leaders.* MA: Addison-Wesley.

Van der Heijden, K. (1996). *Scenarios. The art of strategic conversation.* New York: John Wiley & Sons.

Van Grembergen, W. (2002). Introduction to the Minitrack: IT Governance and its mechanisms. *Proceedings of the 35th Hawaii International Conference on System Sciences (HICSS), IEEE.*

Van Grembergen, W., & Saull, R. (2001). Information Technology governance through the balanced scorecard. *Proceedings of the 34th Hawaii International Conference on System Sciences (HICSS), IEEE.*

Van Grembergen, W., & Van Bruggen, R. (1997). Measuring and improving corporate information technology through the balanced scorecard technique. *Proceedings of the 4th European Conference on the Evaluation of Information Technology, Delft,* (pp. 163-171).

Venkatraman, N. (2000). Five steps to a dot-com strategy: How to find your footing on the web. *Sloan Management Review, 41*(3), 15-28.

Vitale, M. (2001). *The dot.com legacy: Governing IT on internet time.* Working Paper. Information Systems Research Center, University of Houston, Houston, Texas, USA.

Von Bertalanffy, L. (1968). *General System Theory: Foundations, development, applications.* New York: George Braziller.

Weick, K. E., & Roberts, K. H. (1993). Collective mind in organizations: Heedful interrelating on flight decks. *Academy of Science Quarterly, 38,* 357-381.

Weill, P., & Broadbent, M. (1998). *Leveraging the new infrastructure: How market leaders capitalize on information technology.* Boston, MA: Harvard Business School Press.

Weill, P., & Vitale, M. (2001). *Place to space. Migrating to eBusiness models.* Boston, MA: Harvard Business School Press.

Weill, P., & Vitale, M. (2002). What IT infrastructural capabilities are needed to implement e-business models. *MIS Quarterly Executive, 1*(1), 17-34.

Weill, P., & Woodam, R. (2002). *Don't just lead, govern: Implementing effective IT Governance.* CISR Working Paper No. 326. Sloan School of Management, Cambridge, Massachusetts, USA.

Whetherbe, J. (2001). *Achieving the high-performance, information-based networked organization.* Working Paper, Information Systems Research Center, University of Houston, Houston, Texas, USA.

Willcocks, L. (1996). *Investing in information systems: Evaluation and management.* London, Chapman & Hall.

Willcocks, L. P., Feeny, D. F., & Islei, G. (1997). *Managing IT as a strategic resource.* London: McGraw-Hill Companies.

Zachman, J. A. (1987). A framework for information systems architecture. *IBM Systems Journal, 26*(3), 276-292.

Zmud, R. W., Boynton, A. C., & Jacobs, G. C. (1986). The information economy: A new perspective for effective information systems management. *Database,* (Fall), 17-23.

ENDNOTES

1   Two exceptions and noteworthy initiatives undertaken to address the development of a cumulative body of knowledge and skills are the organization of a special 'IT Governance research track' at the 35th Hawaii International Conference on System Sciences (2002), and the IT Governance Institute™, established by the Information Systems Audit and Control Association® ISACA™ in 1998.

2   Theoretically, there are eight distinct federal patterns in IT Governance (excluding the completely centralized and the fully decentralized IT Governance model).

3   IT application decisions address applications prioritization and planning, budgeting, and the delivery of application services.

4   IT development involves blending knowledge of business processes and functions with IT infrastructure capabilities along the complete IT systems development life cycle.

5   In a federal IT Governance model, enterprise IT infrastructure decisions are always allocated to a central corporate IT office.
IT infrastructure decisions address the hardware/software platforms, network and data architectures, and the standards for procurement and deployment of IT resources.

6   The systems view (Ashby, 1956; Von Bertalanffy, 1968) is based on cybernetic principles (Wiener, 1956), which draws upon the Greek word *Kubernesis* — steersmanship, the task of keeping a ship on its course in the midst of unexpected changing circumstances.

7   Organization Design Economics is an interdisciplinary field of study based on the work of, e.g., March & Simon, 1958; Cyert & March, 1963; Lawrence & Lorsch, 1967, 1969; Thompson, 1967; Galbraith, 1973, 1994; Daft & Lengel, 1984; 1986; Williamson, 1996; Malone & Crowston, 1994; Hitt et al., 1998; Nadler & Tushman, 1998.

Chapter III

An Emerging Strategy for E-Business IT Governance

Nandish V. Patel
Brunel University, UK

ABSTRACT

A critical aspect of global e-business information technology (IT) governance is ensuring that it is integrated and that it enables economic viability of a company. Poorly thought through purposes will result in poor IT Governance. The aim is to improve IT Governance and business efficiency and effectiveness. A framework for global e-business IT Governance is developed. It is based on fundamental re-directions in global e-business IT Governance thinking and it applies to companies that seek to integrate Internet, intranet and WWW technologies into their business activities in some form of an e-business model. Such integration is termed the fusion of IT and business into an e-business. The framework explains and elaborates e-business strategies for coping with emergent organisations and planned aspects of IT. The basic premise of the proposed framework is that organisation, especially virtual organisation, is both planned and emergent, diverging from the dominant premise of central control in IT Governance.

INTRODUCTION

In essence, e-business information technology (IT) governance addresses how to design and implement effective organisations by creating flexible IT and information systems (IS) structures and processes. IT Governance in a global context has to cater for intensive competition, cultural diversity, and various fluctuating economic conditions. A static model of IT Governance and organisation cannot adequately address these issues.

The prime aim of IT Governance is to contribute to business activity in terms of lower costs, satisfied customers and better quality products or service provided by a company. Governance assumes accountability, making improving the channels of accountability an important feature of IT Governance, especially accounting for return on investment. Many problems need to be addressed by the IT function: weak planning, rapid business and environmental change, and management involvement are some. The emergent process of IT Governance reveals that managers need to understand that they are neither all-powerful nor powerless to effect change. Rather, they are in partial control of emerging processes that result in new organisational designs. They need to consider the importance of global business management, cultural diversity, ethics and advanced production and information technologies as the boundaries between the Internet and customer strategy continue to merge. Some fundamental re-directions in e-business IT Governance strategy thinking are considered and a framework for global e-business IT Governance and organisational design as both a planned and an emergent process is proposed.

Management strategies are concerned with reaching a specific destination, and in particular with how to reach the destination. Company strategies are unique and difficult to differentiate from a specific company's values, goals and mission. Organisations cannot expect to extrapolate or borrow a strategy from another company. What works strategically for one company may not have the same impact on another organisation. Similarly, e-business IT Governance is affected by an organisation's unique culture and working practices, and should reflect its own goals and ambitions. The proposed framework is not a prescriptive IT Governance package that can be replicated across all organisations or even for all time in a particular organisation. Its purpose is to enable decision-makers to take a holistic and alternative view of IT Governance and to enable them to find their own appropriate mechanisms for devising an IT Governance strategy that fits their particular organisation. This approach is based on the increasing literature on emergent organisations and its corresponding effect on IS development and IT Governance (Pawson et al., 1995; Truex et al., 1999; Patel, 1999). Some authors state that IS development in IT Governance is possible without formal methods (Baskerville, 1992). The proposed framework for global e-business IT Governance supposes that the problem is one of recognising and accommodating emergent activity rather than focusing purely on planned rational governance.

The chapter is organised around the central problem that addresses global e-business IT Governance as combined planned governance with emergent needs. Whilst planning is a vital aspect of IT Governance, the pace of economic change nationally and internationally quickly makes plans outdated. Business needs for IT and IS tend to emerge as a result of organisational and economic factors; thus e-business models, as discussed in the following section, need to encompass emergent activity. The business rationale for e-business requires a broader scope for IT Governance, taking into account both IT and business issues. The e-business IT Governance framework itself is built on radical re-directions from traditional IT Governance, discussed in the section on radical re-directions. The framework section details activities that need to be continuously carried out to ensure plans are relevant to business needs and account for emergent needs. A critical aspect of e-business IT is the development of organisational interfaces, which traditional IT Governance has not had to deal with. These interfaces, for example between customer and organisation or business partner and organisation, are vital for the success of e-business IT Governance. The conclusion reached is that global e-business IT Governance should be regarded both as a systematic and organic approach to IT resource management.

E-BUSINESS MODELS

A distinction between e-commerce and e-business is necessary to appreciate the magnitude of change in the business and operational activities that e-business has created for IT Governance. It may be argued that e-commerce is the use of IT to support business activity. E-commerce equals business plus technology that is limited in scope to transaction-based information flows; for example, EDI. E-business, in contrast, is the complex fusion of IT and business activity that necessitates the governance of IT to consider for the first time the economic aspects of the business. The very economic survival of an e-business company rests on the efficacy of IT and the successful integration of internal and external business processes. E-business equals business plus technology, plus economics, as it brings about a new facet of the economy, namely the e-business economy. The new economics of e-business cover, for example, supply chain management, customer relationship management, and human resources. This new link of IT with economics and the need for business organisations to be agile means that rather than a hard-wired e-business strategy (simple planning and implementation) companies require re-wireable business agility (organic IT Governance and flexible systems) (Allen & Boynton, 1990). This is possible with networked organisations that exist as virtual structures, but only if the corresponding IT Governance strategy is equally flexible.

Global e-business IT Governance cannot be accomplished with the traditional models of aligning IT strategy with business strategy. Earl (1999) succinctly sums up the evolution of IT strategy making by formulating three problems. The first is the perennial problem of aligning business strategy with IT strategy. The second is the periodic problem of securing business opportunity that IT strategy may pose. The third is the paradigm problem of integration, which he calls 'information business strategy'. The new e-business models as defined by Timmers (1999) address the third problem and are a fusion of IT and business activity that make economic viability of a company the central concern of IT Governance. The fundamental difference between the new e-business models and the traditional view of aligning IT support or even transforming business (Venkatraman, 1991) is that in the new fused e-businesses, IT is integrated into business activity. Integration is such that the boundary between pure business activity and pure

IT support is nonexistent. It is thus described as a fusion of business activity and IT that has resulted in unprecedented organisation structures that themselves require a radically different governance and which call for a radical rethinking in IT Governance. The question of managing such novel organisation structures is the single most critical issue that strategists and organisational theorists have to resolve.

An important aspect of emergence is business networks or alliances. Models of business competition that stress efficient use of resources by a firm to differentiate itself from its rivals have given way to models of co-option (Henderson, 1990). Short and Venkatraman (1992) discuss business networks and their value to companies. They show how both business processes and business networks need to be redesigned in co-option models. Consequently, computer networks, especially the Internet and the World Wide Web, have become central to IT Governance. Companies use intranets to support work and distribute information, and commercial-off-the-shelf packages like Lotus Notes need to be managed to avoid problems and cease the opportunities they afford to manage knowledge, for example.

Global IT Governance and IS design in these new organisation structures are inextricably tied with scaling. Scaling is the ability to support larger organisation structures. During the industrial revolution, the need for organisation meant that the craftsman's skills were split into design and subsequent production to enable scaling. Specialists in design and planning addressed the problem of organisation (Groth, 1999). Organisation itself becomes a form of problem solving. Global e-business IT Governance in essence is about scaling. E-business solutions implementations require scale and a global perspective developed through a careful analysis of business rationale. For global companies, IT Governance is concerned with design and planning the application of IT organisation-wide. And yet, it is also concerned with meeting variable and local needs of subsidiary companies and divisions.

We have come a full circle in the 21st century and returned to the reincarnated craftsman — the knowledge worker — in the e-business organisation. Unlike the single craftsman with his craftshop, the knowledge worker has to work in a team and across a distributed organisation. Like the craftsman, the knowledge worker has to specialise, often to very high levels of granularity, but unlike the craftsman, the knowledge worker has to communicate and coordinate within a team and across an organisation, increasingly now across a virtual organisation in many global companies. The medium and mode of this communication is now largely IS, making its design a critical aspect of organisation design and IT Governance itself, and in improvements in effectiveness and internal productivity gains.

Emergent and predicted information and knowledge is communicated and shared in organisations. Available technologies that reflect emergence have given rise to alternative organisational structures (Berners-Lee, 1999) largely based on intranets and extranets, different ways of supporting business processes, and novel ways of working. The new e-business organisational forms are based on information and knowledge assets and seek to facilitate knowledge creation. Deferred system's design (DSD) can be an integral aspect of these new organisations (Patel, 1999), because it seeks to design tools that enable organising virtually and defers IS design decisions to employees to mitigate risk. DSD has the potential to address the problem of emergent information needs as part of balanced IT Governance, balanced between planned activity and emergent needs.

E-BUSINESS IT GOVERNANCE AND BUSINESS RATIONAL

Governance is a multifaceted activity requiring the efficient and effective uses of resources to achieve desired aims. In e-business IT Governance it is the ability to manage IT, develop strategies, and create systems that are relevant to business operations and customers who interface with an organisation. IT Governance involves building a professional IT capability that is able to offer a business strategic advantages. The professional IT executive needs to work closely with business executives to determine how IT can add value. The value contribution of IT can be determined by considering facets of global e-business IT Governance such as:

- Develop an IT strategy, and undertake critical strategic and operational reviews. Strategy formulation requires an imagination to use IT capability to build better relationships with partners, customers and employees.

- Develop and manage the distributed IT/IS systems, e-crm and e-technology infrastructures.

- Ensure that business-critical projects are completed.

- Define methods, tools, and processes.

- Define best practices.

- Manage application development.

- Manage outsourced providers and multi-site procurement policies.

- Ensure effective IT services delivery strategy to business segments that lead to internal productivity gains.

- Develop key performance indicators.

- Critically review current organisation structures and capability and implement cost savings to improve efficiency and effectiveness.

Underlying all the above activities is the aim of meeting operating needs of a company. Any IT Governance mechanism should be rooted in business logic. For global companies with e-business aspirations, three segments of business need to be considered: marketing, human factors, and business-to-business relations. In terms of marketing, a company needs to consider how its e-business strategy supports its overall mission and communications objectives. It needs to develop a one-to-one marketing strategy over the Internet for customers and the extranet business partners. It needs to determine how to relate digitally with its customers. In terms of human factors, a company needs to assess how its customers will respond to digitised interaction. E-crm strategies need to be customer-focused and, as explained in the following section, appropriate customer-organisation interaction models need to be developed. This may require developing easy-to-use interfaces for customers who are simply interested in purchasing items or services. Finally, in terms of business-to-business, a company will need to assess how to develop the interaction between itself and its business partners and suppliers.

Most e-business models tend to overlook the customer as an integral aspect of an e-business. E-business IT Governance needs to be customer-centric. The customer is regarded as an operational aspect of e-business in the framework presented later. No physical boundaries exist between a business and its partners, suppliers or investors, or between a business and its customers in an e-business. Business processes that deliver a product or service now extend virtually to the customer. Dell, the personal computer manufacturer, produces customised products through its corporate portal, linking its operational process directly to the customer. Thus both suppliers and business partners, and critically, customers, now become operational issues in e-business enterprises and e-business IT Governance. Business processes that link directly to customer requirements mean that IT Governance too needs to consider the company's customer in its systems development approaches and strategies. Amazon.com and Yahoo! are examples of companies that operate beyond notions of business transformation; they are truly networked organisations that are superimposed on transient physical and organisational structures. The role of IT Governance in such organisations is beyond the simple management of the IT tool. It involves ensuring the very economic viability of a company.

Some Radical Re-Directions in E-Business IT Governance

The e-business IT Governance framework elaborated in the following section is built on radical re-directions from traditional IT Governance. E-business is the integration of economic, business and technology aspects of business activity. The scope of e-business IT Governance is not now simply inward IT management but outward relations, covering business partners, suppliers, and critically, customers. Traditional IT Governance focused on the technology and its application to business operations, whereas e-business IT Governance is intertwined with business and economic management, with suppliers, business partners and customers. In e-business, IT Governance has thus moved onto a different plane, requiring fundamental re-directions discussed below that need to be considered for effective global e-business IT Governance.

Traditional IT Governance's modus operandi is planning. In global e-business IT Governance it is necessary to consider both planned e-business IT and emergent requirements. Modern organisations cannot be viewed solely as planned and directed entities. Organisational life is about 'being in the process' and not only about definable structures, especially when considering the virtuality of organisation structures. There is evidence that organisation structure is dynamic. In terms of IS development, research reveals that developers need to consider the emergent information and knowledge needs of the organisation (Baskerville, Travis, & Truex, 1992) in such organic structures. Similarly, strategies should be free to appear at any time and in any place in the organisation. There is a 'messy process of informal learning' through which strategies may be formulated. Planning itself needs to be of the rolling wave kind to cater for uncertainty and, possibly, contractual work in systems development.

IS development needs to be re-scoped to include customers, business partners, and suppliers. For e-business, IS development is not simply an 'internal' problem as in traditional IT Governance. In e-business it extends outside the organisation to include business partners and suppliers, but most critically it needs to include customers. Pure e-business organisation is directly linked to its customers through the Internet. Its business processes and operations are driven by this direct interface. As the interface is enabled by IT, its development and the development of associated systems needs to involve all interfaces. Thus the very problem of systems development extends outside the organisation.

Consequently, e-business IT Governance is about developing new interfaces to fundamentally change the way in which an organisation interacts with its customers, partners and suppliers. The new interfaces are between:

- Customer — organisation
- Partner — organisation
- Supplier — organisation

These interfaces are vital for the viability of a company and pose a new problem for global IT Governance. The problem is how to design efficient business processes that extend to interfaces as well as the interface itself. In some virtual organisation forms the customer is a co-producer of the goods or services; for example, where the buyer of cars or personal computers can customise the requirement for a product online. For the customer-organisation interface, one aspect of the problem is how to design interfaces that cater for cultural diversity to be found globally. These interfaces cover both process issues and its fused IT. The customer-organisation interface should be monitored to extract vital business intelligence from customers.

There are various reasons why all systems requirements cannot be known in advance to facilitate detailed IT plans and development. The users may not know what is required, or if they do they may not be able to explain or express the problem in terms that are readily understandable and can be modelled. Therefore global e-business IT Governance needs to develop local information and Knowledge Management tools. Global businesses will need to devise and implement varying marketing strategies for local needs. Web-based marketing systems require incorporating customising or tailoring tools to allow different product promotions or application tailoring (Wolfgang et al., 1998).

Historically, the level of sophistication of tools in a society reflects its intelligent activity. It is not possible to achieve an objective without some kind of tools or devised method. A tool is a 'wholly constructed expression of both knowledge and values' (Groth, 1999). Interestingly, there has been a paucity of tools in IS given its pervasiveness in organisations and, during the last decade, in society generally. E-business tools contribute to organisation structure, its effectiveness and efficiency. Tool building that facilitates the collective experiences of individuals leads to the design of better and effective tools, as it leads to the design of sophisticated and precise tools that solve the problem at hand.

Traditional technology has not had an all-encompassing effect on organisation structure and communication. Traditional IT Governance has not had to deal with questions of organisation structure, except with the notion of business transformation. E-business IT Governance by necessity has to consider the all-encompassing effect that the new networking digital technologies have on organisation. Internet and web technologies enable organising virtually. Policies need to be developed to enable organising virtually, as well as:

- Developing and enable virtual structures, which by definition will change;
- Ensuring economic viability, not simple business 'fit';
- Developing solutions that are valid at corporate and business unit levels.

E-business IT Governance is more complex than the traditional alignment of IT with business or deriving business opportunity from IT. It is about integrating IT into the very business, referred to here as fusing IT with business. An e-business should be regarded as an open-ended organisational network. The notion of open systems (Flood & Jackson, 1991) may be one way of conceptualising such an entity. Another way to think about open-ended organisational networks is as 'webs' (Patel, 2001a). The empirically founded web concept is proposed as a conceptual tool to develop applications better suited for business organisations dealing in information and knowledge with emergent needs. It is consistent with the major content of e-business technology, namely information and knowledge processing, and with the plank of information and knowledge ontology within the proposed framework.

Develop ontologies of information and knowledge that are not simple data/information processing algorithms. A significant aspect of e-business IT Governance that is different from traditional IT Governance concerns business intelligence and models of customers. E-business solutions require intricate models of customer behaviour. The various applications need to be integrated to provide a unified view of customers.

Two other radical considerations are cross-organisational IS development teams and reconceptualising time and space in a virtual e-business organisation. Lee (1999) describes temporal changes of export related work in companies using EDI and how IS create temporal symmetry. International businesses have given rise to global and virtual software development teams. These teams are composed of North American and European corporations and companies from the Indian subcontinent. The management of virtual software development teams is a new challenge for e-business IT Governance.

A FRAMEWORK FOR PLANNED AND EMERGENT IT GOVERNANCE

The proposed framework incorporates the above points and recognises that the organisational changes being brought about by e-business technology are profound in that they have radically altered our centuries-old views of companies. For the first time in the history of computing and its application to business, the customer is a vital consideration in e-business strategy and planning. The digital link into homes now makes the customer a critical aspect of business operation planning and management, which necessarily now includes IT planning and management. As discussed in the previous section, the inclusion of the customer into business operations requires developing models of customers' interaction thorough electronic interfaces with the organisation.

IT Governance consists of designing the governance structures and then implementing and managing them. Critically, it also involves being open to unexpected requirements or emergent information or system needs. This is illustrated in the case presented in the section titled "Planned and Emergent IT Governance in Action". In its broader sense, governance concerns the development of IT in an organisation, its procurement, and its application to business activity. But, as discussed above, e-business IT Governance is more than the simple application of IT to business activity. It is the complex fusion of IT with business activity, business partners, suppliers and customers, and critically it is equally as important as corporate governance.

What model of IT Governance is effective in this environment: planned use of IT, Just do it (JDI) or the emergence model? Planned use of IT is necessary for the known aspects of organisational life, for example, a known merger of companies, and caters for corporate, interdepartmental and company-wide systems. JDI is the empowerment of users to develop systems, made possible with web technologies and caters for local, individual and group needs. The emergent aspects arise in the context of the previous two, and take the form of the unexpected from the point of view of IT planners. Whilst strategic planning can be systematized and governance structures elaborated, their actual implementation is done in real organisational contexts where the unplanned event will necessarily emerge or where the poverty of the plan itself will become evident. The framework contains all three elements, especially given the recent literature on emergent organisations discussed earlier. Emergence and JDI cannot prevail because there will always be a need for corporate level application development, which the IT department can develop.

The essential activities for global e-business IT Governance in the framework encompasses the radical redirections discussed earlier. The inclusion of the customer (3) in IS development is critical in e-businesses (numbers in brackets refer to the list below). In particular, empirical models of customer-organisation interaction that cover process and operational issues need to be developed. The issue of planned versus emergent IT Governance is covered (1,2,6,7). E-business is highly volatile, both technologically and economically. Such a dynamic environment cannot be catered for in e-business IT Governance solely through planned achievement of goals; it also requires mechanisms to respond to unforeseen, opportunistic and emergent events. The essential activities for global e-business IT Governance are:

1. Determine business purpose and strategy to define e-business model. Consider which aspects of the business will be digitised and how the IT strategy will add value to the business. Build top level and local channels of governance accountability.
2. Determine the virtual structure of the organisation; allow for emergent forms and virtual working.
3. Develop customer-organisation interface and interaction models. It is critical to understand how customers, business partners and suppliers interact with the e-business through the business process (digital) interfaces. Developing interaction models will provide such an understanding.
4. Develop customer-organisation systems. This requires taking the radical step of involving customers in systems design decisions. In the past, 'users' were excluded from design decisions but were eventually included when developers realised that systems would be more acceptable if users had a say in their design. Similarly, there is a need to include organisational interfaces such as customers in the e-business design process.

5.   Determine what IS concepts to incorporate. An e-business model can incorporate concepts such as information management, Knowledge Management and decision support. This will ensure that integrated solutions are provided.

6.   Determine the scope for IT outsourcing and the role of consultants. Criteria need to be decided for retaining consultants, and adequate contract and servicing details need to be decided for IT service providers.

7.   Determine which business activity, business process, and business relations to digitise - supplier relations, customer relations, and employee relations. A corollary question concerns how to determine what to digitise. Develop plans for legacy and back-end systems integration with e-business solutions. All digital aspects of the business cannot be predetermined. Some will arise in the course of business activities, and will need ongoing development.

8.   Retain flexibility in IT Governance to allow for emergent aspects of e-business strategy. View design as a process, not as a discrete event in time.

9.   Determine technology-centric ways of working. We know that technology is changing the way work is done. IT Governance has to recognise this fact, and include organisational and work-study in its planning process. This is especially the case with e-business technology planning.

10. Evaluate and procure appropriate e-business technology to enable one through six above.

Each of the above activities needs further elaboration. The verbs that describe the activities should be read as in the present continuous tense, for example 'determining business purpose and strategy....' The activities need to be ongoing to enable the capture of emergent events and opportunities. These activities were partially evident in the example case in the section, "Planned and Emergent IT Governance in Action". The company engaged in activities six, seven, nine and 10, which reflect the planned aspects of IT Governance. It did not consider the emergent aspects that eventually defined its IT/IS infrastructure. For example, activity eight requires planners and systems developers to view design as a process, not as a discrete event or project in time. The company had viewed design as an event in time and treated it as a project to be undertaken by a consultant. The approach led to the demise of the project and wasted monetary expenditure. But the 'process' of design continued, as the company continued to introduce IT/IS by learning from its mistakes.

IS Concepts

The development of the IS field, particularly its interdisciplinary development, has resulted in certain concepts that may be regarded as defining modern business activity. These are integral e-business concepts. The interdisciplinary development of IS has resulted in notions of business operation and organisation, that in turn need to be incorporated into the new e-business models. Table 1 is a list with short descriptions of IS concepts that need to be incorporated into e-business models. The realisation of these concepts in business is a major function of IT Governance.

Selecting appropriate technology to incorporate IS concepts is a significant aspect of the framework (Dijkstra, 1968). For example, if a company wants to manage its customer knowledge effectively, it will need intranet database technologies.

Table 1. Relevance of IS Concepts to E-Business IT Governance

IS Concept	Description	Relation to e-business IT Governance
Data	Facts about organisational transactions	Needs to be integrated and available across the intranet or extranet.
Information	Data that is processed to enable decision-making	This information is not limited solely for internal consumption (executives, planners, problem-solvers), but is to be made available to customers to make informed buying decisions, and to suppliers and business partners.
Knowledge	Knowledge is about ideals to achieve customer satisfaction, product quality, etc.	Disseminate it widely throughout the organisation and enable its creation.
Networked or Virtual organizations	IT enabled organising	Networks are the essence of e-business. The form and content of networks determine business viability.

EMERGENT FORMS OF IT GOVERNANCE

The dominant view of systems development is mechanistic. Jackson states: "To develop software is to build a MACHINE, simply by describing it" (Lee, 1999, p. 1). Such a view is true of mechanical devices, but not of social software that supports and augments human social action. Schuler (1994) defines social computing as software that serves as an intermediary or a focus for a social relation. E-business IT Governance is concerned with the design and development of such social software. Social software has a direct influence on human behaviour; particularly the action humans take in social relations, like business relations such as partners, suppliers and customers.

Most IS design activity happens in projects, and some during the enhancement maintenance phase. Traditional IT Governance makes use of projects to develop systems. E-business systems have been developed using the business project framework, as it affords resource management and goal achievement. However, project-based IS design activity does not deal adequately with a number of issues. Emergent requirements, creeping scope, organisational and business impact on the project are all issues that trouble a project. Yet they are the very essence of IS design activity and that needs to be addressed in emergent organisations. Projects restrict software development to professional developers, but there is a need for 'users' to develop systems for local needs (Patel, 2001a).

The 'software crisis' itself can be viewed as a manifestation of postmodernism. We require a new approach that is distributed and sensitive to context. The era of planned releases of systems projects is past, and business organisations have to extend the functionality of systems over short periods and at low costs in response to various organisational and economic competitive factors, especially in an e-business. Attempts at planning completely all systems requirements in projects have proved unachievable with the usual picture of cost overruns and failed delivery times (Ewusi-Mensah, 1997). New alternative models like the distributed model of open source code development are emerging.

It is difficult for e-business IT Governance to remain centrally controlled through mechanisms such as project-based development. Table 2 is a sketched categorisation of

Table 2. Historical Phases of IT Governance and Development Approaches

Systems Conception and Development Method	Systems Rational & Development Focus (Individual, Coordinated, Central or Distributed) (Aspects of IT Governance)	Reference
Programming as an Art	The practice of programming was originally considered an 'art'. Individual.	Circa 1950s
Structured Programming	Emphasis on how software is partitioned and structured. Reflective developers. Individual.	Goto statement, Dijkstra (1968)
Software Engineering	Apply traditional engineering tools to the 'software crisis'. Coordinated.	NATO Science Committee (Oct. 1968)
Project M anagement	Reflective developers in a business project. Central.	Grindley (1986)
Fourth Generation Languages	Action developers. Local.	Grindley (1986)
Packaged Software (COTS)	Reflective developers and action developers. Central.	Brincklin (2000)
Open Source Code	Reflective developers but in a community of practice (not project bound). Distributed.	Halloween Document (2001)

historical approaches to managing systems development. It charts the move from software development as an 'art' through 'engineering' to the present day 'open source code' conception. The business demands on software are complex and methodologico-project frameworks seem unable to cope. Amethodological decentralised models, like the ones used to develop the Internet, web, and Linux seem to be successful. Global e-business IT Governance requires such a distributed and decentralised model for developing e-business systems.

PLANNED AND EMERGENT IT GOVERNANCE IN ACTION

The value of the proposed framework for management action is illustrated briefly with a case in this section. The case is of a contract supplier of toiletries and pharmaceuticals liquids. In five years a strategic vision became blurred and threatened to send the company's finances into a black hole. The introduction of IT and IS was nontrivial as the company discovered its cost. This case describes the unexpected long introduction, over five years, of IT into the company. It demonstrates the complexities and high costs of realising a global strategic information system.

The company wanted to improve its supply chain management and its associated information. Its high overheads, relative to its competitors, led to a decline in its market share. It wanted to use IT to make its supply chain efficient and automate the information associated with the supply chain. The management believed that IT would help them make their supply chain operations into 'real-time'. This was required because of the increasingly complex processes involved in delivering products to customers on time.

The company introduced its strategic IS in 1997. It was to 'compute' complex planning and scheduling scenarios across some 500-plus Shop Keeping Units. Purchasing was struggling to manage the component range; Material Requirements Planning would enable them to control requirements more systematically. It was sourced from a company that already had a wide customer base, which provided the management with confidence to purchase the software. They were also confident in the consultant who was to install the system into the company.

The supply chain information system was installed in 1997 and failed to deliver the benefits expected. The company abandoned it in 1999. They relaunched a tailored version of it in 2002 with the same objectives as in 1997. The company has lost five years of real progress in IT/IS usage. Five years later, the chairman of the company admitted that it had *vastly* underestimated the costs associated with introducing IT/IS and that they are still suffering the consequences today. Fortunately, the company's cost management was robust for it to 'enjoy good margins', which enabled it to survive 'whereas lesser companies may have gone to the wall'.

The company is a successful example of the value that IT/IS could add to its operations and help it gain competitive advantage. What went wrong? Why did its strategic plans fail? The case illustrates that plans alone were not sufficient. Although the company had not deployed formal strategic IT/IS planning tools such as portfolio analysis, it had a clear vision and plan to introduce IT/IS into its operations. As the production manager observed, planning alone could not account for the complex and emerging environment in which the IT/IS was introduced. The initial introduction and subsequent reintroduction is a case of an emerging strategy for e-business IT Governance.

CUSTOMERS-ORGANISATION INTERFACE AND EMERGENT ORGANISATION

The past application of IT has seen the automation, support, and re-design (re-engineering) and transformation of business activities. IT Governance became progressively complex through these phases of IT application to business. The current move is towards new 'models' of e-business. The new e-business models require strategy formulation and careful IT Governance through prescribed methods, for example, IT Balanced Scorecard (Van Grembergen, 2000), but there are also fundamental aspects of e-business that IT Governance needs to consider that prescribed methods cannot cover.

Corporate design, information and knowledge are intertwined. Information and knowledge are a prime element of organisation design, and e-business technology has enabled the complex integration of all three in e-business models such as e-shop, e-mall or third-party marketplace (Timmers, 1999). Organisation theorists assert that information processing and coordination of work tasks are central features of an organisation (Gailbraith, 1977; Minzberg, 1979; Groth, 1999). Following the history of industrial design, the premise in IT Governance and IS development is that computer-based information processing requires central design. The use of IT for information processing makes central IT Governance and designs an invalid proposition in the 21st century organisation.

The various interfaces between a company and its customers, partners and employees need to be both functionally relevant and easy to use. Certain interfaces such as customers cannot be trained to use e-business IT systems. The design of these interfaces is critical to the success of an e-business. The new e-business organisation requires a multidisciplinary team to deliver relevant and effective solutions. Designers, creatives, psychologists and developers all can contribute to the novel e-business systems. IT Governance and design need to be local and in actual-time (when it is required). So modern information processing in organisation requires an amethodological or distributed governance too. In methodological approaches the analysis, design, and implementation of IS solutions to organisational problems are separated and controlled centrally. An amethodological approach proposed by Patel (1999), deferred system's design (DSD), enables organisations to delay design decision making to mitigate risk, and permit procedural, operational or policy problems to be resolved locally. E-business systems in banking incorporate DSD (Theotokis et al., 1997) to allow emergent and tailorable information processing needs to be facilitated locally.

The System's Environment

A system that is not impacted highly by the environment remains constant. Its architecture and functionality remain stable with minor changes because the human or organisational force for change is nil or minimal. It is difficult to find examples of such systems in e-business systems. A system that has a high environmental impact on it, for example a web-based marketing system, needs to constantly change. The forces for change are high and constant on such systems. In general, e-business systems are of the latter type.

The complexity of a customised order processing system such as for personal computers or cars or an electronic bidding system such as for auctions increases with the degree of their embeddedness in the environment. Generally, such e-business systems have a high correlation with the organisational (and economic) environment in which they function. Figure 1 is an organisation environment impact analysis of e-business systems that need to cater for organisational emergence. When the correlation with the environment is high, e-business systems need to be developed using DSD, as shown in the top right quadrant. Over time, environment impact on systems requires most systems to move clockwise from the bottom left quadrant to the top right quadrant.

Emergent ways of IT Governance do not seek to specify fixed systems architecture, so that the future use of IS and its flexibility can be accommodated. Information and

Figure 1. Organisation Environment Impact Analysis

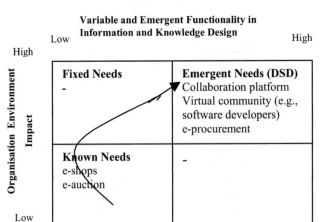

knowledge systems that are built on emergent principles such as DSD would be capable of accommodating the complexity of organisational phenomena and change increasingly evident in an e-business economy.

Information and knowledge ontologies are not an aspect of current IT Governance, especially that concerned with aligning IT with business strategy. This aspect is critical given the inclusion of the customer in the operations of the business. E-business companies will have to develop deeper understandings of their products, customers, and partners through better information and knowledge creation, sharing and analysis in a shifting environment.

CONCLUSION

E-business IT Governance has been conceptualised as encompassing both systematic planned activities and organic emergent needs to ensure successful e-business applications development. E-business models need to cater for emergent requirements and regard suppliers, business partners, and especially customers as integral. Global e-business governance needs a radically different perspective on IS development, organisation interfaces, organisation structures, and ontologies of information and knowledge. The radical alternative theme is to understand the technological, managerial and organisational (including interfaces) influences that both define e-business and its eventual success.

The framework proposes various activities in IT Governance that can cater for the new challenges of global e-business IT Governance. The development of electronic supply chains, networks, and customer-facing practices are vital but the question of measuring the success of these developments has yet to be resolved. It is an interesting research question, especially in organisations that permit emergent activity and appropriate IT responses to it, both corporate and local.

Planning is a vital aspect of IT Governance, but the pace of economic change nationally and internationally quickly makes plans outdated. Business needs for IT and IS tend to emerge as a result of organisational and economic factors; thus e-business models need to encompass emergent activity. The business rationale for e-business requires a broader scope for IT Governance, taking into account both IT and business issues. Organisations need to implement activities that need to be continuously carried out to ensure plans are relevant to business needs and account for emergent needs. A critical aspect of e-business IT is the development of organisational interfaces, which traditional IT Governance has not had to deal with. These interfaces are vital for the success of e-business. Global e-business IT Governance should be regarded both as a systematic and organic approach to IT resource management.

REFERENCES

Allen, B., & Boynton, A (1990). *Restructuring information systems: The high road and the low road*. In A. Boynton & R. Zmud (Eds.), *Management Information Systems: Readings and Cases. A Managerial Perspective* (pp. 316-329). Glenview, IL: Scott, Foresman.

Baskerville, R., Travis, J., & Truex, D. (1992). Systems without methods: The impact of new technologies on information systems development projects. In Kendell et al. (Eds.), *The Impact of Computer Supported Technologies in Information Systems Development* (pp. 241-269). North Holland: Elsevier Science Publishers BV.

Berners-Lee, T. (1999). With Mark Fischetti. *Weaving the Web*. HarperSanFrancisco.

Brincklin, D. (2000). www.brincklin.com.

Dijkstra, E. (1968). GOTO statement considered harmful. *Communications of the ACM, 11*(3), 147-148.

Earl, M. J. (1999). Strategy-making in the Information Age. In W.L. Curry & B. Galliers, B. (Eds.), *Rethinking Management Information Systems*. Oxford University Press.

Ewusi-Mensah, K. (1997). Critical issues in abandoned information systems development projects. *Communications of the ACM, 40*(9), 74-80.

Flood, R. L., & Jackson, M. C. (1991). *Creative problem solving*. Chichester, UK: John Wiley & Sons.

Gailbraith, J. R. (1977). *Organisation design*. Reading, MA: Addison-Wesley.

Grindley. (1986). *Fourth generation languages: Volume 1, A survey of best practice*. London: IDPM Publications.

Groth, L. (1999). *Future organisation design*. Chichester, UK: John Wiley & Sons.

Halloween Document. (2001). www.opensource.org/halloween/halloween1.html.

Henderson, J. C. (1990). Plugging into strategic partnerships: The critical IS connection. *Sloan Management Review, 30*(3), 7-18.

Jackson, M. (1998). *Software requirements & specifications*. Harlow, UK: Addison-Wesley.

Lee, H. (1999). Time and information technology: Monochronicity, polychronicity and temporal symmetry. *European Journal of Information Systems, 8,* 16-26.

Minzberg, H. (1979). *The structure of organisations*. Englewood Cliffs, NJ: Prentice Hall.

NATO Science Committee. (1968). *Conference on software engineering.* Germische, Oct.

Patel, N. V. (1999). The spiral of change model for coping with changing and ongoing requirements. *Requirements Engineering, 4,* 77-84.

Patel, N. V. (2001a). The structure of information and knowledge in a market research company: Systems or webs? *Proceedings of the 9th European Conference on Information Systems*, (June 27-29), Bled, Slovenia.

Patel, N. V. (2001b). Towards a tailorable system architecture for corporate and local information and knowledge management. *Proceedings of the 6th UK Academy for Information Systems Conference,* (April 18-20), University of Portsmouth.

Pawson R., Bravard J-L., & Cameron, L. (1995). The case for expressive systems. *Sloan Management Review,* (Winter), 41-48.

Schuler, D. (1994). Social computing. *Communications of the ACM, 37*(1), 29.

Short, J., & Venkatraman, N (1992). Beyond business process redesign: Redefining Baxter's business network. *Sloan Management Review,* (Autumn), 7-21.

Theotokis, D., Gyftodimos, G., Geogiadis, P., & Philokyprou, G. (1997). Atoma: A component object oriented framework for computer based learning. In G. M. Chapman (Ed.), *Proceedings of the Third International Conference on Computer Based Learning in Science (CBLIS' 97)* (July 4-8 1997, pp. B1(15)), De Montford University, Leicester, UK.

Timmers, P. (1999). *Electronic commerce*. Chichester, UK: John Wiley & Sons.

Truex, D. P., Baskerville, R., & Klein, H (1999). Growing systems in emergent organisations. *Communications of the ACM, 42*(8).

Van Grembergen, W. (2000). The balanced scorecard and IT Governance. *Information Systems Control Journal, 2*, 40-43.

Venkatraman, N. (1991). IT-Induced business reconfiguration. In M. S. Scott-Morton (Ed.), *The Corporation of the 1990s*. Oxford University Press.

Wolfgang, A., Hinrich, E., & Woetzel, G. (1998). Effectiveness and efficiency: The need for tailorable user interface on the Web. *Computer Networks and ISDN Systems, 30*, 499-508.

SECTION II:

PERFORMANCE MANAGEMENT AS IT GOVERNANCE MECHANISM

Chapter IV

Assessing Business-IT Alignment Maturity

Jerry Luftman
Stevens Institute of Technology, USA

ABSTRACT

Strategic alignment focuses on the activities that management performs to achieve cohesive goals across the IT (Information Technology) and other functional organizations (e.g., finance, marketing, H/R, R&D, manufacturing). Therefore, alignment addresses both how IT is in harmony with the business, and how the business should, or could, be in harmony with IT. Alignment evolves into a relationship where the function of IT and other business functions adapt their strategies together. Achieving alignment is evolutionary and dynamic. It requires strong support from senior management, good working relationships, strong leadership, appropriate prioritization, trust, and effective communication, as well as a thorough understanding of the business and technical environments. The strategic alignment maturity assessment provides organizations with a vehicle to evaluate these activities. Knowing the maturity of its strategic choices and alignment practices make it possible for a firm to see where it stands and how it can improve. This chapter discusses an approach for assessing the maturity of the business-IT alignment. Once maturity is understood, an organization can identify opportunities for enhancing the harmonious relationship of business and IT.

INTRODUCTION

Business-IT alignment refers to applying IT in an appropriate and timely way, in harmony with business strategies, goals and needs. It has remained a fundamental concern of business and IT executives for more than 15 years. This definition of alignment addresses how:

1.   IT is aligned with the business
2.   The business should or could be aligned with IT.

Mature alignment evolves into a relationship where IT and other business functions adapt their strategies together. When discussing business-IT alignment, terms like harmony, linkage, fusion, fit, match, and integration are frequently used synonymously with the term alignment. It does not matter whether one considers business-IT alignment or IT-business alignment; the objective is to ensure that the organizational strategies adapt harmoniously.

The evidence that IT has the power to transform whole industries and markets is strong (e.g., King, 1995; Luftman, 1996; Earl, 1993; Earl, 1996; Luftman et al., 1993; Goff, 1993; Liebs, 1992; Robson, 1994; Luftman, Papp, & Brier, 1999; Luftman & Brier, 1999). Important questions that need to be addressed include the following:

- How can organizations assess alignment?
- How can organizations improve alignment?
- How can organizations achieve mature alignment?

The purpose of this chapter is to present an approach for assessing the maturity of a firm's business-IT alignment. Until now, none was available. The alignment maturity assessment approach described in this chapter provides a comprehensive vehicle for organizations to evaluate business-IT alignment in terms of where they are and what they can do to improve alignment. The maturity assessment applies the previous research that identified enablers/inhibitors to achieving alignment (Luftman, Papp, & Brier, 1995; Luftman & Brier, 1999), and the author's consulting experience that applied the methodology that leverages the most important enablers and inhibitors as building blocks for the evaluation. The maturity assessment also applies concepts included in the popular work done by the Software Engineering Institute (Humphrey, 1988), Keen's reach and range (Keen, 1996) and an evolution of the Nolan and Gibson stages of growth (Nolan, 1979).

Alignment's importance has been well known and well documented since the late 1970s (e.g., McLean & Soden, 1977; IBM, 1981; Mills, 1986; Parker & Benson, 1988; Brancheau & Whetherbe, 1987; Dixon & Little, 1989; Niederman et al., 1991; Chan & Huff, 1993; Henderson & Venkatraman, 1996; Luftman & Brier, 1999). Over the years, it persisted among the top-ranked concerns of business executives. Alignment seems to grow in importance as companies strive to link technology and business in light of dynamic business strategies and continuously evolving technologies (Papp, 1995; Luftman, 1996). Importance aside, what is not clear is how to achieve and sustain this harmony relating business and IT, how to assess the maturity of alignment, and what the impact of misalignment might be on the firm (Papp & Luftman, 1995). The ability to achieve and sustain this synergistic relationship is anything but easy. Identifying an organization's alignment maturity provides an excellent vehicle for understanding and improving the business-IT relationship. It is the maturity of the IT-business alignment relationship that is the focus of this chapter.

This chapter, after this Introduction, is divided into five sections. They are:

1. The Strategic Alignment Maturity Assessment Description — explains the essential components of the maturity assessment.

2. The Six Strategic Alignment Maturity Criteria — illustrates each of the six criteria that are evaluated in deriving the level of strategic alignment maturity. Examples from many of the previously conducted assessments are included.

3. Conducting a Strategic Alignment Maturity Assessment — describes the process applied in carrying out an evaluation. This section ties the respective assessment metrics together. Along with the examples in the Appendix, the last section serves as the vehicle for validating the model.

4. Conclusions — summarizes the strategic alignment maturity assessment research, to date.

5. Appendices

    A - Strategic Alignment Maturity Assessment Experiences — highlights the experiences with 25 Fortune 500 companies that participated in the initial strategic alignment maturity assessments. It also includes summaries of six assessments of Fortune 200 companies and a large university.

    B - The Five Levels of Strategic Alignment Maturity — describes each of the five levels of strategic alignment maturity.

STRATEGIC ALIGNMENT MATURITY ASSESSMENT

As the summary of the maturity assessment in Figure 1 illustrates, the model involves the following five levels of strategic alignment maturity:

1. Initial/Ad Hoc Process
2. Committed Process
3. Established Focused Process
4. Improved/Managed Process
5. Optimized Process

Each of the five levels of alignment maturity focuses, in turn, on a set of six criteria based on practice validated with an evaluation of 25 Fortune 500 companies. A summary of the evaluations is presented in Appendix A. There are now over 60 organizations that have taken the assessment. Research and benchmarking are continuing under the sponsorship of The Conference Board and Society for Information Management (SIM). The five levels of maturity are described in detail in Appendix B. The same criteria are used for each level of maturity.

The six IT-business alignment criteria are illustrated in Figure 2 and are described in the following section of this chapter. These six criteria are:

1. Communications Maturity
2. Competency/Value Measurement Maturity

Figure 1. Strategic Alignment Maturity Summary

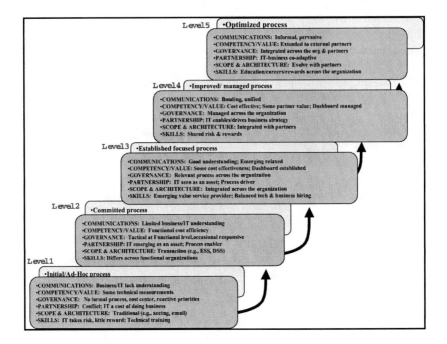

3.   Governance Maturity
4.   Partnership Maturity
5.   Scope & Architecture Maturity
6.   Skills Maturity

The procedure for assessing maturity is as follows:

1.   Each of the criteria is assessed individually by a team of IT and business unit executives to determine the firm's level of strategic maturity on this criterion. In other words, each of the six criteria is found to be at Level 1, Level 2, Level 3, Level 4, or Level 5.

2.   The evaluation team converges on a single assessment level for each of the six criteria. The discussions that ensue are extremely valuable in understanding both the current state of the organization's alignment maturity and how the organization can best proceed to improve the maturity.

3.   The evaluation team, after assessing each of the six criteria from level one to five, uses the results to converge on an overall assessment level of the maturity for the firm. They apply the next higher level of maturity as a roadmap to identify what they should do next.

This conceptual framework (qualities and attributes) is described in Appendix B. The process of conducting a Strategic Alignment Maturity Assessment is described in more detail later in this chapter.

Figure 2. Alignment Maturity Criteria

THE SIX STRATEGIC ALIGNMENT MATURITY CRITERIA

This section describes each of the six criteria (illustrated in Figure 2) that are evaluated in deriving the level of strategic alignment maturity. Examples taken from actual assessment summaries illustrate the kinds of insights that can be identified. Appendix A includes a more complete description of seven of these studies, as well as the benchmark data attained thus far.

Most organizations today are at a Level 2 with some attributes of Level 3. This is similar to what has been found by the Carnegie software models that identify the comparable stage of application development. Naturally, the objective of the Strategic Alignment Maturity model is to assess the organization at a higher stratum.

Communications

Effective exchange of ideas and a clear understanding of what it takes to ensure successful strategies are high on the list of enablers and inhibitors to alignment. Too often there is little business awareness on the part of IT or little IT appreciation on the part of the business. Given the dynamic environment in which most organizations find themselves, ensuring ongoing knowledge sharing across organizations is paramount.

Many firms choose to draw on liaisons to facilitate this knowledge sharing. The key word here is facilitate. Often the author has seen facilitators whose role is to serve as the sole conduit of interaction among the different organizations. This approach tends to stifle rather than foster effective communications. Rigid protocols that impede discussions and the sharing of ideas should be avoided.

For example, a large aerospace company assessed its communications alignment maturity at Level 2. Business-IT understanding is sporadic. The relationship between IT and the business function could be improved. Improving communication should focus on how to create the understanding of IT as a strategic business partner by the businesses it supports rather than simply a service provider. The firm's CIO made the comment that there is "no constructive partnership". However, in an interview with the firm's Director of Engineering & Infrastructure, he stated that he views his organization as a "strategic business partner". One way to improve communications and, more important, understanding, would be to establish effective business function/IT liaisons that facilitate sharing of knowledge and ideas.

In a second case, a large financial services company's communication alignment maturity placed it in Level 2 with some attributes of Level 1. Business awareness within IT is through specialized IT business analysts, who understand and translate the business needs to other IT staff (i.e., there is limited awareness of business by general IT staff). Awareness of IT by the firm's business functions is also limited, although senior and mid-level management are aware of IT's potential. Communications are achieved through bi-weekly priority meetings attended by the senior and middle level management from both groups, where they discuss requirements, priorities and IT implementation.

In a third example, a large utility company's communication alignment maturity places it at a Level 2. Communications are not open until circumstances force the business to identify specific needs. There is a lack of trust and openness between some business units and their IT team. IT business partners tend to be bottlenecks in meeting commitments. Its poor performance in previous years left scars that have not healed.

Competency/Value Measurements

Too many IT organizations cannot demonstrate their value to the business in terms that the business understands. Frequently business and IT metrics of value differ. A balanced "dashboard" that demonstrates the value of IT in terms of contribution to the business is needed.

Service levels that assess IT's commitments to the business often help. However, the service levels have to be expressed in terms that the business understands and accepts. The service levels should be tied to criteria (see subsection, Partnership) that clearly define the rewards and penalties for surpassing or missing the objectives.

Frequently organizations devote significant resources to measuring performance factors. However, they spend much less of their resources on taking action based on these measurements. For example, an organization that requires an ROI before a project begins, but that does not review how well objectives were met after the project was deployed provides little to the organization. It is important to assess these criteria to understand (1) the factors that lead to missing the criteria and (2) what can be learned to improve the environment continuously.

For example, a large aerospace company assessed its competency/value measurement maturity to be at a Level 2. IT operates as cost center. IT metrics are focused at the functional level, and Service Level Agreements (SLAs) are technical in nature. One area that could help to improve maturity would be to add more business-related metrics to SLAs to help form more of a partnership between IT and the business units. Periodic formal assessments and reviews in support of continuous improvement would also be beneficial.

A large software development company assessed its competency/value measurement maturity at Level 3. Established metrics evaluate the extent of service provided to the business functions. These metrics go beyond basic service availability and help desk responsiveness, evaluating such issues as end-user satisfaction and application development effectiveness. The metrics are consolidated onto an overall dashboard. However, because no formal feedback mechanisms are in place to react to a metric, the dashboard cannot be considered to be managed.

At a large financial services company, IT competency/value was assessed at a Level 2 because they use cost efficiency methods within the business and functional organizations. Balanced metrics are emerging through linked business and IT metrics, and a Balanced Scorecard is provided to senior management. Service level agreements are technical at the functional level. Benchmarking is not generally practiced and is informal in the few areas where it is practiced. Formal assessments are done typically for problems and minimum measurements are taken after the assessment of failures.

Governance

The considerations for IT Governance are described in Luftman and Brier (1999). Ensuring that the appropriate business and IT participants formally discuss and review the priorities and allocation of IT resources is among the most important enablers/inhibitors of alignment. This decision-making authority needs to be clearly defined.

For example, IT Governance in a large aerospace company is tactical at the core business level and not consistent across the enterprise. For this reason, they reported a Level 2 maturity assessment. IT can be characterized as reactive to CEO direction. Developing an integrated enterprise-wide strategic business plan for IT would facilitate better partnering within the firm and would lay the groundwork for external partnerships with customers and suppliers.

A large communications manufacturing company assessed its governance maturity at a level falling between one and two. IT does little strategic planning because it operates as a cost center and, therefore, cost reduction is a key objective. In addition, priorities are reactive to business needs as business managers request services.

A large computing services company assessed their governance maturity at a level 1+. A strategic planning committee meets twice a year. The committee consists of corporate top management with regional representation. Topics or results are not discussed nor published to all employees. The reporting structure is federated with the CIO reporting to a COO. IT investments are traditionally made to support operations and maintenance. Regional or corporate sponsors are involved with some projects. Prioritization is occasionally responsive.

Partnership

The relationship that exists between the business and IT organizations is another criterion that ranks high among the enablers and inhibitors. Giving the IT function the opportunity to have an equal role in defining business strategies is obviously important. However, how each organization perceives the contribution of the other, the trust that develops among the participants, ensuring appropriate business sponsors and champions of IT endeavors, and the sharing of risks and rewards are all major contributors to mature alignment. This partnership should evolve to a point where IT both enables AND drives changes to both business processes and strategies. Naturally, this demands having a good business design where the CIO and CEO share a clearly defined vision.

For example, a large software development company assessed their partnership maturity at a level of two. The IT function is mainly an enabler for the company. IT does not have a seat at the business table, either with the enterprise or with the business function that is making a decision. In the majority of cases, there are no shared risks because only the business will fail. Indications are that the partnership criterion will rise from a Level 2 to 3 as top management sees IT as an asset, and because of the very high enforcement of standards at the company.

Partnership for a large communications manufacturing company was assessed at Level 1. IT is perceived as a cost of being in the communications business. Little value is placed on the IT function. IT is perceived only as help desk support and network maintenance.

For a large utility company, partnership maturity was assessed at a level of 1+. IT charges back all expenses to the business. Most business executives see IT as a cost of doing business. There is heightened awareness that IT can be a critical enabler to success, but there is minimal acceptance of IT as a partner.

Partnership for a large computing services company was assessed at Level 2. Since the business executives pursued e-commerce, IT is seen as a business process enabler as demonstrated by the Web development. Unfortunately, the business now assigns IT with the risks of the project. Most IT projects have an IT sponsor.

Scope and Architecture

This set of criteria tends to assess information technology maturity. The extent to which IT is able to:

- go beyond the back office and the front office of the organization
- assume a role supporting a flexible infrastructure that is transparent to all business partners and customers
- evaluate and apply emerging technologies effectively
- enable or drive business processes and strategies as a true standard
- provide solutions customizable to customer needs

Scope and Architecture was assessed at a level of 2+ at a large software development company. This is another area where the company is moving from a Level 2 to a Level 3. ERP systems are installed and all projects are monitored at an enterprise level. Standards are integrated across the organization and enterprise architecture is integrated. It is only in the area of Inter-enterprise that there is no formal integration.

A large financial services company assessed their scope and architecture at Level 1. Although standards are defined, there is no formal integration across the enterprise. At best, only functional integration exists.

Skills

Skills include all of the human resource considerations for the organization. Going beyond the traditional considerations such as training, salary, performance feedback, and career opportunities are factors that include the organization's cultural and social environment. Is the organization ready for change in this dynamic environment? Do individuals feel personally responsible for business innovation? Can individuals and organizations learn quickly from their experience? Does the organization leverage innovative ideas and the spirit of entrepreneurship? These are some of the important conditions of mature organizations.

For example, a large aerospace company assesses their skills maturity at a Level 2. A definite command and control management style exists within IT and the businesses. Power resides within certain operating companies. Diverse business cultures abound. Getting to a non-political, trusting environment between the businesses and IT, where risks are shared and innovation and entrepreneurship thrive, is essential to achieving improvements in each of the other maturity tenets.

Skills maturity at a large computing services company is assessed at a level of one. Career crossover is not encouraged outside of top management. Innovation is dependent on the business unit, but in general is frowned upon. Management style is dependent on the business unit, but is usually command and control. Training is encouraged but left up to the individual employee.

CONDUCTING A STRATEGIC ALIGNMENT MATURITY ASSESSMENT

An essential part of the assessment process is recognizing that it must be done with a team including both business and IT executives. The convergence on a consensus of the maturity levels and the discussions that ensue are extremely valuable in understanding the problems and opportunities that need to be addressed to improve business-IT alignment. The most important part of the process is the creation of recommendations addressing the problems and opportunities identified. The most difficult step, of course, is actually carrying out the recommendations. This section ties the assessment metrics together. The examples and experiences provided in Appendix A, together with the procedure described here, serves as the vehicle for validating the model.

Each of the criteria and levels are described by a set of attributes that allow a particular dimension to be assessed using a 1 to 5 Likert scale, where:

   1 = this does not fit the organization, or the organization is very ineffective
   2 = low level of fit for the organization
   3 = moderate fit for the organization, or the organization is moderately effective
   4 = this fits most of the organization
   5 = strong level of fit throughout the organization, or the organization is very effective

Different scales can be applied to perform the assessment (e.g., good, fair, poor; 1, 2, 3). However, whatever the scale, it is important to evaluate each of the six criteria with both business and IT executives to obtain an accurate assessment. The intent is to have the team of IT and business executives converge on a maturity level. Typically, the initial review will produce divergent results. This outcome is indicative of the problems/ opportunities being addressed.

The relative importance of each of the attributes within the criteria may differ among organizations. For example, in some organizations the use of SLAs (Service Level Agreements) might not be considered as important to alignment as the effectiveness of liaisons. Hence, giving SLAs a low maturity assessment should not significantly impact the overall rating in this case. However, it would be valuable if the group discusses why the organization does not consider a particular attribute (in this example, SLAs) to be significant.

Using a Delphi approach with a Group Decision Support Tool (Luftman, 1997) often helps in attaining the convergence. The author's experience suggests that "discussions" among the different team members helps to ensure a clearer understanding of the problems and opportunities that need to be addressed.

Keep in mind that the primary objective of the assessment is to identify specific recommendations to improve the alignment of IT and the business. The evaluation team, after assessing each of the six criteria from Level 1 to 5, uses the results to converge on an overall assessment level of the maturity for the firm. They apply the next higher level of maturity as a roadmap to identify what they should do next. A trained facilitator is typically needed for these sessions.

Experience with the initial 25 Fortune 500 companies indicates that more than 80% of the organizations are at Level 2 maturity with some characteristics of Level 3 maturity. Figure 3 (including parts A through F) in Appendix A illustrates the "average" results of the Strategic Alignment Maturity assessments for these 25 companies. These results are the start of a Strategic Alignment Maturity Assessment benchmark repository. As the sample grows, it is anticipated that exemplar benchmarks based on factors such as industry, company age, and company size will be available. The figure shows the maturity attributes for each of the six maturity components. Figure 3 (without the average numbers) can be used as the basis for determining an organization's maturity level.

The specific results of the maturity assessment for seven firms are also included in Figure 3. Keep in mind that the results of these maturity assessments were not the principal objective of this exercise. Rather, the goal is to provide the firm with specific insights regarding what it can do to improve the maturity level and thereby improve IT- business strategic alignment.

Strategic Alignment as a Process

The approach applied to attain and sustain business-IT alignment focuses on understanding the alignment maturity, and on maximizing alignment enablers and minimizing inhibitors. The process (Luftman & Brier, 1999) includes the following six steps:

1. *Set the goals and establish a team.* Ensure that there is an executive business sponsor and champion for the assessment. Next, assign a team of both business and IT leaders. Obtaining appropriate representatives from the major business

functional organizations (e.g., Marketing, Finance, R&D, Engineering) is critical to the success of the assessment. The purpose of the team is to evaluate the maturity of the business-IT alignment. Once the maturity is understood, the team is expected to define opportunities for enhancing the harmonious relationship of business and IT. Assessments range from three to twelve half-day sessions. The time demanded depends on the number of participants, the degree of consensus required, and the detail of the recommendations to carry out.

2.   *Understand the business-IT linkage.* The Strategic Alignment Maturity Assessment is an important tool in understanding the business-IT linkage. The team evaluates each of the six criteria. A trained facilitator can be valuable in guiding the important discussions.

3.   *Analyze and prioritize gaps.* Recognize that the different opinions raised by the participants are indicative of the alignment opportunities that exist. Once understood, the group needs to converge on a maturity level. The team must remember that the purpose of this step is to understand the activities necessary to improve the business-IT linkage. The gap between where the organization is today and where the team believes it needs to be are the gaps that need to be prioritized. Apply the next higher level of maturity as a roadmap to identify what can be done next.

4.   *Specify the actions (project management).* Naturally, knowing where the organization is with regards to alignment maturity will drive what specific actions are appropriate to enhance IT-business alignment. Assign specific remedial tasks with clearly defined deliverables, ownership, timeframes, resources, risks, and measurements to each of the prioritized gaps.

5.   *Choose and evaluate success criteria.* This step necessitates revisiting the goals and regularly discussing the measurement criteria identified to evaluate the implementation of the project plans. The review of the measurements should serve as a learning vehicle to understand how and why the objectives are or are not being met.

6.   *Sustain alignment.* Some problems just won't go away. Why are so many of the inhibitors IT related? Obtaining IT-business alignment is a difficult task. This last step in the process is often the most difficult. To sustain the benefit from IT, an "alignment behavior" must be developed and cultivated. The criteria described to assess alignment maturity provide characteristics of organizations that link IT and business strategies. By adopting these behaviors, companies can increase their potential for a more mature alignment assessment and improve their ability to gain business value from investments in IT. Hence, the continued focus on understanding the alignment maturity for an organization and taking the necessary action to improve the IT-business harmony is key.

CONCLUSION

Achieving and sustaining IT-business alignment continues to be a major issue. Experience shows that no single activity will enable a firm to attain and sustain alignment. There are too many variables. The technology and business environments are too dynamic.

The strategic alignment maturity assessment provides a vehicle to evaluate where an organization is and where it needs to go to attain and sustain business-IT alignment. The careful assessment of a firm's alignment maturity is an important step in identifying the specific actions necessary to ensure IT is being used to appropriately enable or drive the business strategy.

The research to derive the business-IT alignment maturity assessment has just begun. The Conference Board and Society of Information Management (SIM) is sponsoring research to assess and benchmark alignment maturity. To date, more than 50 organizations have participated in the research. The appendices discuss the results of the initial 25 organizations that conducted formal assessments. The author would appreciate hearing from practitioners, researchers, and consultants, as the strategic alignment process and the alignment maturity assessment are applied. The intent is to enhance the alignment assessment tool and provide a vehicle to benchmark exemplar organizations. The journey continues.

REFERENCES

Brancheau, J., & Wetherbe, J. (1987). Issues in information systems management. *MIS Quarterly, 11*(1), 23-45.

Dixon, P., & John, D. (1989). Technology issues facing corporate management in the 1990s. *MIS Quarterly, 13*(3), 247-255.

Earl, M. J. (1996). Experience in strategic information systems planning. *MIS Quarterly, 17*(1), 1-24.

Faltermayer, E. (1994). Competitiveness: How U.S. companies stack up now. *Fortune, 129*(8), (April 18), 52-64.

Goff, L. (1993). You say tomayto, I say tomahto. *Computerworld*, (November 1), 129.

Henderson, J., & Venkatraman, N. (1990). *Strategic alignment: A model for organizational transformation via Information Technology*. Working Paper 3223-90, Cambridge, MA: Sloan School of Management, Massachusetts Institute of Technology.

Henderson, J., & Venkatraman, N. (1996). Aligning business and IT strategies. In J. N. Luftman (Ed.), *Competing in the Information Age: Practical Applications of the Strategic Alignment Model*. New York: Oxford University Press.

Humphrey, W.S. (1988). Characterizing the software process: A maturity framework. *IEEE Software, 5*(2), 73-79.

IBM. (1981). *Business Systems Planning, Planning Guide*, GE20-0527. White Plains, NY: IBM Corporation.

Keen, P. (1996). Do you need an IT strategy? In J. N. Luftman (Ed.), *Competing in the Information Age*. New York: Oxford University Press.

King, J. (1995). Re-engineering focus slips. *Computerworld*, (March 13).

Liebs, S. (1992). We're all in this together. *Information Week*, (October 26).

Luftman, J. (1996). *Competing in the Information Age: Practical applications of the strategic alignment model*. New York: Oxford University Press.

Luftman, J. (1997). Align in the sand. *Computerworld*, (February 17).

Luftman, J., & Brier, T. (1999). Achieving and sustaining business-IT alignment. *California Management Review, 1*(Fall), 109-122.

Luftman, J., Lewis, P., & Oldach, S. (1993). Transforming the enterprise: The alignment of business and Information Technology strategies. *IBM Systems Journal, 32*(1), 198-221.

Luftman, J., Papp, R., & Brier, T. (1995). The strategic alignment model: Assessment and validation. *Proceedings of the Information Technology Management Group of the Association of Management (AoM) 13th Annual International Conference,* Vancouver, British Columbia, Canada (August 2-5, pp. 57-66).

Luftman, J., Papp, R., & Brier, T. (1999). Enablers and inhibitors of business-IT alignment. *Communications of the Association for Information Systems, 1*(11).

McLean, E., & Soden, J. (1977). *Strategic planning for MIS.* New York: John Wiley & Sons.

Mills, P. (1986). *Managing service industries.* New York: Ballinger.

Niederman, F., Brancheau, J., & Wetherbe, J. (1991). Information systems management issues for the 1990s. *MIS Quarterly, 15*(4), 475-95.

Nolan, R. L. (1979). Managing the crises in data processing. *Harvard Business Review,* (March 1).

Papp, R. (1995). *Determinants of strategically aligned organizations: A multi-industry, multi-perspective analysis* (PhD Dissertation). Hoboken, NJ: Stevens Institute of Technology.

Papp, R., & Luftman, J. (1995). Business and IT strategic alignment: New perspectives and assessments. *Proceedings of the Association for Information Systems, Inaugural Americas Conference on Information Systems,* Pittsburgh, PA (August 25-27).

Parker, M., & Benson, R. (1988). *Information economics.* Englewood Cliffs, NJ: Prentice-Hall.

Watson, R., & Brancheau, J. (1991). Key issues in information systems management: An international perspective. *Information & Management, 20,* 213-23.

APPENDIX A
Strategic Alignment Maturity Assessment Experiences

In 2000, pilot studies were conducted using formal assessments of 25 Fortune 500 firms. The last column in Figure 3 (A, B, C, D, E, and F) in this appendix illustrates the "average" evaluations (rated using a Likert scale) for the six criteria of the Strategic Alignment Maturity assessments for these 25 firms. The numbers are the average responses from all participants (e.g., IT, Finance, Marketing from all 25 firms) for each of the respective components of the six criteria. These results are the start of a Strategic Alignment Maturity Assessment benchmark repository. Future assessments will be included to provide exemplar benchmarks based on decisive factors such as industry and company size.

Figure 3 (A, B, C, D, E, and F) in this appendix also includes the responses from six actual assessments of Fortune 200 companies and a large university. These seven assessments represent the average evaluations (rated using a Likert scale) that the multi-functional group (e.g., IT, Finance, Marketing) from each of the firms identified. They are a subset of the 25 firms.

Typically, after getting the individual responses from the participants for their perception of the level of maturity for each of the six criteria, a discussion was facilitated to obtain consensus on the respective maturity level for each of the six criteria. In one case, a Delphi was used to derive the consensus. The maturity level at the bottom of each column represents the consensus for the respective group. Most of the examples used in the main part of this chapter, especially in the section The Six Alignment Maturity Criteria, come from these seven firms. Figure 3 (without the average numbers) can be used as the basis for determining an organization's maturity level.

Figure 3A. Communications

		7 Assessments Summarized							Initial
		1	2	3	4	5	6	7	25 Firms
Understanding of Business by IT									
1.	IT Management not aware	3	3	1	3	2	1	0	2
2.	Limited IT awareness	4	3	3	2	2	4	2	4
3.	Senior and mid-management	2	1	3	1	1	2	5	3
4.	Pushed down through organization	0	0	1	0	0	0	1	1
5.	Pervasive	0	0	0	0	0	0	0	1
Understanding of IT by Business									
1.	Business Management not aware	2	3	2	3	2	2	2	3
2.	Limited Business awareness	4	3	4	2	3	4	3	4
3.	Emerging business awareness	1	1	1	0	0	1	2	1
4.	Business aware of potential	0	0	0	0	0	0	0	1
5.	Pervasive	0	0	0	0	0	0	0	0
Inter/Intra-organizational learning/education									
1.	Casual, ad-hoc	3	2	2	3	2	1	0	4
2.	Informal	3	4	3	2	3	4	5	4
3.	Regular, clear	0	1	2	0	0	0	0	1
4.	Unified, bonded	0	0	0	0	0	0	0	1
5.	Strong and structured	0	0	0	0	0	0	0	0
Protocol Rigidity									
1.	Command and Control	4	3	2	4	4	4	5	4
2.	Limited relaxed	2	2	4	2	2	2	0	3
3.	Emerging relaxed	0	0	1	0	0	1	0	1
4.	Relaxed, informal	0	0	0	0	0	0	0	1
5.	Informal	0	0	0	0	0	0	0	0
Knowledge (Intellectual Capital) Sharing									
1.	Ad-hoc	1	2	1	2	1	0	1	0
2.	Semi structured	2	2	3	3	2	4	5	5
3.	Structured around key processes	2	4	3	1	1	3	0	3
4.	Institutionalized	0	0	0	0	0	1	0	1
5.	Extra-enterprise	0	0	0	0	0	0	0	0
Liaison(s) Breadth/Effectiveness									
1.	None or Ad-hoc	4	2	1	2	1	1	1	2
2.	Limited tactical technology based	1	2	3	3	4	4	4	4
3.	Formalized, regular meetings	0	0	4	0	1	2	2	3
4.	Bonded, effective at all internal levels	0	0	1	0	0	0	0	1
5.	Extra-enterprise	0	0	0	0	0	0	0	0
MATURITY LEVEL		**2**	**2**	**2**	**2**	**2**	**2**	**2**	**2+**

Figure 3B. Competency/Value Measurements

		7 Actual Assessments Summarized							Initial
		1	2	3	4	5	6	7	25 Firms
IT Metrics									
1.	Technical; not related to business	4	2	1	5	3	4	5	5
2.	Cost efficiency	3	3	4	4	4	4	5	4
3.	Traditional financial	3	4	4	3	4	3	3	3
4.	Cost effectiveness	1	3	2	1	1	0	0	2
5.	Extended to external partners	0	1	1	0	0	0	0	1
Business Metrics									
1.	Ad-hoc; not related to IT	4	2	2	4	4	2	5	4
2.	At the functional organization	3	3	3	4	4	4	4	4
3.	Traditional financial	2	4	4	4	4	4	5	4
4.	Customer based	0	3	1	1	1	1	0	3
5.	Extended to external partners	0	1	1	1	0	0	0	2
Balanced Metrics									
1.	Ad-hoc metrics unlinked	3	2	0	1	3	3	4	3
2.	Business and IT metrics unlinked	4	3	2	5	5	4	4	4
3.	Emerging business & IT metrics linked	0	2	4	0	0	2	0	3
4.	Business & IT metrics linked	0	0	1	0	0	0	0	1
5.	Business, partners & IT metrics linked	0	0	0	0	0	0	0	1
Service Level Agreements									
1.	Sporadically present	1	2	2	0	3	4	4	3
2.	Technical at the functional level	5	3	5	5	4	2	3	4
3.	Emerging across the enterprise	1	4	1	2	1	0	0	2
4.	Enterprise wide	0	1	1	1	0	0	0	1
5.	Extended to external partners	0	0	1	0	0	0	0	1
Benchmarking									
1.	Not generally practiced	2	1	1	1	1	3	2	2
2.	Informal	4	2	4	3	2	4	5	4
3.	Focused on specific processes	2	4	3	4	4	3	1	3
4.	Routinely performed	2	3	1	2	3	1	0	2
5.	Routinely performed with partners	1	1	0	1	1	0	0	1
Formal Assessments/Reviews									
1.	None	1	0	0	1	0	2	2	2
2.	Some; typically for problems	4	2	4	4	4	5	4	4
3.	Emerging formality	2	3	2	3	4	1	0	2
4.	Formally performed	2	3	1	1	2	0	0	1
5.	Routinely performed	0	0	0	1	1	0	0	0
Continuous Improvement									
1.	None	1	0	2	0	2	2	3	2
2.	Minimum	3	2	3	3	3	3	3	3
3.	Emerging	1	4	3	3	3	1	2	3
4.	Frequently	1	1	1	2	2	0	0	2
5.	Routinely performed	1	1	1	1	1	0	0	1
MATURITY LEVEL		**2**	**3**	**2**	**2**	**2**	**2**	**2**	**2+**

Figure 3C. Governance

	7 Assessments Summarized							Initial
	1	2	3	4	5	6	7	25 Firms
Business Strategic Planning								
1. Ad-hoc	3	1	1	4	2	1	2	3
2. Basic planning at the functional level	5	3	5	3	4	5	3	5
3. Some inter-organizational planning	2	2	2	1	2	3	1	2
4. Managed across the enterprise	0	1	1	0	1	0	0	1
5. Integrated across & outside the enterprise	0	0	0	0	0	0	0	0
IT Strategic Planning								
1. Ad-hoc	3	1	1	4	5	4	4	3
2. Functional tactical planning	5	4	5	2	2	2	5	4
3. Focused planning, some inter-organizational	2	4	4	1	1	1	1	4
4. Managed across the enterprise	0	1	1	1	0	0	0	1
5. Integrated across & outside the enterprise	0	0	0	0	0	0	0	0
Reporting/Organization Structure								
1. Central/Decentral; CIO reports to CFO	2	0	3	5	3	5	4	4
2. Central/Decentral, some co-location; CIO reports to CFO	5	4	5	2	4	1	4	4
3. Central/Decentral, some federation; CIO reports to COO	1	3	0	1	0	0	0	3
4. Federated; CIO reports to COO or CEO	0	4	0	0	0	0	0	2
5. Federated; CIO reports to CEO	0	2	0	0	0	0	0	2
Budgetary Control								
1. Cost Center; Erratic/inconsistent spending	2	2	3	3	4	4	5	3
2. Cost Center by functional organization	5	5	5	5	3	1	3	5
3. Cost Center; some investments	1	4	1	1	1	0	1	3
4. Investment Center	0	0	0	0	0	0	0	1
5. Investment Center; Profit Center	0	0	0	0	0	0	0	1
IT Investment Management								
1. Cost based; Erratic spending	4	2	3	5	5	5	5	4
2. Cost based; Operations, maintenance focus	4	2	5	4	4	4	5	5
3. Traditional; Process enabler	1	4	2	2	1	1	4	3
4. Cost effectiveness; Process driver	0	0	1	0	0	0	0	1
5. Business value; Extended to business partners	0	0	0	0	0	0	0	0
Steering Committee(s)								
1. Not formal/regular	2	2	2	4	4	4	2	2
2. Periodic organized communication	5	4	3	3	2	3	5	4
3. Regular clear communication	0	2	1	0	0	0	1	1
4. Formal, effective committees	0	0	1	0	0	0	0	1
5. Partnership	0	0	0	0	0	0	0	0
Prioritization Process								
1. Reactive	4	2	3	5	4	4	5	4
2. Occasional responsive	4	4	5	3	2	2	2	4
3. Mostly responsive	1	4	2	0	0	0	0	3
4. Value add, Responsive	0	1	0	0	0	0	0	1
5. Value added partner	0	1	0	0	0	0	0	0
MATURITY LEVEL	**2**	**3**	**2**	**1+**	**1+**	**1**	**2**	**2+**

Figure 3D. Partnership

	7 Assessments Summarized							Initial
	1	2	3	4	5	6	7	25 Firms
Business Perception of IT Value								
1. IT perceived as a cost of business	4	4	3	5	5	4	5	4
2. IT emerging as an asset	5	5	5	1	1	5	3	5
3. IT is seen as an asset	2	1	2	0	0	2	0	2
4. IT is part of the business strategy	1	0	1	0	0	0	0	1
5. IT-business co-adaptive	0	0	0	0	0	0	0	0
Role of IT in Strategic Business Planning								
1. No seat at the business table	2	5	3	5	5	5	5	4
2. Business process enabler	5	5	5	2	2	5	4	5
3. Business process driver	0	0	0	0	0	0	0	1
4. Business strategy enabler/driver	0	0	0	0	0	0	0	0
5. IT-business co-adaptive	0	0	0	0	0	0	0	0
Shared Goals, Risk, Rewards/Penalties								
1. IT takes risk with little reward	5	5	3	5	5	5	4	4
2. IT takes most of risk with little reward	4	5	5	2	2	4	3	5
3. Risk tolerant; IT some reward	1	0	1	0	0	0	3	1
4. Risk acceptance & rewards shared	0	0	0	0	0	0	0	0
5. Risk & rewards shared	0	0	0	0	0	0	0	0
IT Program Management								
1. Ad-hoc	2	1	1	1	2	2	4	2
2. Standards defined	5	5	5	4	4	3	4	4
3. Standards adhered	2	4	3	2	2	2	2	2
4. Standards evolve	2	3	3	2	0	0	0	2
5. Continuous improvement	0	0	1	0	0	0	0	0
Relationship/Trust Style								
1. Conflict/Minimum	3	3	3	4	4	4	4	3
2. Primarily transactional	4	4	5	3	3	4	5	4
3. Emerging valued service provider	2	3	3	0	0	0	0	2
4. Valued service provider	1	1	1	0	0	0	0	0
5. Valued Partnership	1	0	0	0	0	0	0	0
Business Sponsor/Champion								
1. None	2	4	3	5	4	3	4	4
2. Limited at the functional organization	2	4	4	2	4	3	4	4
3. At the functional organization	4	2	3	0	0	0	4	3
4. At the HQ level	1	1	1	0	0	0	0	1
5. At the CEO level	1	1	0	0	0	0	0	1
MATURITY LEVEL	**2**	**2**	**2**	**1**	**1+**	**2**	**2**	**2+**

Figure 3E. Scope and Architecture

	7 Assessments Summarized							Initial
	1	2	3	4	5	6	7	25 Firms
Traditional, Enabler/Driver, External								
1. Traditional (e.g., accounting, e-mail)	2	2	3	4	2	4	5	2
2. Transaction (e.g., ESS, DSS)	2	3	4	3	3	2	2	3
3. Expanded scope (e.g., business process enabler)	5	4	3	2	4	0	0	4
4. Redefined scope (business process driver)	1	0	0	1	0	0	0	1
5. External scope; Business strategy driver/enabler	0	0	0	0	0	0	0	0
Standards Articulation								
1. None or ad-hoc	0	0	4	4	0	4	3	2
2. Standards defined	5	4	3	2	4	1	4	4
3. Emerging enterprise standards	4	3	1	3	4	0	1	3
4. Enterprise standards	3	3	0	0	3	0	0	1
5. Inter-enterprise standards	0	0	0	0	0	0	0	0
Architectural Integration:								
Functional Organization								
1. No formal integration	0	0	5	4	1	4	5	2
2. Early attempts at integration	3	3	2	2	2	3	1	5
3. Integrated across the organization	4	4	0	0	4	0	0	1
4. Integrated with partners	1	0	0	0	0	0	0	0
5. Evolve with partners	0	0	0	0	0	0	0	0
Enterprise								
1. No formal integration	1	2	5	5	1	4	4	3
2. Early attempts at integration	3	4	3	3	3	3	1	4
3. Standard enterprise architecture	4	3	2	1	4	1	1	3
4. Integrated with partners	1	1	0	0	0	0	0	0
5. Evolve with partners	0	0	0	0	0	0	0	0
Inter-Enterprise								
1. No formal integration	2	3	5	4	3	4	4	3
2. Early concept testing	4	3	2	2	3	1	0	3
3. Emerging with key partners	3	1	0	0	1	0	0	2
4. Integrated with key partners	2	0	0	0	0	0	0	1
5. Evolve with all partners	0	0	0	0	0	0	0	0
Architectural Transparency, Agility, Flexibility								
1. None	2	2	3	5	4	4	3	4
2. Limited	4	4	5	2	4	4	4	4
3. Focused on communications	5	3	2	3	3	1	2	3
4. Effective emerging technology manag.	3	2	0	2	2	0	0	2
5. Across the infrastructure	2	1	0	0	2	0	0	2
MATURITY LEVEL	**3**	**2+**	**1**	**1**	**2+**	**1**	**1**	**2+**

Figure 3F. Skills

		7 Assessments Summarized							Initial
		1	2	3	4	5	6	7	25 Firms
Innovation, Entrepreneurs hip									
1.	Discouraged	3	3	4	3	4	5	3	4
2.	Dependent on functional organization	4	5	5	4	5	3	4	3
3.	Risk tolerant	1	2	0	0	1	1	2	2
4.	Enterprise, partners, and IT managers	0	0	0	0	0	0	0	1
5.	The norm	0	0	0	0	0	0	0	0
Cultural Locus of Power									
1.	In the business	3	2	4	2	2	5	3	3
2.	Functional organization	4	4	2	4	4	2	4	4
3.	Emerging across the organization	4	2	0	1	1	1	1	2
4.	Across the organization	0	0	0	0	0	0	0	1
5.	All executives, including CIO & partners	0	0	0	0	0	0	0	0
Management Style									
1.	Command and control	5	3	4	3	4	4	3	4
2.	Consensus-based	2	4	2	3	3	1	2	3
3.	Results based	1	2	2	2	2	1	3	2
4.	Profit/value based	0	0	0	1	0	0	0	1
5.	Relationship based	0	0	0	0	0	0	0	0
Change Readiness									
1.	Resistant to change	4	4	5	3	4	4	3	4
2.	Dependent on functional organization	4	5	1	5	4	3	4	4
3.	Recognized need for change	2	2	1	2	2	2	4	2
4.	High, focused	0	0	0	0	0	0	1	0
5.	High, focused	0	0	0	0	0	0	0	0
Career Crossover									
1.	None	2	1	5	2	1	4	3	3
2.	Minimum	5	5	3	5	4	2	4	4
3.	Dependent on functional organization	1	3	2	1	3	2	1	2
4.	Across the functional organization	0	0	0	0	0	0	0	0
5.	Across the enterprise	0	0	0	0	0	0	0	0
Education, Cross-Training									
1.	None	3	2	1	1	3	4	4	3
2.	Minimum	4	4	5	4	4	2	4	4
3.	Dependent on functional organization	4	4	4	2	4	2	3	3
4.	At the functional organization	0	0	0	0	0	0	1	1
5.	Across the organization	0	0	0	0	0	0	0	0
Social, Political, Trusting Interpersonal Environment									
1.	Minimum	3	3	4	2	2	4	3	4
2.	Primarily transactional	4	4	3	3	3	1	4	3
3.	Emerging among IT & business	3	3	1	0	2	0	3	3
4.	Achieved among IT & business	0	0	0	0	0	0	0	1
5.	Extended to customers & partners	0	0	0	0	0	0	0	0
MATURITY LEVEL		**2**	**2**	**1**	**2**	**2**	**1**	**2+**	**2**

APPENDIX B
The Five Levels of Strategic Alignment Maturity

This appendix describes each of the five levels of strategic alignment maturity summarized in Figure 1. Each of the six criteria described in the main part of this chapter are evaluated in deriving the level of strategic alignment maturity.

Level 1: Initial/Ad Hoc Process

Organizations that meet many of the characteristics of the attributes in the six Strategic Alignment Maturity criteria for Level 1 can be characterized as having the lowest level of Strategic Alignment Maturity. It is highly improbable that these organizations will be able to achieve an aligned IT business strategy, leaving their investment in IT significantly unleveraged.

COMMUNICATIONS

ATTRIBUTE	CHARACTERISTICS
Understanding of Business by IT	Minimum
Understanding of IT by Business	Minimum
Inter/Intra-organizational learning	Casual, Ad-hoc
Protocol Rigidity	Command and Control
Knowledge Sharing	Ad-hoc
Liaison(s) Breadth/Effectiveness	None or Ad-hoc

COMPETENCY/VALUE MEASUREMENTS

ATTRIBUTE	CHARACTERISTICS
IT Metrics	Technical; Not related to business
Business Metrics	Ad-hoc; Not related to IT
Balanced Metrics	Ad-hoc unlinked
Service Level Agreements	Sporadically present
Benchmarking	Not generally practiced
Formal Assessments/Reviews	None
Continuous Improvement	None

GOVERNANCE

ATTRIBUTE	CHARACTERISTICS
Business Strategic Planning	Ad-hoc
IT Strategic Planning	Ad-hoc
Reporting/Organization Structure	Central/Decentral; CIO reports to CFO
Budgetary Control	Cost Center; Erratic spending
IT Investment Management	Cost based; Erratic spending
Steering Committee(s)	Not formal/regular
Prioritization Process	Reactive

PARTNERSHIP

ATTRIBUTE	CHARACTERISTICS
• Business Perception of IT Value	IT Perceived as a cost of business
• Role of IT in Strategic Business Planning	No seat at the business table
• Shared Goals, Risk, Rewards/Penalties	IT takes risk with little reward
• IT Program Management	Ad-hoc
• Relationship/Trust Style	Conflict/Minimum
• Business Sponsor/Champion	None

SCOPE & ARCHITECTURE

ATTRIBUTE	CHARACTERISTICS
• Traditional, Enabler/Driver, External	Traditional (e.g., accounting, e-mail)
• Standards Articulation	None or ad-hoc
• Architectural Integration: - Functional Organization - Enterprise - Inter-enterprise	No formal integration
• Architectural Transparency, Flexibility	None

SKILLS

ATTRIBUTE	CHARACTERISTICS
• Innovation, Entrepreneurship	Discouraged
• Locus of Power	In the business
• Management Style	Command and control
• Change Readiness	Resistant to change
• Career crossover	None
• Education, Cross-Training	None
• Social, Political, Trusting Interpersonal Environment	Minimum
• Attract & Retain best talent	No program

Level 2: Committed Process

Organizations that meet many of the characteristics of the attributes in the six Strategic Alignment Maturity criteria for Level 2 can be characterized as having committed to begin the process for Strategic Alignment Maturity. This level of Strategic Alignment Maturity tends to be directed at local situations or functional organizations (e.g., Marketing, Finance, Manufacturing, H/R) within the overall enterprise. However, due to limited awareness by the business and IT communities of the different functional organizations use of IT, alignment can be difficult to achieve. Any business-IT alignment at the local level is typically not leveraged by the enterprise. However, the potential opportunities are beginning to be recognized.

COMMUNICATIONS

ATTRIBUTE	CHARACTERISTICS
• Understanding of Business by IT	Limited IT awareness
• Understanding of IT by Business	Limited Business awareness
• Inter/Intra-organizational learning	Informal
• Protocol Rigidity	Limited relaxed
• Knowledge Sharing	Semi structured
• Liaison(s) Breadth/Effectiveness	Limited tactical technology based

COMPETENCY/VALUE MEASUREMENTS

ATTRIBUTE	CHARACTERISTICS
• IT Metrics	Cost efficiency
• Business Metrics	At the functional organization
• Balanced Metrics	Business and IT metrics unlinked
• Service Level Agreements	Technical at the functional level
• Benchmarking	Informal
• Formal Assessments/Reviews	Some, typically for problems
• Continuous Improvement	Minimum

GOVERNANCE

ATTRIBUTE	CHARACTERISTICS
• Business Strategic Planning	Basic planning at the functional level
• IT Strategic Planning	Functional tactical planning
• Reporting/Organization Structure	Central/Decentral, some co-location; CIO reports to CFO
• Budgetary Control	Cost Center by functional organization
• IT Investment Management Cost based;	Operations & maintenance focus
• Steering Committee(s)	Periodic organized communication
• Prioritization Process	Occasional responsive

PARTNERSHIP

ATTRIBUTE	CHARACTERISTICS
• Business Perception of IT Value	IT emerging as an asset
• Role of IT in Strategic Business Planning	Business process enabler
• Shared Goals, Risk, Rewards/Penalties	IT takes most of the risk with little reward
• IT Program Management	Standards defined
• Relationship/Trust Style	Primarily transactional
• Business Sponsor/Champion	Limited at the functional organization

SCOPE & ARCHITECTURE

ATTRIBUTE	CHARACTERISTICS
• Traditional, Enabler/Driver, External	Transaction (e.g., ESS, DSS)
• Standards Articulation	Standards defined
• Architectural Integration:	
- Functional Organization	Early attempts at integration
- Enterprise	Early attempts at integration
- Inter-enterprise	Early concept testing
• Architectural Transparency, Flexibility	Limited

SKILLS

ATTRIBUTE	CHARACTERISTICS
• Innovation, Entrepreneurship	Dependent on functional organization
• Locus of Power	Functional organization
• Management Style	Results based
• Change Readiness	Dependent on functional organization
• Career crossover	Minimum
• Education, Cross-Training	Minimum
• Social, Political, Trusting Interpersonal Environment	Primarily transactional
• Attract & Retain best talent	Technology focused

Level 3: Established Focused Process

Organizations that meet many of the characteristics of the attributes in the six Strategic Alignment Maturity criteria for Level 3 can be characterized as having established a focused Strategic Alignment Maturity. This level of Strategic Alignment Maturity concentrates governance, processes and communications towards specific business objectives. IT is becoming embedded in the business. Level 3 leverages IT assets on an enterprise-wide basis and applications systems demonstrate planned, managed direction away from traditional transaction processing to systems that use information to make business decisions. The IT extrastructure (leveraging the inter-organizational infrastructure) is evolving with key partners.

COMMUNICATIONS

ATTRIBUTE	CHARACTERISTICS
• Understanding of Business by IT	Senior and mid-management
• Understanding of IT by Business	Emerging business awareness
• Inter/Intra-organizational learning	Regular, clear
• Protocol Rigidity	Emerging relaxed
• Knowledge Sharing	Structured around key processes
• Liaison(s) Breadth/Effectiveness	Formalized, regular meetings

COMPETENCY/VALUE MEASUREMENTS

ATTRIBUTE	CHARACTERISTICS
• IT Metrics	Traditional Financial
• Business Metrics	Traditional Financial
• Balanced Metrics	Emerging business and IT metrics linked
• Service Level Agreements	Emerging across the enterprise
• Benchmarking	Emerging
• Formal Assessments/Reviews	Emerging formality
• Continuous Improvement	Emerging

GOVERNANCE

ATTRIBUTE	CHARACTERISTICS
• Business Strategic Planning	Some inter-organizational planning
• IT Strategic Planning	Focused planning, some inter-organizational
• Reporting/Organization Structure	Central/ Decentral, some federation; CIO reports to COO
• Budgetary Control	Cost Center; some investments
• IT Investment Management	Traditional; Process enabler
• Steering Committee(s)	Regular clear communication
• Prioritization Process	Mostly responsive

PARTNERSHIP

ATTRIBUTE	CHARACTERISTICS
• Business Perception of IT Value	IT seen as an asset
• Role of IT in Strategic Business Planning	Business process enabler
• Shared Goals, Risk, Rewards/Penalties	Risk tolerant; IT some reward
• IT Program Management	Standards adhered
• Relationship/Trust Style	Emerging valued service provider
• Business Sponsor/Champion	At the functional organization

SCOPE & ARCHITECTURE

ATTRIBUTE	CHARACTERISTICS
• Traditional, Enabler/Driver, External	Expanded scope (e.g., business process enabler)
• Standards Articulation	Emerging enterprise standards
• Architectural Integration:	Integrated across the organization
- Functional Organization	Integrated for key processes
- Enterprise	Emerging enterprise architecture
- Inter-enterprise	Emerging with key partners
• Architectural Transparency, Flexibility	Focused on communications

SKILLS

ATTRIBUTE	CHARACTERISTICS
• Innovation, Entrepreneurship	Risk tolerant
• Locus of Power	Emerging across the organization
• Management Style	Consensus based
• Change Readiness	Recognized need for change
• Career crossover	Dependent on functional organization
• Education, Cross-Training	Dependent on functional organization
• Social, Political, Trusting Interpersonal Environment	Emerging among IT & business
• Attract & Retain best talent	Tech & business focus; Retention program

Level 4: Improved/Managed Process

Organizations that meet many of the characteristics of the attributes in the six Strategic Alignment Maturity criteria for Level 4 can be characterized as having a managed Strategic Alignment Maturity. This level of Strategic Alignment Maturity demonstrates effective governance and services that reinforce the concept of IT as a value center. Organizations at Level 4 leverage IT assets on an enterprise-wide basis and the focus of applications systems is on driving business process enhancements to obtain sustainable competitive advantage. A Level 4 organization views IT as an innovative and imaginative strategic contributor to success.

COMMUNICATIONS

ATTRIBUTE	CHARACTERISTICS
• Understanding of Business by IT	Pushed down through organization
• Understanding of IT by Business	Business aware of potential
• Inter/Intra-organizational learning	Unified, bonded
• Protocol Rigidity	Relaxed, informal
• Knowledge Sharing	Institutionalized
• Liaison(s) Breadth/Effectiveness	Bonded, effective at all internal levels

COMPETENCY/VALUE MEASUREMENTS

ATTRIBUTE	CHARACTERISTICS
• IT Metrics	Cost effectiveness
• Business Metrics	Customer based
• Balanced Metrics	Business and IT metrics linked
• Service Level Agreements	Enterprise wide
• Benchmarking	Routinely performed
• Formal Assessments/Reviews	Formally performed
• Continuous Improvement	Frequently

GOVERNANCE

ATTRIBUTE	CHARACTERISTICS
• Business Strategic Planning	Managed across the enterprise
• IT Strategic Planning	Managed across the enterprise
• Organizational Reporting Structure	Federated; CIO reports to COO or CEO
• Budgetary Control	Investment Center
• IT Investment Management	Cost effectiveness; Process driver
• Steering Committee(s)	Formal, effective committees
• Prioritization Process	Value add, responsive

PARTNERSHIP

ATTRIBUTE	CHARACTERISTICS
• Business Perception of IT Value	IT is seen as a driver/enabler
• Role of IT in Strategic Business Planning	Business strategy enabler/driver
• Shared Goals, Risk, Rewards/Penalties	Risk acceptance & rewards shared
• IT Program Management	Standards evolve
• Relationship/Trust Style	Valued service provider
• Business Sponsor/Champion	At the HQ level

SCOPE & ARCHITECURE

ATTRIBUTE	CHARACTERISTICS
• Traditional, Enabler/Driver, External	Redefined scope (business process driver)
• Standards Articulation	Enterprise standards
• Architectural Integration:	Integrated with partners
- Functional Organization	Integrated
- Enterprise	Standard enterprise architecture
- Inter-enterprise	With key partners
• Architectural Transparency, Flexibility	Emerging across the organizations

SKILLS

ATTRIBUTE	CHARACTERISTICS
• Innovation, Entrepreneurship	Enterprise, partners, and IT managers
• Locus of Power	Across the organization
• Management Style	Profit/value based
• Change Readiness	High, focused
• Career crossover	Across the functional organization
• Education, Cross-Training	At the functional organization
• Social, Political, Trusting Interpersonal Environment	Achieved among IT & business
• Attract & Retain best talent	Formal program for hiring & retaining

Level 5: Optimized Process

Organizations that meet the characteristics of the attributes in the six Strategic Alignment Maturity criteria for Level 5 can be characterized as having an optimally aligned Strategic Alignment Maturity. A sustained governance process integrates the IT strategic planning process with the strategic business process. Organizations at Level 5 leverage IT assets on an enterprise-wide basis to extend the reach (the IT extrastructure) of the organization into the supply chains of customers and suppliers. It is often difficult to determine if a Level 5 organization is more a technology company than it is a securities, insurance, travel, retail company.

COMMUNICATIONS

ATTRIBUTE	CHARACTERISTICS
Understanding of Business by IT	Pervasive
Understanding of IT by Business	Pervasive
Inter/Intra-organizational learning	Strong and structured
Protocol Rigidity	Informal
Knowledge Sharing	Extra-enterprise
Liaison(s) Breadth/Effectiveness	Extra-enterprise

COMPETENCY/VALUE MEASUREMENTS

ATTRIBUTE	CHARACTERISTICS
IT Metrics	Extended to external partners
Business Metrics	Extended to external partners
Balanced Metrics	Business, partner, & IT metrics
Service Level Agreements	Extended to external partners
Benchmarking	Routinely performed with partners
Formal Assessments/Reviews	Routinely performed
Continuous Improvement	Routinely performed

GOVERNANCE

ATTRIBUTE	CHARACTERISTICS
Business Strategic Planning	Integrated across & outside the enterprise
IT Strategic Planning	Integrated across & outside the enterprise
Organizational Reporting Structure	Federated; CIO reports to CEO
Budgetary Control	Investment Center; Profit Center
IT Investment Management	Business value; Extended to business partners
Steering Committee(s)	Partnership
Prioritization Process	Value added partner

PARTNERSHIP

ATTRIBUTE	CHARACTERISTICS
• Business Perception of IT Value	IT co-adapts with the business
• Role of IT in Strategic Business Planning	Co-adaptive with the business
• Shared Goals, Risk, Rewards/Penalties	Risk & rewards shared
• IT Program Management	Continuous improvement
• Relationship/Trust Style	Valued Partnership
• Business Sponsor/Champion	At the CEO level

SCOPE & ARCHITECTURE

ATTRIBUTE	CHARACTERISTICS
• Traditional, Enabler/Driver, External	External scope; Business strategy driver/enabler
• Standards Articulation	Inter-Enterprise standards
• Architectural Integration:	Evolve with partners
- Functional Organization	Integrated
- Enterprise	Standard enterprise architecture
- Inter-enterprise	With all partners
• Architectural Transparency, Flexibility	Across the infrastructure

SKILLS

ATTRIBUTE	CHARACTERISTICS
• Innovation, Entrepreneurship	The norm
• Locus of Power	All executives, including CIO & partners
• Management Style	Relationship based
• Change Readiness	High, focused
• Career crossover	Across the enterprise
• Education, Cross-Training	Across the enterprise
• Social, Political, Trusting Interpersonal Environment	Extended to external customers & partners
• Attract & Retain best talent	Effective program for hiring & retaining

Chapter V

Linking the IT Balanced Scorecard to the Business Objectives at a Major Canadian Financial Group[*]

Win Van Grembergen
University of Antwerp, Belgium

Ronald Saull
London Life, Canada

Steven De Haes
University of Antwerp Management School, Belgium

ABSTRACT

The Balanced Scorecard (BSC) initially developed by Kaplan and Norton is a performance management system that enables businesses to drive strategies based on measurement and follow-up. In recent years, the BSC has been applied to information technology (IT). The IT BSC is becoming a popular tool, with its concepts widely supported and dispersed by international consultant groups such as Gartner Group, Renaissance Systems, Nolan Norton Institute, and others. As a result of this interest, the first real-life applications are starting to emerge. In this paper, the development and implementation of a departmental BSC within an Information Services Division (ISD) serving a Canadian financial group will be described and discussed. We use an IT BSC maturity model to determine the maturity level of the IT BSC under review.

INTRODUCTION

Kaplan and Norton (1992, 1993, 1996a, 1996b) introduced the Balanced Scorecard (BSC) at an enterprise level. Their fundamental premise is that the evaluation of a firm should not be restricted to a traditional financial evaluation but should be supplemented with measures concerning customer satisfaction, internal processes and the ability to innovate. Results achieved within these additional perspective areas should assure future financial results and drive the organization towards its strategic goals while keeping all four perspectives in balance. For each of the four perspectives they propose a three layered structure: (1) mission (e.g., to become the customers' most preferred supplier), (2) objectives (e.g., to provide the customers with new products), and (3) measures (e.g., percentage of turnover generated by new products). The Balanced Scorecard can be applied to the IT function and its processes as Gold (1992, 1994) and Willcocks (1995) have conceptually described and has been further developed by Van Grembergen and Van Bruggen (1997), Van Grembergen and Timmerman (1998) and Van Grembergen (2000).

In this chapter, the development and implementation of an IT BSC within the Information Services Division (ISD) of a Canadian tri-company financial group consisting of Great-West Life, London Life and Investors Group (hereafter named The Group) is described and discussed. We use an IT BSC maturity model (adapted from the capability maturity model developed by the Software Engineering Institute) to determine the maturity level of the IT BSC under review. An important conclusion of the paper is that an IT BSC must go beyond the operational level and must be integrated across the enterprise in order to generate business value. This can be realized through establishing a linkage between the business Balanced Scorecard and different levels of IT Balanced Scorecards and through the definition of clear cause-and-effect relationships between outcome measures and performance drivers throughout the whole scorecard.

IT BALANCED SCORECARD CONCEPTS

In Figure 1, a generic IT Balanced Scorecard is shown (Van Grembergen & Van Bruggen, 1998). The *User Orientation* perspective represents the user evaluation of IT. The *Operational Excellence* perspective represents the IT processes employed to develop and deliver the applications. The *Future Orientation* perspective represents the human and technology resources needed by IT to deliver its services over time. The *Business Contribution* perspective captures the business value created from the IT investments.

Each of these perspectives has to be translated into corresponding metrics and measures that assess the current situation. These assessments need to be repeated periodically and aligned with pre-established goals and benchmarks. Essential components of the IT BSC are the cause-and-effect relationships between measures. These relationships are articulated by two key types of measures: outcome measures and performance drivers. A well developed IT scorecard contains a good mix of these two types of measures. Outcome measures such as programmers' productivity *(e.g., number of function points per person per month)* without performance drivers such as IT staff education *(e.g., number of educational days per person per year)* do not communicate

Figure 1. Generic IT Balanced Scorecard

USER ORIENTATION	BUSINESS CONTRIBUTION
How do users view the IT department? **Mission** To be the preferred supplier of information systems. **Objectives** • Preferred supplier of applications • Preferred supplier of operations vs. proposer of best solution, from whatever source • Partnership with users • User satisfaction	How does management view the IT department? **Mission** To obtain a reasonable business contribution from IT investments. **Objectives** • Control of IT expenses • Business value of IT projects • Provision of new business capabilities
OPERATIONAL EXCELLENCE	**FUTURE ORIENTATION**
How effective and efficient are the IT processes? **Mission** To deliver effective and efficient IT applications and services. **Objectives** • Efficient and effective developments • Efficient and effective operations	How well is IT positioned to meet future needs? **Mission** To develop opportunities to answer future challenges. **Objectives** • Training and education of IT staff • Expertise of IT staff • Research into emerging technologies • Age of application portfolio

how the outcomes are to be achieved. And performance drivers without outcome measures may lead to significant investment without a measurement indicating whether the chosen strategy is effective. These cause-and-effect relationships have to be defined throughout the whole scorecard (Figure 2): More and better education of IT staff (future orientation) is an enabler (performance driver) for a better quality of developed systems (operational excellence perspective) that in turn is an enabler for increased user satisfaction (user perspective), that eventually will lead to higher business value of IT (business contribution).

The proposed standard IT BSC links with business through the business contribution. The relationship between IT and business can be more explicitly expressed through a cascade of Balanced Scorecards (Van der Zee, 1999; Van Grembergen, 2000). In Figure 3, the relationship between IT scorecards and the business scorecard is illustrated. The IT Development BSC and the IT Operational BSC both are enablers of the IT Strategic BSC that in turn is the enabler of the Business BSC. This cascade of scorecards becomes a linked set of measures that will be instrumental in aligning IT and business strategy and will help to determine how business value is created through information technology.

RESEARCH METHODOLOGY

Case research is particularly appropriate for research within the IT area because researchers in this field often lag behind practitioners in discovering and explaining new methods and techniques (Benbasat et al., 1987). This is certainly true for the Balanced Scorecard and its application to IT. The Balanced Scorecard is becoming a popular

Figure 2. Cause-and-Effect Relationships

IF	
IT employees' expertise is improved	(future orientation)
THEN	
this may result in a better quality of developed systems	(operational excellence)
THEN	
this may better meet user expectations	(user orientation)
THEN	
this may enhance the support of business processes	(business contribution)

technique, with its concepts supported and dispersed by consultants. A single case design is appropriate when "the investigator has access to a situation previously inaccessible to scientific observation" (Yin, 1994). Like Benbasat et al. (1987) we believe "that the case research strategy is well-suited to capturing the knowledge of practitioners and developing theories from it."

A case study research approach is used to study the phenomenon of the IT BSC and its development and implementation in a single organization. In case study research, the researcher is an observer/investigator rather than a participant (Benbasat et al., 1987). The Chief Information Officer (CIO) of the case company (also the second author of the article) applied the Balanced Scorecard technique to his IT organization. The other co-authors conducted all interviews (including interviews with the CIO) to gather data for this study. Their role was purely the role of observers who were interested in investigating how the IT BSC concepts they and other researchers developed in earlier publications were applied by practitioners and how the experience and knowledge of practitioners could help to improve the earlier proposed IT BSC frameworks. Although the CIO/author used one of the leading author's publications (Van Grembergen & Van Bruggen, 1997) to build his first scorecard, the leading author/researcher was never involved as an advisor in the further developments and implementations.

Figure 3. Balanced Scorecards Cascade

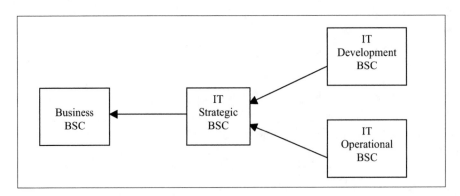

The initial research took place from the end of 1999 until mid 2000 (Period-1). The IT BSC project within the case company is still an ongoing project. During the revision period of August-November 2002 (Period-2), the article has been updated.

In both research periods, the data were collected through in-depth interviews with the CIO by means of multiple e-mail conversations and also through some casual face-to-face conversations when the authors met during international conferences on IT performance measurement. During the second research period in 2002, additional in-depth interviews were conducted with the project manager of the IT Balanced Scorecard project. Also, six individuals who have key roles and accountabilities for scorecard deliverables at the Group were interviewed (including the Vice President Information Services, the Financial Control Director, the Operations & Technical Support Technology Services Director, the Mainframe Technical Support Manager, the Career Centers Director and the Project Management Career Center Leader). These interviews were done by means of e-mail and telephone conversations and an intensive workshop at the headquarters of the company. Data from other sources such as internal reports and slides from the CIO's presentations for his management were used to develop and complete an understanding of the case company, its processes, its technology, its IT organization and its development and implementation of the IT BSC.

CASE COMPANY INTRODUCTION: A TRI-COMPANY

The Great-West Life Assurance Company, London Life and Investors Group are members of the Power Financial Corporation group of companies, with London Life as a wholly owned subsidiary of The Great-West Life Assurance Company. In 2001, MacKenzie Financial was also acquired by the Power Financial Corporation Group, but as the IT Balanced Scorecard project does not cover this company, MacKenzie's organization and IT division will not be taken into account in this article.

The Great-West Life Assurance Company is an international corporation offering life insurance, health insurance, retirement savings, specialty reinsurance and general insurance, primarily in Canada and the United States. Great-West serves the financial security needs of more than 13 million people in Canada and the United States. Great-West has more than $86.9 billion (all figures in this article are in Canadian dollars) in assets under administration and $477 billion of life insurance in force. Founded in Winnipeg in 1891, Great-West is now a leading life and health insurer in the Canadian market in terms of market share.

London Life was founded in Ontario in 1874 and has the leading market share of individual life insurance in Canada. London Life markets life insurance, disability insurance and retirement savings and investment products through its exclusive sales force. The company is a supplier of reinsurance, primarily in the US and Europe, and is a 39% participant in a joint venture life insurance company, Shin Fu, in Taiwan. London Life has more than $30 billion assets under administration and $142.6 billion of life insurance in force.

Investors Group, with its corporate headquarters in Winnipeg, was founded more than 70 years ago. Investors Group is Canada's leading provider of mutual funds, offering a wide spectrum of funds, including those created through strategic partnerships with some of the best known Canadian and international investment management firms. It also offers a wide range of insurance and mortgage options, and currently has $17.1 billion of life insurance coverage in force through three different carriers, and administers with more than $7.6 billion of primarily residential mortgages. Investors Group manages assets of $40.5 billion.

THE TRI-COMPANY IT MERGER

The trend in financial services industry consolidation was a motivating factor behind the acquisition of London Life by Great-West Life and the merger of the IT divisions of the three companies in November 1997. At that time, the tri-company IT expenditures had exceeded $200 million. The ability to reduce these costs and to achieve true synergies and economies of scale within the IT operations was clearly a driver and opportunity for the companies to realize. The merger enabled single systems solutions across all three companies to be explored and implemented as well as single operational processes. Forming a tri-company shared services organization positioned management to:

- achieve world-class status as an information services group,
- maximize purchasing power and operating efficiency,
- leverage technology investments,
- optimize technical infrastructure and application support costs.

Figure 4 depicts the current IT organizational structure of the merged IT division, which employed 812 full-time/part-time employees at the time of the second research period. Also, the position of the IT division relative to the higher reporting levels is indicated. *Application Delivery* and *Technology Services* are respectively the traditional IT department's Systems Development and Operations of the combined organizations. Application Delivery is separated from account management and people management in order to focus on continuous improvement of delivery performance. *Account Management* is the linkage with the clients/users. This component ensures effective communication and translation of business needs into IT processes and educates users on the IT corporate agendas. Account Management employs IT generalists who provide IT insights into business strategy and decision making. *Career Centers* are focused on the professional development of IT people and ensure attention to people issues in order to reduce turnover of talented IT employees. *Corporate Technology* enables the development of a common architecture and provides technology directions. The *eBusiness Solution Center* works on the introduction of new technologies that enable eBusiness solutions for The Group. *Management Services* focuses on running IT as a business and ensures effective financial management and management reporting, including IT scorecard reporting.

Figure 4. Organization Chart of the Merged IT Division

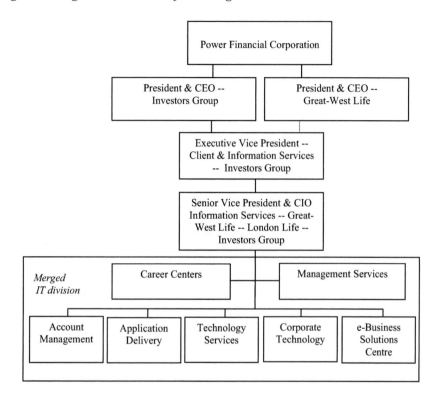

IT BSC PROJECT AND ITS ORGANIZATION

Before the merger, the CIO of Great-West Life (who is the present CIO of the merged IT division) began focusing on the scorecard as a (potentially) effective measurement tool. His objective was to ensure that IT was fairly evaluated. In his own words: "Through the Balanced Scorecard I would know what was important to the business and I would not fall victim to the early termination syndrome. Or at least I would have a better chance of survival." However, once the three companies came together through the acquisition and merger of the IT groups, the stakes were raised considerably. Now, the IT division had exposures on multiple fronts with stakeholders who were concerned about the perceived loss of control over their vital IT services. This prompted an executive request for a formal measure of factors to measure IT success. The response of the merged IT division was to formalize the criteria into a new and extended IT scorecard based on the experiences gained within Great-West Life.

Senior management of all the three companies questioned the benefits of huge investments in IT and how more value might be achieved through better alignment of business strategy and IT strategy. Within The Group the specific concerns for the different stakeholders are shown in Figure 5.

Figure 5. IT Concerns of the Different Stakeholders

Stakeholders	Key questions
Board of Directors Executive Management Committee	Does IT support the achievement of business objectives? What value does the expenditure on IT deliver? Are IT costs being managed effectively? Are IT risks being identified and managed? Are targeted inter-company IT synergies being achieved?
Business Unit Executives	Are IT's services delivered at a competitive cost? Does IT deliver on its service level commitments? Do IT investments positively affect business productivity or the customer experience? Does IT contribute to the achievement of our business strategies?
Corporate Compliance Internal Audit	Are the organization's assets and operations protected? Are the key business and technology risks being managed? Are proper processes, practices and controls in place?
IT Organization	Are we developing the professional competencies needed for successful service delivery? Are we creating a positive workplace environment? Do we effectively measure and reward individual and team performance? Do we capture organizational knowledge to continuously improve performance? Can we attract/retain the talent we need to support the business?

The concepts of the Balanced Scorecard and its application to information technology were discovered through an Internet search, primarily through the web site of the IT Governance Institute (www.itgi.org). Departing from this web site, relevant publications on the IT Balanced Scorecard from academics and practitioners were identified and consulted. It was believed that the scorecard could provide an answer to the key questions of the different stakeholders.

The formal development of the IT Balanced Scorecard began in 1998, and from the start the objectives were clearly stated:

- align IT plans and activities with business goals and needs,
- align employees' efforts toward IT objectives,
- establish measures for evaluating the effectiveness of the IT organization,

- stimulate and sustain improved IT performance,
- achieve balanced results across stakeholder groups.

At the beginning of the initial research period (December 1999), the situation was that the scorecard effort was not yet approached as a formal project and as a result, progress had been somewhat limited. In 2000 the formality of the project was increased and the CIO (Information Services Executive) was appointed as sponsor. In 2001, a project manager/analyst was formally assigned to the IT Balanced Scorecard project. Status to-date (end of 2002) is that the case company is still completing the scorecard: 66% of the measures are completed, 29% are in progress and 5% are not yet started.

BUILDING THE IT BSC

It was recognized by the CIO that building an IT BSC was meaningful under two conditions which required (1) a clearly articulated business strategy, and (2) the new Information Services Division moving from a commodity service provider to a strategic partner, as illustrated by Venkatraman (1999) (Figure 6).

The newly constructed ISD is viewed as a strategic partner. During several meetings between IT and executive management, the vision, strategy, measures of success and value of IT were jointly created. Typically, pure business objectives were used as the standard to assess IT. The vision and strategy of ISD were defined as:

- ISD is a single IT organization focused on developing world-class capabilities to serve the distinct customer needs of its three sponsoring companies,
- ISD operates as a separate professional services business on a full recovery, non-profit basis,
- ISD supports the achievement of company strategies and goals through the industry consolidation period,
- ISD becomes the "supplier of choice" of information services,
- ISD establishes a forward looking enterprise architecture strategy which enables the use of technology as a competitive edge in the financial service market place,
- ISD becomes the "employer of choice" for career-oriented IT professionals in the markets in which ISD and The Group operate.

Figure 6. IT Division as a Service Provider or Strategic Partner

Service provider	Strategic partner
• IT is for efficiency	• IT for business growth
• Budgets are driven by external benchmarks	• Budgets are driven by business strategy
• IT is separable from the business	• IT is inseparable from the business
• IT is seen as an expense to control	• IT is seen as an investment to manage
• IT managers are technical experts	• IT managers are business problem solvers

Figure 7. Perspective Questions and Mission Statements of the IT Strategic Scorecard

CUSTOMER ORIENTATION	CORPORATE CONTRIBUTION
Perspective question How should IT appear to business unit executives to be considered effective in delivering its services? **Mission** To be the supplier of choice for all information services, either directly or indirectly through supplier relationships.	**Perspective question** How should IT appear to the company executive and its corporate functions to be considered a significant contributor to company success? **Mission** To enable and contribute to the achievement of business objectives through effective delivery of value added information services.
OPERATIONAL EXCELLENCE	FUTURE ORIENTATION
Perspective question At which services and processes must IT excel to satisfy the stakeholders and customers? **Mission** To deliver timely and effective IT services at targeted service levels and costs.	**Perspective question** How will IT develop the ability to deliver effectively and to continuously learn and improve its performance? **Mission** To develop the internal capabilities to continuously improve performance through innovation, learning and personal organizational growth.

These issues go to the heart of the relationship between IT and the business and will be reflected in the IT strategic Balanced Scorecard, as is illustrated in Figures 7 and 8. Figure 7 shows the perspective questions and mission statements for the four quadrants: corporate contribution, customer orientation, operational excellence and future orientation. Figure 8 displays the measures for each perspective. The details regarding the individual perspectives and their measures are in the annex.

MATURITY OF THE DEVELOPED IT BSC

At the beginning of the project, the IT BSC was primarily focused on the operational level of the IT department. It was acknowledged from the beginning that this could not be the end result. Therefore, actions were started to go beyond the operational IT BSC and to measure the true value of IT at the business level. The organization established two ways to demonstrate the business value, one at service delivery level and one at the IT strategy level. As will be illustrated hereafter, the goal is to evolve to an IT strategic BSC that shows how the business objectives are enabled by IT.

A cascade of Balanced Scorecards has been established to create a link between the scorecards at the unit level and the overall business objectives (see Figure 9). A link between the IT BSC and the Business BSC is not yet implemented, as there is currently no formal Business BSC for the Group. The scorecards at the unit level are classified into three groups: operational services scorecards (e.g., IT service desk scorecard), gover-

Figure 8. IT Strategic Scorecard Framework

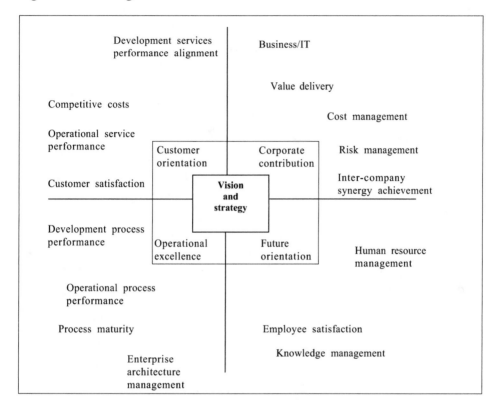

nance services scorecards (e.g., career center scorecard), and development services scorecards (e.g., application development scorecard). The measures of these unit scorecards are *rolled-up* or *aggregated* in the IT strategic Balanced Scorecard. This in turn is fed into and evaluated against the business objectives. In this way, the service (and value) delivered by IT is directly measured against the objectives of the overall business. Further, on an annual basis, the IT strategic BSC is reviewed by business and IT management and the result is fed back into the next annual planning cycle. This planning cycle defines what the business needs are and what IT must do to accomplish those needs.

For example, from the IT service desk scorecard (i.e., a unit scorecard which is situated in the operational services scorecard group), metrics such as average speed of answer, overall resolution rate at initial call and call abandonment rate (all three customer orientation metrics) are *rolled-up* to service level performance metrics in the IT strategic Balanced Scorecard. Other metrics of this unit scorecard, such as expense management (corporate contribution perspective), client satisfaction (customer orientation perspective), process maturity of incident management (operational excellence perspective) and staff turnover (future orientation perspective), will *aggregate* as part of the IT strategic scorecard. The overall view of the IT strategic Balanced Scorecard is then fed into and evaluated against the defined business objectives.

Figure 9. Cascade of Scorecards to Link Unit Scorecards, IT Strategic Scorecard and Business Objectives

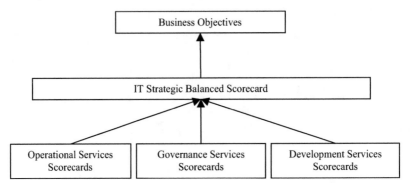

The second way to demonstrate business value is situated within the IT strategic Balanced Scorecard. The cause-and-effect relationships between performance drivers and outcome measures of the four quadrants are established as indicated in Figure 10. These connections help to understand how the contribution of IT towards the business will be realized: building the foundation for delivery and continuous learning and growth (future orientation perspective) is an enabler for carrying out the roles of the IT division's mission (operational excellence perspective) that is in turn an enabler for measuring up to business expectations (customer expectations perspective), that eventually must lead to ensuring effective IT Governance (corporate contribution perspective).

Figure 10. Cause-and-Effect Relationships within the IT Strategic Balanced Scorecard

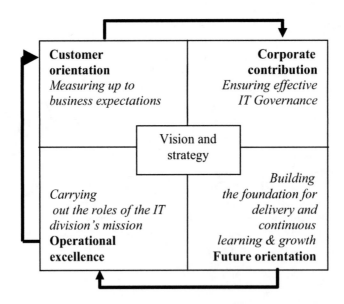

Figure 11. Maturity Levels for the IT Balanced Scorecard

Level 1 Initial
There is evidence that the organization has recognized there is a need for *a measurement system* for its information technology division. There are ad hoc approaches to measure IT with respect to the two main IT processes, i.e., operations and systems development. This measurement process is often an individual effort in response to specific issues.

Level 2 Repeatable
Management is aware of the concept of *the IT Balanced Scorecard* and has communicated its intent to define appropriate measures. Measures are collected and presented to management in a scorecard. Linkages between outcome measures and performance drivers are generally defined but are not yet precise, documented or integrated into strategic and operational planning processes. Processes for scorecard training and review are informal and there is no compliance process in place.

Level 3 Defined
Management has standardized, documented and communicated the IT BSC through formal training. The scorecard process has been structured and **linked to business planning cycle**. The need for compliance has been communicated but compliance is inconsistent. Management understands and accepts the need to integrate the IT BSC within the alignment process of business and IT. Efforts are underway to change the alignment process accordingly.

Level 4 Managed
The IT BSC is fully integrated into the strategic and operational planning and review systems of the business and IT. Linkages between outcome measures and performance drivers are systematically reviewed and revised based upon the analysis of results. There is a full understanding of the issues at all levels of the organization that is supported by formal training. Long term stretch targets and priorities for IT investment projects are set and linked to the IT scorecard. A business scorecard and **a cascade of IT scorecards** are in place and are communicated to all employees. Individual objectives of IT employees are connected with the scorecards and incentive systems are linked to the IT BSC measures. The compliance process is well established and levels of compliance are high.

Level 5 Optimized
The IT BSC is fully aligned with the business strategic management framework and vision is frequently reviewed, updated and improved. Internal and external experts are engaged to ensure industry best practices are developed and adopted. The **measurements** and results are part of management reporting and **are systematically acted upon** by senior and IT management. Monitoring, self-assessment and communication are pervasive within the organization and there is optimal use of technology to support measurement, analysis, communication and training.

Establishing the link with the business objectives through a cascade of scorecards and defining the cause-and-effect relationships within the scorecards are important steps in determining the maturity of the IT Balanced Scorecard. This maturity can be assessed through a maturity model. We therefore used an IT Maturity Model (MM) to match the case company's scorecard level against the levels of the IT MM. The IT MM we used is based on the Software Engineering Institute's Capability Maturity Model CMM (Paulk et al., 1993). Our IT BSC Maturity Model highlights five maturity levels with the characteristics found in Figure 11.

According to this IT BSC maturity model the case company is at the "Repeatable" stage (Level 2). Parts of the Level 3 maturity are achieved, but the basic principle of maturity assessment states that all conditions have to be fulfilled before moving to a higher maturity level. The challenge is to reach stage 4, the "Managed" level, within two to three years. It is understood that major milestones in this further development will be:

- the detailed cause-and-effect relationships between the output measures and performance drivers have to be further elaborated,

- short and long term targets have to be further defined,
- individual and group objectives of IT employees have to be further linked to the IT BSC,
- the scorecards have to be further integrated in the strategic and operational management processes.

The construction of cause-and-effect relationships is a critical issue in the further development of the IT strategic BSC. These relationships have not yet been explicitly defined, although they are implicit in the existing scorecard. E.g., the *Professional development days per staff member* measure (Figure 16 in annex) can be identified as a performance driver for the outcome measures *Development process performance* (Figure 15). The Corporate contribution perspective of Figure 13 is an enabler (performance driver) of the (generic) business objectives of the financial Group with its specific measures such as *Business/IT alignment, Value delivery, Cost management, Risk management*, and *Inter-company synergy achievement.* The CIO and its executive management are aware that an explicit articulation of these relationships has to be done and that it may help to improve the IT strategic BSC and its link with the business objectives, later on with the implementation of a Business BSC.

IT management is now in the process of determining how they might progress in terms of maturity level over time. The ultimate goal is to reach the "Managed" Level 4. Figure 12 displays the concrete improvements plans for the further development of the scorecard as articulated by IT management.

The improvement in unit cost measurements should enable the organization to break down the IT activities' costs in a more detailed level. The second improvement for 2002 refers to the approval processes of both the overall enterprise architecture and the systems level architectures delivered through major projects. The enterprise architecture dictates certain architectural and technical standards for application and technical systems and is reviewed and reapproved on a regular basis. In 2002, operational services baselines and targets and unit scorecards will be developed and a measurement process for personal development will be implemented. For 2003 and the short-term future, the development priority is set on the establishment of a risk management measure. The goal

Figure 12. Improvement Plans for the IT BSC

Improvements for 2002 (in progress)
• improvement in unit cost measurements
• target state and project architecture approval process
• development of operational service baselines and targets, and unit scorecards
• implementation of personal development days measurement process
Improvements for 2003 and later (in order of priority)
• risk management measure
• customer satisfaction
• 'state of the infrastructure' assessment
• 'lessons learned' sharing process
• explicit articulation of cause-and-effect relationships

is to develop an overall risk management strategy and measure our attainment of the defined target state risk level. Next, a regular survey process using generic questions needs to be developed to measure customer satisfaction and a process for assessing the "state of the infrastructure" will be implemented. This assessment compares the status-to-date of the existing infrastructure against the "to-be" position. Lowest priority but scheduled on the short- and midterm improvements plan is the sharing process of "lessons learned" on development projects.

Based on our maturity model shown in Figure 11, actions for the subsequent years should put the case company close to Level 4. It is the belief of the CIO that these plans are realistic but that "this desired timeline is probably quite optimistic and it may well take twice as long to accomplish these changes." However, the most important aspect is that all stakeholders in the process were engaged by the end of 2001 and that progress is made each subsequent year.

LESSONS LEARNED

The following lessons can be attributed to this IT BSC case:

1. *Start simultaneously constructing a business and IT scorecard.* The IT BSC within the case company was started within the IT organization primarily with the objective to ensure that IT is fairly evaluated by the business. This is a rather defensive approach and focuses merely on the internal IT processes. Although it is clearly recognized within the case company that a more explicit linkage with the business (with a business Balanced Scorecard) has to be developed and supported, the question still remains whether it is more appropriate (a) to start with a business Balanced Scorecard followed by the subsequent creation of the corresponding IT scorecards or (b) to develop both scorecards simultaneously. It is now our conclusion that it is probably more ideal to start simultaneously with both scorecards which requires both IT and senior management to discuss the opportunities of information technologies, which supports the IT/business alignment and IT Governance process.

2. *Consider the scorecard technique as a supportive mechanism for IT/business alignment and IT Governance.* Recurring issues in IT practice and IT academic publications focus on how to align IT and business and how to control IT. It is our strong belief that a cascade of business and IT Balanced Scorecards may support both processes. However, as is shown in this case study, the Balanced Scorecard is only a technique that can only be successful if the business and IT work together and act upon the measurements of the scorecards. The Balanced Scorecard approach will only have results when other mechanisms such as a well functioning Board and IT Steering Committee are in place.

3. *Consider the construction and implementation of an IT Balanced Scorecard as an evolutionary project.* Constructing an IT Balanced Scorecard is not a one week project. It requires considerable time and other major resources. Moreover, it is a project that is to be matured over time and that is characterized by different stages, as is illustrated by the IT BSC Maturity Model introduced in this paper. This iterative approach is confirmed by this case. The described IT BSC began at a lower level with actions currently in place to reach a higher level where a more explicit

connection exists between outcome measures and performance drivers, and where an explicit linkage is established with business requirements.

4. *Provide a formal project organization.* Good project management is a critical success factor for effective construction and implementation of an IT Balanced Scorecard. IT management of the case company confronted with the question of how the IT BSC project was organized had to admit that in the beginning, there was no real formal organization in place and that this delayed the progress of its implementation. Currently, the sponsor of the IT BSC is the CIO, and one full-time project manager is assigned to the project. A group of 15 individuals have key roles and accountabilities for scorecard deliverables.

5. *Provide best IT practices.* Introducing an IT Balanced Scorecard in an IT environment with poor management and IT practices is too large a challenge. The implementation of the IT BSC within the case company was certainly supported by practices already in place, such as ROI-evaluation of IT projects, the existence of IT steering committees, Service Level Agreement practices, etc. If it is decided to implement, e.g., the Information Economics approach to score and evaluate projects and to integrate this method within the IT BSC, this will take considerable time and is to be seen as a separate project.

6. *Revisit the dynamic measures.* The implementation of the IT Balanced Scorecard requires the establishment and definition of a large number of metrics. As business requirements change, these metrics are dynamic and should be reevaluated on a regular basis. Most important in this regard is that it should always remain very clear why a certain issue is measured, i.e., what the value is of measuring it. When this value cannot be demonstrated any more, a measure should be challenged and changed or replaced by another one.

CONCLUSION

In this chapter, the development and implementation of an IT Balanced Scorecard within a large Canadian insurance group is described and discussed. It was shown that building and implementing such a scorecard is a project that needs substantial human and financial resources. Furthermore, setting up an IT BSC is a project that is characterized by different phases in time. The current status of the case scorecard is Level 2 of the IT BSC Maturity Model that is introduced in this paper. This implies that the case IT scorecard to date has to be linked with the business scorecard or at least the business objectives to support the IT/business alignment process and the IT Governance process. Currently, a plan for the next two years has been developed with the objective to build a mature IT BSC explicitly linked to the business. It is recognized within the case company that this will be a great challenge for both IT and business people.

The case under review illustrated one of the most crucial issues in building and implementing an IT strategic Balanced Scorecard: its required linkage with the business objectives. To create this link, a cascade of Balanced Scorecards has been established with the lower level unit scorecards for the operational and development services. The measures of these unit scorecards are rolled-up or aggregated in the IT strategic scorecard that ultimately realizes the link with the business objectives through its corporate contribution perspective. The precise articulation of the cause-and-effect relationships through the identification of outcome measures and their corresponding performance drivers seemed to be a critical success factor. These relationships are implicit in the current IT strategic Balanced Scorecard but are to be defined more explicitly.

REFERENCES

Bensabat, I., Goldstein, D., & Mead, M. (1987). The case research strategy in studies of information systems. *MIS Quarterly, September*, 369-386.

Gold, C. (1992). *Total quality management in information services – IS measures: A balancing act.* Boston, MA: Research Note Ernst & Young Center for Information Technology and Strategy.

Gold, C. (1994). *U.S. measures – A balancing act.* Boston, MA: Ernst & Young Center for Business Innovation.

IT Governance Institute. (2000). *CobiT (Control Objectives for IT and related Technologies) 3e edition.* Available online: www.isaca.org/cobit.htm.

Kaplan, R., & Norton, D. (1992). The Balanced Scorecard – measures that drive performance. *Harvard Business Review,* (January/Febraury), 71-79.

Kaplan, R., & Norton, D. (1993). Putting the Balanced Scorecard to work. *Harvard Business Review,* (September/October), 134-142.

Kaplan, R., & Norton, D. (1996a). Using the Balanced Scorecard as a strategic management system. *Harvard Business Review,* (January/February), 75-85.

Kaplan, R., & Norton, D. (1996b). *The Balanced Scorecard: Translating vision into action.* Boston, MA: Harvard Business School Press.

Paulk, M., Curtis, B., Chrissis, M.B., & Weber, C. (1993). Capability maturity model for software, version 1.1. *Technical Report Software Engineering Institute,* CMU/SEI-93-TR-024, ESC-TR-93-177.

Saull, R. (2000). The IT Balanced Scorecard – A roadmap to effective governance of a shared services IT organization. *Information Systems Control Journal* (previously *IS Audit and Control Journal*), *2,* 31-38.

Van der Zee, J. (1999). Alignment is not enough: Integrating business and IT management with the Balanced Scorecard. *Proceedings of the 1st Conference on the IT Balanced Scorecard, Antwerp* (March, pp. 1-221).

Van Grembergen, W. (2000). The Balanced Scorecard and IT governance. *Information Systems Control Journal* (previously *IS Audit & Control Journal*), *2,* 40-43.

Van Grembergen, W., & Timmerman, D. (1998). Monitoring the IT process through the balanced score card. *Proceedings of the 9th Information Resources Management (IRMA) International Conference, Boston, MA* (May, pp. 105-116).

Van Grembergen, W., & Van Bruggen, R. (1997). Measuring and improving corporate information technology through the Balanced Scorecard technique. *Proceedings of the Fourth European Conference on the Evaluation of Information technology, Deflt,* (October, pp. 163-171).

Venkatraman, N. (1999). *Valuing the IS contribution to the business.* Computer Sciences Corporation.

Willcocks, L. (1995). *Information management. The evaluation of information systems investments.* London: Chapman & Hall.

Yin, K. (1994). *Case study research. Design and methods.* Thousand Oaks, CA: Sage
    Publications.

** A version of this chapter also appeared in the Journal of Information Technology Cases and
Applications, 5(1), 2003.*

APPENDIX:
DETAILS ON THE MEASURES OF
THE IT STRATEGIC SCORECARD

In this annex, the IT strategic Balanced Scorecard is discussed in more detail (see
also Saull, 2000). In each of the four quadrants, the objectives, measures and benchmarks
will be elaborated. Many of the measures are rolled-up or aggregated from the unit
scorecards (e.g., data center scorecard) to metrics in the IT strategic scorecard. Some of
the measures defined in the IT strategic scorecard are high-level but cover specific
concrete metrics. At this moment, the collecting process of the data is often very labor-
intensive, but it is the belief of management that first the correct measures have to be
defined before implementing tools that can automate the data collecting process.

Corporate Contribution Scorecard

The Corporate contribution perspective evaluates the performance of the IT
organization from the viewpoint of executive management, the Board of Directors and the
shareholders, and provides answers to the key questions of these stakeholders concerning IT Governance (cf., Figure 5). The key issues, as depicted by Figure 13, are business/
IT alignment, value delivery, cost management, risk management and inter-company
synergy achievement. Benchmarks have been used where an objective standard was
available or could be determined in most cases from external sources.

The main measurement challenges are with the areas of business/IT alignment and
the value delivery.

Currently, *business/IT alignment* is measured by the approval of the IT operational
plan and budget. Although not a discrete measure of alignment, the approval process
within the Group is particularly thorough and as a result is accepted by business
executives as a good indicator. All aspects of development, operations and governance/
support services are examined and challenged to ensure they are essential to achieving
business objectives or supporting the enabling IT strategy.

In the *value delivery* area, the performance of a specific IT services group delivering
to a specific business unit (e.g., 'group insurance' services) is measured. For each
business unit, specific metrics are and/or will be defined. The ultimate responsibility for
achieving and measuring the business value of IT rests with the business and is reflected
in the business results of the individual lines of business in different ways, depending
on the nature of value being sought.

Cost management is a traditional financial objective and is in the first place
measured through the attainment of expense and recovery targets. The expenses refer
to the costs that the IT organization has made for the business, and the recovery refers
to the allocation of costs to IT services and the internal charge back to the business. All

IT costs are fully loaded (no profit margin) and recovered from the lines of business on a fair and equitable basis as agreed to by the companies' CFOs. Comparisons with similar industries will be drawn to benchmark these metrics. Next to this, IT unit costs (e.g., application development) will be measured and compared to the 'top performing levels' benchmark provided by Compass.

The development of the *risk management* metrics are the priority for the upcoming year. At this moment, the results of the internal audits are used and benchmarked against criteria provided by OSFI, the Canadian federal regulator in the financial services sector. The execution of the Security Initiative and the delivery of a Disaster Recovery Assessment need to be accomplished in the upcoming year. This will enable the business to get an insight on how well it is prepared to respond to different disaster scenarios.

Synergy achievement is measured through the achievement of single system solutions, targeted cost reductions and the integration of the IT organizations. This measure is very crucial in the context of the merger of the three IT organizations in the sense that it enables a post-evaluation of this merger and demonstrates to management whether the new IT organization is effective and efficient. The selection of single system solutions was a cooperative effort between business leaders and IT staff, resulting in a "Target State Architecture" depicting the target applications architecture. The synergy targets were heavily influenced by the consulting firm (Bain & Co.) that was used to assist in evaluating the London Life acquisition and the tri-company IT merger potential. The consultants suggested specific dollar reduction targets for technology services (IT operations) and application delivery services (IT development), largely based on norms they had developed from their previous merger and acquisition work. The approval of the Target State Architecture plan and the attainment of the targeted integration cost reductions will be measured. The IT organization integration metric refers to the synergies within the IT organization, e.g., is there one single service desk for the three companies or are there three different ones?

Customer Orientation Scorecard

The Customer orientation perspective evaluates the performance of IT from the viewpoint of internal business users (customers of IT) and, by extension, the customers of the business units. It provides answers to the key questions of these stakeholders concerning IT service quality (cf., Figure 5). As shown in Figure 14, the issues this perspective focuses on are competitive costs, development services performance, operational services performance and customer satisfaction.

In the *Customer satisfaction* area, the IT BSC of the merged IT organization is relying on annual interviews with key business managers. It is the intent to set up one generic survey, which can be re-used, with relevant questions that cover the topics mentioned in Figure 14.

Insight into the *competitive costs* area can demonstrate to the business how cost competitive the IT organization is compared to other (e.g., external) parties. This insight is realized by measuring the attainment of IT unit cost targets and the blended labor rate. This rate model provides an overall single rate for any IT professional who is appointed to the business. The competitive costs measures are benchmarked against Compass's operational 'Top Performing level' and against the offerings of commercial IT service vendors (market comparisons).

Figure 13.   Corporate Contribution Scorecard

Objective	Measures	Benchmarks
Business/IT alignment	• Operational plan/budget approval	• Not applicable
Value delivery	• Measured in business unit performance	• Not applicable
Cost management	• Attainment of expense and recovery targets • Attainment of unit cost targets	• Industry expenditure comparisons • Compass operational 'top performance' levels
Risk management	• Results of internal audits • Execution of Security Initiative • Delivery of disaster recovery assessment	• OSFI sound business practices • Not applicable • Not applicable
Inter-company synergy achievement	• Single system solutions • Target state architecture approval • Attainment of targeted integration cost reductions • IT organization integration	• Merger & Acquisition guidelines • Not applicable • Not applicable • Not applicable

Development services performance measures are project-oriented, using attributes such as goal attainment, sponsor satisfaction and project governance (i.e., the way the project is managed). These data are mostly captured by interviews with key managers. The most effective time to establish the basis for these (project) development measures is at the point where business cases are being prepared and projects are evaluated. Each IT project initiative will be evaluated by the IS Executive Committee in which IT and business managers determine — based on the business drivers, budget and state architecture compliance — which projects need to be executed. When a project is approved, the project manager defines clear targets for cost, schedule, quality, scope and governance. The quantitative data (e.g., budget) are reported throughout the lifecycle of the project. After completion of the project, the quantitative and qualitative data are evaluated during the major project review and the main success drivers, delivery issues and lessons learned are documented.

In terms of *operational service performance,* IT management measures achievement against targeted service levels. For each operational unit (e.g., data center), average response time, service availability and resolution time for incidents are rolled-up to these service performance metrics in the strategic Balanced Scorecard. The results are benchmarked against the performance of competitors.

Operational Excellence Scorecard

The operational excellence scorecard provides the performance of IT from the viewpoint of IT management (process owners and service delivery managers) and the

Figure 14. Customer Orientation Scorecard

Objective	Measures	Benchmarks

Customer satisfaction	• Business unit survey ratings:
	   • Cost transparency and levels
	   • Service quality and responsiveness
	   • Value of IT advice and support
	   • Contribution to business objectives
		• Not applicable

Competitive costs	• Attainment of unit cost targets • Blended labor rates	• Compass operational 'Top Level Performing' levels • Market comparisons

Development services performance	• Major project success scores
	   • Recorded goal attainment
	   • Sponsor satisfaction ratings
	   • Project governance rating
		• Not applicable

Operational services performance	• Attainment of targeted service levels	• Competitor comparisons

audit and regulatory bodies. The operational excellence perspective copes with the key questions of these stakeholders and provides answers to questions of maturity, productivity and reliability of IT processes (cf., Figure 5). The issues that are of focus here, as displayed in Figure 15, are development process performance, operational process performance, process maturity and enterprise architecture management.

In relation to *development process performance*, function point-based measures of productivity, quality and delivery rate such as number of faults per 100 installed function points and delivery rate of function points per month are defined. Benchmark data on industry performance will be gathered from a third party (e.g., Compass). In the operational process performance area, measures of productivity, responsiveness, change management effectiveness and incident occurrence level are benchmarked against selected Compass studies (e.g., on data centers, client server, etc.).

The *process maturity* is assessed using the CobiT (Control Objectives for IT and related Technology) framework and maturity models (ITGI, 2000). CobiT identifies 34 IT processes within four different domains (see Figure 15) and describes detailed maturity levels for each of these processes. The Group has identified 15 out the 34 priority processes that should have a maturity assessment in 2003 and the other processes will be measured later.

Enterprise architecture management deals with the IT responsibility to define an enterprise architecture which supports long-term business strategy and objectives and to act as a steward on behalf of business executives to protect the integrity of that architecture. Major project architecture approval measures the compliance of net new systems as they are proposed, developed and implemented. Product acquisition compliance technology standards measure our adherence to detailed technology standards which are at the heart of minimizing technology diversity and maximizing inter-company technology synergies. The "State of the Infrastructure" assessment measures the degree to which IT has been able to maintain a robust and reliable infrastructure as required to deliver effectively to business needs. It does so by comparing each platform area against risk-based criteria for potential impact to business continuity, security and/or compliance.

Future Orientation Perspective

The future orientation perspective shows the performance of IT from the viewpoint of the IT organization itself: process owners, practitioners and support professionals. The future orientation perspective provides answers to stakeholder questions regarding IT's readiness for future challenges (cf., Figure 5). The issues focused on, as depicted

Figure 15. Operational Excellence Scorecard

Objective	Measures	Benchmarks

Development process performance	• Function point measures of:
	   • Productivity
	   • Quality
	   • Delivery rate
		• To be determined

Operational process performance	• Benchmark based measures of:
	   • Productivity
	   • Responsiveness
	   • Change management effectiveness
	   • Incident occurrence levels
		• Selected Compass benchmark studies

Process maturity	• Assessed level of maturity and compliance in priority processes within:
	   • Planning and organization
	   • Acquisition and implementation
	   • Delivery and support
	   • Monitoring
		• To be defined

Enterprise architecture management	• Major project architecture approval • Product acquisition compliance to technology standards • 'State of the infrastructure' assessment	• Not applicable

Figure 16. Future Orientation Scorecard

Objective	Measures	Benchmarks

Human resource management	• Results against targets:
	   • Staff complement by skill type
	   • Staff turnover
	   • Staff 'billable' ratio
	   • Professional development days per staff member
		• Not applicable
		
		• Market comparison
		• Industry standard
		• Industry standard

Employee satisfaction	• Employee satisfaction survey scores in:
	   • Compensation
	   • Work climate
	   • Feedback
	   • Personal growth
	   • Vision and purpose
		• North American technology-dependent companies

Knowledge management	• Delivery of internal process improvements to 'Cybrary' • Implementation of 'lessons learned' sharing process	• Not applicable • Not applicable

in Figure 16, are human resources management, employee satisfaction and Knowledge Management. The metrics that will appear in the future orientation quadrant of the IT strategic Balanced Scorecard are in many cases the aggregated results of measures used in the unit scorecards (e.g., career center).

Human resource management is an objective that is tracked by comparing measures as described in Figure 16 against predefined targets: the staff complement by skill type (number of people with a certain profile, e.g., systems analyst), staff turnover, staff 'billable' ratio (i.e., hours billed/total hours salary paid; if this ratio can be increased, the IT organization can charge lower rates to the business for the IT assigned people), and professional development days per staff member.

Employee satisfaction is measured by using surveys with questions relating to compensation, work climate, feedback, personal growth, and vision and purpose. Benchmark data of North American technology-dependent companies are provided by a third party.

In the *Knowledge Management* area, the delivery of internal process improvements to the 'Cybrary' is very important. The 'Cybrary' refers to the intranet that all employees can access for seeking and sharing knowledge. To measure improvements, metrics (e.g., number of hits per day on the Cybrary) still need to be developed. Closely linked to this, Knowledge Management is also measured by the implementation of the 'lessons learned' sharing process. Here too, specific metrics still need to be developed.

Chapter VI

Measuring and Managing E-Business Initiatives through the Balanced Scorecard

Wim Van Grembergen
University of Antwerp, Belgium

Isabelle Amelinckx
University of Antwerp, Belgium

ABSTRACT

The Balanced Scorecard (BSC) initially developed by Kaplan and Norton is a performance measurement system that supplements traditional financial measures with the criteria that measure performance from three additional perspectives: customer perspective, internal business perspective, and innovation and learning perspective. In recent years, the Balanced Scorecard has been applied to information technology in order to ensure that IT is fairly evaluated. The proposed methodology can also be applied to e-business initiatives. In this chapter, it is illustrated how the BSC can be used to measure and manage e-business initiatives. A generic e-business Balanced Scorecard is proposed and its development and implementation is discussed.

INTRODUCTION

As we enter the new millennium, the internet-based way of doing business is certainly going to change whole industries and markets and will therefore have a great impact on consumers and businesses. Electronic business (e-business) can be described as the process of buying and selling or the exchanging of products, services, and information; generating demand for them through marketing and advertising; servicing customers; collaborating with business partners; and conducting electronic transactions within an organization via computer networks including the internet (Turban et al., 2000). Similar definitions exist (see e.g., Hartman et al., 2000): "An e-business initiative is any internet initiative — tactical or strategic — that transforms business relationships, whether those relationships are business-to-consumer, business-to-business, intrabusiness or consumer-to-consumer." Electronic business will change all aspects of our lives — how we work, play, learn and shop. It will transform our economic infrastructure in the sense that new methods of supply, distribution, marketing, service, and management will emerge. E-business will improve business performance through low cost and open connectivity by introduction of new technologies in the value chain and connecting value chains across businesses in order to improve service, reduce costs, open new channels, and transform the competitive landscape.

Companies are becoming increasingly aware of the many potential benefits provided by e-business. Some of the e-business potential benefits for organizations are (Turban et al., 2000; Amor, 2000):

- supporting Business Process Reengineering (BPR) efforts;
- expanding the marketplace;
- strengthening the relationships with customers and suppliers;
- reducing costs through the deployment of electronic internal and external business processes;
- lowering telecommunications costs as a result of the inexpensive internet infrastructure.

Because of the intangible nature of some of these benefits, it is difficult to measure the contribution of e-business initiatives to business performance and to manage these initiatives to ensure that real profits are realized (Giaglis et al., 1999). In practice, when starting an e-business initiative, organizations focus too much on the technology (Rifkin & Kurtzman, 2002; Barua et al., 2001). An example is the Belgian online grocery store *Ready.be* that used its web storefront to take customers' orders. It relied heavily on manual processes to fulfill the orders. In less than two years, *Ready.be* set up a centralized warehouse and fifty points of distribution where customers could pick up their purchases they made through the internet. Besides this, *Ready.be* renewed its web site three times and even started a WAP (Wireless Application Protocol) project that would allow customers to mail their shopping list via their mobile phone. In the year 2000 the losses of *Ready.be* amounted to 12 million Euros and the online grocery had to stop its business. This mini case shows that too much attention was paid to the technology and that this e-business initiative should have been monitored. In the grocery case, the use of a monitoring instrument could easily have shown that too many costs were made that could be avoided by just using the existing warehouses and shops of their traditional grocery chain and by not starting yet the WAP project (this pervasive computing project was clearly technically driven). Therefore, in this chapter a recent developed monitoring instrument, the Balanced Scorecard, will be presented and applied to e-business initiatives.

The need for measuring e-business performance is confirmed by a study conducted by the consulting firm Accenture (formerly Andersen Consulting) and the Cranfield School of Management (Adams et al., 2001). Senior managers from more than 70 bricks-and-mortar, clicks-and-mortar, and dot.com firms were surveyed regarding their performance management systems. One of the major findings is that dot.coms appear to measure more than the two other types of businesses, but that may be misguided because numerous publications reveal that e-businesses are failing to deliver the expected service and even go bankrupt. We agree with one of the study's conclusions that "even if they do have the data, they would appear to be failing to act on it." Another observation is that too many dotcoms are "obsessed with measurement rather than management." The deployment of an e-business Balanced Scorecard may overcome these problems if it is implemented as a measurement and management system.

IT EVALUATION

Evaluation is often considered as a process to diagnose malfunctions and to suggest appropriate planning and treatment by providing feedback information and contributing to organizational planning. It is generally aimed at the identification and quantification of costs and benefits. There are different monitoring instruments available and these are dependent on the features of the costs and benefits.

When both costs and benefits can be easily quantified and assigned a monetary value, traditional financial performance measures work well. These "hard" evaluation techniques draw upon the traditional skills of the financial analyst. Whatever features are looked at, they have to be given numeric values. Figure 1 shows an overview of the most popular IT evaluation techniques.

Return On Investment (ROI): Return on investment is the ratio of average annual net income of the project divided by the internal investment in the project. To find the ROI, first the average net benefit has to be calculated. This net benefit is divided by the total initial investment to arrive at ROI. The weakness of ROI is that it can ignore the time value of money.

Net Present Value (NPV): Evaluating a capital project requires that the cost of an investment be compared with the net cash inflows that occur many years later. But these two kinds of inflows are not directly comparable because of the time value of money. Thus, to compare the investment with future savings, you need to discount the earnings to their present value and then calculate the net present value of the investment. To calculate the NPV, the opportunity cost of capital is used as discount rate. The net present value is the amount of money an investment is worth, taking into account its costs, earnings and other time value of money.

Internal Rate of Return (IRR): The IRR is defined as the rate of return or profit that an investment is expected to earn. It is a variation of the NPV method but calculated using an interest rate that will cause the NPV to equal zero. This is also called the yield of the investment and is often used to define a hurdle rate. The IRR intends

Figure 1. IT Evaluation

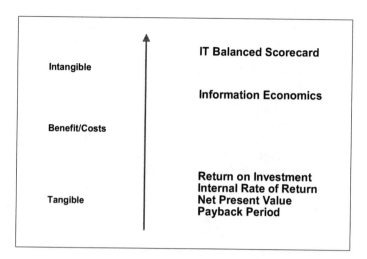

to make projects more comparable by showing what interest would have to be received on the sum of the investment to get the same return offered.

Payback Method: The payback method is a measure of the time required to pay back the initial investment of a project. It is the period between the moment that an IT investment gets funded and the moment that the total sum of the investment is recovered through the net incoming cash flows. A short payback period reduces the risk of the project and many organizations will select projects on the basis of short payback periods, even at the cost of lower net benefit levels. The payback period calculations are easy and hence popular but they suffer from the fact that they take no account of the time value of money, the amount of cash flow after the payback period, the disposal value (usually zero with computer systems) and the profitability of the investment.

Because they need monetary values for benefits and costs, the traditional financial performance measures such as ROI, NPV, IRR, and the payback method are problematic in measuring IT and e-business investments. Multi-criteria methods may solve this problem because they account for tangibles as well as for intangible impacts (Berghout & Renkema, 2001). One of the best known multi-criteria methods is Information Economics (Parker, 1996). This method uses an extended form of the ROI, which includes an assessment of intangible elements. The greatest weakness of this technique is the fact that it is a scoring technique with the difficulty that one has to agree on scores (Robson, 2001).

The Information Economics method is a scoring technique whereby the value and risk categories are attributed scores between 0 and 5. For a value category, 0 means "no positive contribution" and a 5 refers to "a large positive contribution". For a risk category, 0 means "no risk" and a 5 means "large risk". Each of these categories is assigned a weight. By adding the weighted scores of the value categories and subtracting the weighted scores of the risk categories, one can calculate the weighted scores of each

IT project. The categories of Table 1 have an indicative meaning, so when applying this method one has to adapt these categories to his own needs and specifications.

Although most categories of Table 1 are clear, a few of them might require some explanation. *Value linking* incorporates, e.g., interest savings due to an accelerated cashing of invoices realized through electronic payment. *Value acceleration* contains the additional cash flows due to reduced time scale for e-business operations. *Value restructuring* refers to the efficiency and effectiveness of the employees. *Innovation valuation* incorporates the additional cash flows arising from the innovating aspects of the e-business investment. *Strategic IT architecture* assesses the degree of value to which the project fits into the e-business plan. *Business strategy risk* and *IT strategy risk* respectively refer to the degree of risk in terms of how well the company or IT department succeeds in achieving its strategic objectives. *Definitional uncertainty* indicates the degree of risk in terms of how clearly the functional requirements and specifications have been agreed upon. The bigger the change, the bigger the risk and the larger the negative score for the project. *Technical risk* assesses the skills, hardware and software innovations required by the project to attach a score to the uncertainty. The more innovative, the bigger the risk and the bigger the negative score for the project.

Although Information Economics do consider non-quantifiable elements such as flexibility, however, no rational economic value can be assigned.

The solution for our measurement problem seems to be the Balanced Scorecard technique developed in the 90s, which is a measurement system that also takes into account intangible elements. Moreover, it enables businesses to derive strategies based on measurement and follow-up. In recent years, the BSC has also been applied to information technology.

Table 1. Information Economics

Traditional ROI (+)		
+ value linking (+) + value acceleration (+)	+ value restructuring (+) + innovation (+)	
= Adjusted ROI	+ Business Value	+ IT Value
	• strategic match (+) • competitive advantage (+) • competitive response (+) • management information (+) • service and quality (+) • environmental quality (+) • empowerment (+) • cycle time (+) • mass customization (+)	• strategic IT architecture (+)
	- Business Risk	- IT Risk
	• business strategy risk (-) • business organization risk (-)	• IT strategy risk (-) • Definitional uncertainty (-) • Technical risk (-) • IT service delivery risk (-)
= VALUE (business contribution)		

BALANCED SCORECARD (BSC)

In the 1990s, Kaplan and Norton developed the Balanced Scorecard. Their idea is that the evaluation of a company should not be restricted to the traditional financial performance measures but should be supplemented with measures concerning customer satisfaction, internal processes, and the ability to innovate. Results achieved within the additional perspectives should assure future financial results (Kaplan & Norton, 1992, 1993, 1996a, 1996b).

Kaplan and Norton propose a three layered structure for the four perspectives: mission (to become the customers' most preferred supplier), objectives (to provide the customers with new products), and measures (percentage of turnover generated by new products). To put the BSC to work, companies should translate each of the perspectives into corresponding metrics and measures that assess the current situation. These assessments have to be repeated periodically and have to be confronted with the goals that have to be set beforehand. At first, the BSC was used as a performance measurement system and a planning and control device. Later on, some companies moved beyond this early vision of the scorecard. They discovered that the measures on a Balanced Scorecard can be used as the cornerstone of a management system that communicates strategy, aligns individuals and teams to the strategy, establishes long-term strategic targets, aligns initiatives, allocates long- and short term resources and finally, provides feedback and learning about the strategy (Kaplan & Norton, 1992, 1993, 1996a, 1996b).

GENERIC IT BALANCED SCORECARD

Different market situations, product strategies, business units, and competitive environments require different scorecards to fit their mission, strategy, technology, and culture. The general BSC-framework can be translated to the more specific needs of the monitoring and evaluation of the IT function, and recently the IT BSC has emerged in practice (Graeser et al., 1998; Van Grembergen & Saull, 2001). In Van Grembergen and Van Bruggen (1997) and Van Grembergen and Timmerman (1998) a generic IT scorecard is proposed consisting of four perspectives: business contribution, user orientation, operational excellence, and future orientation (Table 2). This IT scorecard differs from the company-wide BSC because it is a departmental scorecard for an internal service supplier (IT): the customers are the computer users, the business contribution is to be considered from management's point of view, the internal processes under consideration are the IT processes (systems development and operations), and the ability to innovate is measuring the use of new technologies and the human IT resources.

A detailed version of the IT BSC model is depicted in Table 2.

GENERIC E-BUSINESS
BALANCED SCORECARD

The costs that go with the development, implementation, and the maintenance of an e-business initiative can be very high. Therefore, e-business projects, like any other projects, need to be evaluated and monitored to find out if the project delivers what it

Table 2. Balanced Scorecard Applied to IT

USER ORIENTATION **How do the users view the IT department?**	BUSINESS CONTRIBUTION **How does management view the IT department?**
Mission To be the preferred supplier of information systems and to exploit business opportunities maximally through information technology *Objectives* • preferred supplier of applications • preferred supplier of operations • partnership with users • user satisfaction	*Mission* To obtain a reasonable business contribution of investment in IT *Objectives* • control of IT expenses • sell IT products and services to third parties • business value of new IT projects • business value of the IT function
OPERATIONAL EXCELLENCE **How effective and efficient are the IT processes?**	FUTURE ORIENTATION **Is IT positioned to meet future needs?**
Mission To deliver efficiently IT products and services *Objectives* • efficient software development • efficient operations • acquisition PCs and PC software • problem management • user education • managing IT staff • use of communication software	*Mission* To develop opportunities to answer future challenges *Objectives* • permanent training and education of IT staff • expertise of IT staff • age of application portfolio • research into emerging information technologies

was supposed to deliver (Giaglis et al., 1999). This means that performance measurement is one of the most important activities that occur once an e-business initiative is started. However, many of the e-business benefits such as better customer service, increased responsiveness, and faster deliveries are intangibles that are difficult to translate into monetary benefits. Raisinghani (2001) reports that three-fourths of information systems investments, ranging from data centers to web sites, offer no calculable business value.

Within an IT scorecard, we certainly will have some contribution measures regarding e-business initiatives, but within the e-business scorecard we will have more detailed and in-depth e-business initiative measures that will allow us to monitor and manage these emerging initiatives. Similar to Rosemann (2001) who applies the BSC to ERP (Enterprise Resource Planning) projects, the scorecard concepts are applied to another type of projects, in casu e-business projects (see also, e.g., Hasan & Tibbits, 2000).

In Table 3 a generic e-business Balanced Scorecard is proposed. The *Customer Orientation* perspective represents the evaluation of the consumer and business clients of the web site and in this way also the supporting back office systems. The *Operational Excellence* perspective represents the e-business processes employed to deliver the demanded services and the e-business applications. The *Future Orientation* perspective represents the human and technology resources needed by the e-business initiative to deliver its services over time. The *Business Contribution* perspective captures the

Table 3. Generic E-Business Balanced Scorecard

CUSTOMER ORIENTATION	BUSINESS CONTRIBUTION
What is the company's success in acquiring and retaining customers through the web site?	**How should the e-business initiative appear to the Board in order to be considered as significant contribution to company success?**
Mission To be the preferred supplier through the Internet *Objectives* • customer satisfaction • customer retention • acquiring new customers • effective internet marketing	*Mission* To enable and contribute to the achievement of business strategies through effective application of e-business *Objectives* • e-business strategic plan achievements • business value of e-business initiative • compliance with budget
OPERATIONAL EXCELLENCE	FUTURE ORIENTATION
At which services and processes must the e-business application excel to satisfy the stakeholders and customers?	**How will IT develop the ability to change and improve in order to better achieve the company's strategy through e-business application?**
Mission To deliver timely and effective e-business services at targeted service levels *Objectives* • fulfillment process • availability of the e-business system • improvement of system development • security and safety	*Mission* To enable and contribute to the achievement of business strategies through effective application of e-business *Objectives* • e-business expertise of developers • e-business staff management effectiveness • independence of consultants • reliability of software vendors

business value created from the e-business investments. In building this generic e-business scorecard, performance measures defined in Van Grembergen and Saull (2001), Rosemann (2001), and Chaffey et al. (2000) are integrated into this scorecard.

Measures for Business Contribution

The ultimate goal of e-business initiatives, as any other IT initiative, is to satisfy the Board of Directors and consequently the shareholders. Surprisingly, the Adams' survey (2001) revealed that only 56% of the dot.com companies tracked shareholder satisfaction, which is low if one takes into consideration their reliance on the stock exchange and investors (for the bricks-and-mortar and the clicks-and-mortar firms percentages were even lower, with respectively 36% and 44%). We suppose that this is caused by the fact that many e-business initiatives are technically driven and that not always is the business evaluation taken too seriously. Therefore, in a *Balanced Scorecard*, the business contribution perspective is as important — not to say the most important — as the other three perspectives. Motivated, trained, and experienced IT employees (future orientation) should improve the delivery of excellent e-business processes and applications (operational excellence), that in turn should enhance customer satisfaction (customer orientation), and finally should by all means result in financial profits (business contribution).

The key issues, as depicted by Table 4, are e-business strategic plan achievements, business value of e-business initiatives, and compliance with budget.

E-Business Strategic Plan Achievements

E-business initiatives often are deployed on the basis of a step-by-step approach. A well-known Gartner-model describes four levels of e-business: (1) the publishing level, focusing on showing information of the company on a web site, (2) the prospecting level with customer oriented information, (3) the business integration level, which is transaction-centric and can be defined as e-commerce, and (4) the business transformation level, which is the mature level of e-business that includes supplier and customer integration. *Completion of steps of the e-business project plan* will represent this evaluation item.

Business Value of E-Business Initiatives

The business value of e-business initiatives shows how the e-business initiatives are affecting the performance of the whole business. It gives an idea of the contribution of the initiative measured through the standard financial measures that are used to determine the health of the business. Typical measures are the *profitability of the web site* that can be defined as the direct revenue of the web site minus the operational costs of the site, and a combination of *Return on Investment* and *Information Economics* to capture the tangible and intangible benefits (cf., paragraphs on IT evaluation).

The *direct online contribution* measures the extent to which the internet contributes to sales and refers to the sales that are actually placed on the web site. It does not include the amount of revenue indirectly achieved due to the internet influencing buying decisions, although the internet has also in this case made a real contribution to sales (Lee, 1998).

Operational cost reductions, cost reductions of acquiring a new customer, cost reductions of customer relationship management, and cost reductions of promotional

Table 4. Measures for Business Contribution

E-Business strategic plan achievements
• completion of steps of the E-business project plan
Business value of E-business initiatives
• profitability of the web site
• Return on Investment (ROI) or Information Economics
• direct online contribution to revenue
• operational cost reductions
• cost reductions of acquiring a new customer
• cost reductions of customer relationship management
• cost reductions of promotional material
Compliance with budget
• actual versus budgeted expenses (ongoing development and maintenance)

material are measures that explain how the web site is helping to reduce costs. Cost reductions of promotional material, e.g., are lower printing and distribution costs. In direct mail, money has to be spent for every additional person that a company wants to reach, whereas with a web site there is no extra cost. E-business is supporting the customer on an ongoing basis through interactive online user groups, online technical support, frequently asked questions and answers, newsletters, and online renewal and subscriptions. Therefore, the cost of supporting customers can be reduced because some of these functions can be partly or totally automated through specific e-business software such as customer relationship management products (e.g., Siebel), customer-facing e-mail products to manage large volumes of incoming e-mails, collaborative filtering packages to derive what products or services individuals will be likely to purchase based on their similarity to other individuals or groups, etc.

Compliance with Budget

An e-business initiative represents a capital investment that entails expenses as well as revenues. The start of an e-business initiative is also the initiation of a permanent commitment to resource demands because of ongoing expenses that are often difficult to predict. Therefore, a financial evaluation is needed that compares the actual costs with the *budgeted expenses*. Important cost categories include the costs associated with the development and — not to forget — the maintenance of a web site and its back office systems.

Measures for Customer Orientation

The customer orientation perspective evaluates the performance of the e-business initiative from the viewpoint of the business and/or consumer clients (Table 5). The measures in this perspective describe the company's success in acquiring and retaining customers and sales. This perspective also represents customer satisfaction and effective internet marketing that both are performance drivers for outcome measures regarding acquiring and retaining clients.

Customer Satisfaction

Customer satisfaction is the feeling that a product or service meets the customer expectations and determines whether a customer will repeat his or her web purchases or not. Since service quality is an antecedent of user satisfaction in web environments, customer satisfaction will be enhanced by increasing customer service (Zeithaml et al., 1993). Typical measures for customer satisfaction are *web site satisfaction scores* calculated on the basis of offline or online surveys, and the *number of customer complaints* eventually compared to *the resolutions within a reasonable time*. Customer satisfaction is perceived by senior managers involved in e-business initiatives as an important measure. All respondents of the Adams et al. (2001) survey said that they measured customer satisfaction or that they should measure it.

Customer Retention

Besides measuring the customer satisfaction, it is also crucial to know how to increase the *customer retention* rate — the degree to which a customer will stay with a

Table 5. Measures for Customer Orientation

Customer satisfaction
- score on online customer satisfaction surveys
- # of customer complaints/resolutions

Customer retention
- retention rates of clients who use the internet compared with those who do not
- % customers placing repeat orders

Acquiring new customers
- customer acquisition or new leads generated by the web site
- sales generated directly and indirectly by the web site

Effective internet marketing
- # of hits
- # of page impressions
- # of site visits
- # of visitors

specific vendor or brand. After all, there is no point having fairly satisfied customers who do not come back to buy again or who are not prepared to recommend the product or service to others. Customers will be likely to return to the merchant's web site if they have had good experience with the firm in the past. Relationships and repeat business happen when customers feel a connection to a site and believe that the site is their best option for doing business. Connection to a site is a function of web design, as customers see and interact with a company via the web interface. Belief in a site or in a company's ability to fulfill its commitments is a function of service design. Both are required for relationships and repeat business. Furthermore, by retaining customers, a company can increase its profits because customers will buy more and sales will grow. In the end, a company's market position is strengthened because customers are kept away from the competitors. Besides this, it is important that they have developed trust. Trust can be developed through reputation, relationship, and knowledge of the other party's business (Stewart, 1998). Measurements are *retention rates of online clients* and loyalty measures such as *percent of online customers placing repeat orders.*

Acquiring New Customers

Acquiring new customers is measured through the *number of new customers* and/ or *new leads* generated via the web site. A more general measure is the *number of sales generated by the web site* eventually compared to the usual business.

Effective Internet Marketing

Internet marketing measures indicate the effectiveness of internet marketing activities in meeting customer, business, and marketing objectives and can be collected online or offline. Although traditional offline metrics are still important, online web metrics are more often used for the assessment of internet marketing effectiveness. Online web metrics are those that are collected automatically on a web server and enable marketers to detect which parts of their web sites are working well and which are not. Before some examples of online metrics will be given, some remarks have to be made.

Measuring and tracking visits and usage of commercial web sites and measuring consumer response to advertisements are hindered by the lack of standardization. Moreover, the complexity of the medium hampers the standardization process. In addition to it, most measures try to prove that a web site is successful when it attracts large numbers of visitors, while it is not so much the quantity of people visiting your web site as the quality of their experience that is important (Schwartz, 1998).

An example of an online metric is a *hit*. A hit is each element, including graphics, text, and interactive items of a requested page recorded by the web server. A single page with multiple graphics can be counted as multiple hits since each graphic is counted as a separate hit. Although the term *hit* is frequently used to indicate how successful a particular site is, it is misleading because a hit includes all units of content sent by a web server when a particular URL is assessed. Merely hitting on a site does not mean that the user did anything with the information. A simple count of the number of web browsers requesting a data transfer from a web server is not enough to explain the browser's interest because it does not account for the user's ability to access information, how effectively information is organized and structured for comprehension, or the appropriateness of the information to the user. A *page impression* is a more reliable measure and equals the number of times a particular web page has been presented to visitors. When customers visit a site, they can visit the home page, browse through the site or visit many pages. When the marketer is not interested in this distinction a useful measure may be the *number of site visits*. A site visit is a series of consecutive web page requests from a visitor to a web site. Measuring the number of site visits can be misleading. A user might enter a new site, then go to lunch leaving his browser on this site to continue clicking around the site when he returns from lunch one hour later. Surprisingly, this would be counted as two visits because there was a period of 30 minutes of inactivity between clicks. Another problem in counting visits is when a user enters a site, and clicks on a link which takes him away for a few minutes but then returns to the site. This will also be counted as two visits even though it might make more sense to count it as one. To gauge performance it is also possible to use the *number of visitors*. A visitor is a unique individual who visits a web site (Alpar et al., 2001). The primary problem in identifying unique visitors is that the web logs only the internet protocol address of the user visiting a site. The problem with IP addresses is that corporate firewalls and internet service providers may allow multiple users to share a single IP address or may assign the same user a different IP address every time he or she connects to the web (Amor, 2000). None of the measures mentioned are able to identify the individual user. Requiring a user to login with a password or using cookies are means of identification that can overcome this. Web site self-registration can be used to identify the individual user and to collect the users' perception of the web site, the company, and its products and services (Chaffey et al., 2000).

Measures for Operational Excellence

The operational excellence perspective (Table 6) focuses on the internal conditions for satisfying the customer expectations as mentioned above. The fulfillment process measures the back office processes, whereas the other measures availability, improvement of systems development and security, represent the IT processes.

Table 6. Measures for Operational Excellence

Fulfillment process
- on-time delivery of products and electronic services
- level of stock-outs
- level of shipping errors
- # problems with customer order processing
- # problems with warehouse processing

Availability of the e-business system
- average system availability
- average downtime
- maximum downtime

Improvement of systems development
- punctuality index of e-business systems delivery
- e-business systems development process excellence
- average time to upgrade the e-business system

Security and safety
- absence of major e-business issues in Internal/External audit reports
- absence of major unrecoverable e-business failures or security breaches

---

Fulfillment Process

The *fulfillment process* measures are very important because it may happen that the back office systems are not adapted to the new e-conomy requirements and/or are not integrated with the front office. Online clients, by the nature of the medium, expect almost immediate delivery of goods and services that can be delayed by *stock-outs, shipping errors,* and *processing problems*. Major critical success factors are: a full integration between front and back office and sufficient stock levels.

Availability of the E-Business system

The *availability of the e-business systems* is crucial. The 24 hours a day, all year round availability is necessary for remaining competitive because companies have a market reach beyond national boundaries or offer e-commerce not restricted to limited opening hours, because customers no longer want to be tied to these opening hours of stores and bank branch offices. Therefore, *average downtime* and *maximum downtime* must be restricted to a minimum.

Improvement of Systems Development

In relation to the measurement of the *e-business systems development process*, significant use can be made of the Capability Maturity Model of the Software Engineering Institute (Paulk et al., 1993). This model defines five maturity levels for systems development that can be easily adapted to the more specific e-business development process. A reasonable level is Level Three which can be defined as (adapted for e-business): The e-business systems development process is documented, standardized, and integrated into a standard software process for the organization. All e-business initiatives use an approved, tailored version of the organization's standard software process for developing and maintaining e-business software. Needless to say that in practice, because of the time pressure on delivering e-business systems, most e-business developments do not reach this level.

Security and Safety

Almost every day publications claim that people's willingness to engage in online transactions requires a perception of security. So, one of the main e-business challenges is to guarantee security and safety, and focus must be on the *occurrence and management of major e-business incidents* and the responses to audit reports on the security of e-business systems.

Measures for Future Orientation

The future orientation perspective (Table 7) examines the company's ability to effectively make use of as well as to improve the e-business project functions nowadays and in the future. Because this ability depends on the know-how and experience of personnel, IT employee-centered measures *(e-business expertise of developers)* such as number of training days, expertise of developers, and their knowledge of new e-business solutions are required. Education and training, and involvement in new e-business technologies are important motivators for IT employees. In light of the shortage of these professionals, it is important to have a follow-up of these measures. Surprisingly, the Adams' (2001) survey shows that only 50% of the dotcoms claim to measure employee satisfaction.

Human resources may be limited, which is often the reason for not delivering e-business systems within time and not delivering the quality that is required. The *e-business staff management effectiveness* measures capture the workload per developer, their rate of absence leave that may indicate problems if this rate is too high, developer backup for the e-business applications, and the satisfaction of the project members and related IT staff.

Table 7. Measures for Future Orientation

E-business expertise of developers
• # of training days per developer
• expertise of developers
• acquaintance with emerging new e-business software and technologies
E-business staff management effectiveness
• rate of absence leave per developer
• average workload per developer
• % of e-business modules covered by more than two developers
• e-business project members satisfaction rate
Independence of consultants
• # of consultant days per module in use more than two years
• # of consultant days per module in use less than two years
Reliability of software vendors
• # releases per year
• # of functional additions
• # of new customers

Other useful measures are these relating to *the dependence on external consultants*. E-business initiatives are often implemented with the assistance of external consultants. However, companies want a quick transfer of the e-business initiatives' know-how to their own staff in order to reduce the need for expensive consultants and to be able to cope with problems. The evolution over time of the number of consultant days spent within the firm can assess the success of such a transfer.

As most e-business initiatives rely on *external software providers*, it is essential to monitor them in order to foresee future problems. Relevant measures are their number of releases per year, the functional enhancements within their packages and their number of new customers.

BUILDING AN E-BUSINESS BALANCED SCORECARD

In building a company-specific e-business Balanced Scorecard, a number of steps need to be followed. First, the concept of the e-business Balanced Scorecard technique has to be presented to senior management, IT management, and e-business project management, and an e-business BSC project team has to be established. Secondly, during the data-gathering phase, information is collected on e-business metrics. The metrics identified have to be specific, measurable, actionable, relevant, and timely (SMART). In this manner, one avoids developing metrics for which accurate or complete data cannot be collected and that lead to actions contrary to the best interests of the business (Chaffey et al., 2000). Thereafter online and offline techniques are introduced to collect metrics. Finally, the organization-specific e-business scorecard, based on the principles of Kaplan and Norton, is developed (Van Grembergen & Timmerman, 1998).

The most important development principle is that an e-business Balanced Scorecard should contain cause-and-effect relationships. The measures selected for the e-business scorecard should be elements in a chain of cause-and-effect relationships. This means that an e-business scorecard needs a good mix of outcomes and performance drivers. While outcome measures reflect the common goals of many strategies as well as similar structures across industries and companies, the performance drivers are the measures that are unique for a particular business unit, or in our case, for an e-business initiative. On the one hand, outcome measures without performance drivers do not indicate how the outcomes are to be achieved. On the other hand, performance drivers without outcome measures may enable the e-business initiative to achieve short term operational improvements but will fail to reveal whether the operational improvements have been translated into expanded business with existing and new customers and enhanced financial performance (Kaplan & Norton, 1992). An example may clarify the difference between outcome measures (sometimes referred to as key goal indicators) and performance drivers (key performance indicators). Improvement of the e-business expertise of the developers (performance driver) may enhance and shorten the delivery of e-business applications (outcome measure). In other words, the developer's experience is an enabler of the fast delivery of e-business applications. This also means that the future orientation and the operational excellence measures are enablers of, respectively, the customer orientation and business contribution goals.

SUMMARY CONCLUSION

In this chapter, the Balanced Scorecard concepts are applied to e-business initiatives. A generic e-business Balanced Scorecard is developed and presented as a measuring and management instrument. The proposed e-business scorecard consists of four perspectives: the Customer perspective representing the evaluation of the consumer and business clients, the Operational perspective focusing on the business and IT processes, the Future perspective showing the human and technology resources needed to deliver the e-business application, and the Contribution perspective capturing the e-business benefits. It is argued that a monitoring instrument such as the proposed e-business scorecard is a must when building, implementing, and maintaining an e-business system because these initiatives are often too technically managed and are often started without a clear business case.

A major pitfall when introducing and using an e-business Balanced Scorecard is that the focus is too much on measurement and that management fails to act on the performance measures. It is our conjecture that the many e-business initiatives within bricks-and-mortar firms and within dotcoms fail because there is no real monitoring of the e-business initiative. At least, a well-developed performance management system may give signals to senior management that something is wrong and that the e-business strategy has to be reconsidered.

REFERENCES

Adams, C., Kapashi, N., Neely, A., & Marr, B. (2001). *Measuring eBusiness performance.* Cranfield University School of Management/Accenture. Available online: http://www.accenture.com.

Alpar, P., Porembski, M., & Pickerodt, S. (2001). Measuring the efficiency of web site traffic generation. *International Journal of Electronic Commerce, 6*(1), 53-74.

Amor, D. (2000). *The E-business revolution: Living and working in an interconnected world.* Upper Saddle River, NJ: Prentice Hall.

Barua, A., Konana, P., Whinston, A., & Yin, F. (2001). Driving e-business excellence. *MIT Sloan Management Review, 43*(1), 36-44.

Berghout, E., & Renkema, T. (2001). Methodologies for IT investment evaluation: A review and assessment. In W. Van Grembergen, (Ed.), *Information Technology Evaluation Methods and Management* (pp. 78-97). Hershey, PA: Idea Group Publishing.

Chaffey, D., Mayer, R., Johnston, K., & Ellis-Chadwick, F. (2000). *Internet marketing.* Harlow: Pearson Education.

Giaglis, G., Paul, R., & Doukidis, G. (1999). Assessing the impact of electronic commerce on business performance: A simulation experiment. *Electronic Markets, 9*(1/2), 25-31.

Graeser, V., Willcocks, L., & Pisanias, N. (1998). *Developing the IT scorecard.* London: Business Intelligence.

Hartman, A., Sifonis J., & Kador, J. (2000). *Net ready – Strategies for success in the E-conomy.* New York: McGraw-Hill.

Hasan, H., & Tibbits, H. (2000). Strategic management of electronic commerce: An adaption of the Balanced Scorecard. *Internet Research, 10*(5), 439-450.

Kaplan, R., & Norton, D. (1992). The Balanced Scorecard – Measures that drive performance. *Harvard Business Review, 70*(1), 71-79.

Kaplan, R., & Norton, D. (1993). Putting the Balanced Scorecard to work. *Harvard Business Review, 71*(5), 134-142.

Kaplan, R., & Norton, D. (1996a). Linking the Balanced Scorecard to strategy. *California Management Review, 39*(1), 53-79.

Kaplan, R., & Norton, D. (1996b). *The Balanced Scorecard translating strategy into action.* Boston, MA: Harvard Business School Press.

Lee, S. (1998). The business value of transaction - based web sites. *Proceedings of the 9th Information Resources Management (IRMA) International Conference, Boston* (pp. 749-751).

Parker, M. (1996). *Strategic transformation and information technology.* Upper Saddle River, NJ: Prentice Hall.

Paulk, M., Curtis, B., Chrissis, M.B., & Weber, C. (1993). *Capability maturity model for software, version 1.1,* Technical Report, CMU/SEI-93-TR024, ESC-TR-93-177, Software Engineering Institute.

Raisinghani, M. (2001). A balanced analytic approach to strategic electronic commerce decisions: A framework of the evaluation method. In W. Van Grembergen, (Ed.), *Information Technology Evaluation Methods and Management* (pp. 185-197). Hershey, PA: Idea Group Publishing.

Rifkin, G., & Kurtzman, J. (2002). Is your e-business plan radical enough? *MIT Sloan Management Review, 43*(3), 91-95.

Robson, W. (2001). Information value and IS investment. In W. Robson (Ed.), *Strategic Management and Information Systems* (pp. 347-394). London: Pitman Publishing.

Rosemann, M. (2001). Evaluating the management of enterprise systems with the Balanced Scorecard. In W. Van Grembergen, *Information Technology Evaluation Methods and Management* (pp. 171-184). Hershey, PA: Idea Group Publishing.

Schwartz, E. (1998). *Webeconomics.* New York: Broadway Books.

Stewart, T. (1998). *The e-business tidal wave.* Available online: http://www.deloitte.com/tidalwave/.

Turban, E., Lee, J., & King, D. (2000). *Electronic commerce: A managerial perspective.* Upper Saddle River, NJ: Prentice Hall.

Van Grembergen, W., & Saull, R. (2001). Information Technology governance through the Balanced Scorecard. *Proceedings of the 34th Hawaii International Conference on System Sciences (HICSS), Maui, Hawaii.* CD-ROM.

Van Grembergen, W., & Timmerman, D. (1998). Monitoring the IT process through the Balanced Scorecard. *Proceedings of the 9th Information Resources Management (IRMA) International Conference* (pp. 105-116). Boston, MA.

Van Grembergen, W., & Van Bruggen, R. (1997). Measuring and improving corporate information technology through the Balanced Scorecard technique. *Proceedings of the 4th European Conference on the Evaluation of Information Technology, Delft* (pp. 163-171). Delft.

Zeithaml, V., Berry, L., & Parasuraman, A. (1993). The nature and determinants of customer expectations of service. *Journal of the Academy of Marketing Science, 21*(Winter), 1-12.

Chapter VII

A View on Knowledge Management: Utilizing a Balanced Scorecard Methodology for Analyzing Knowledge Metrics

Alea Fairchild

Vesalius College/Vrije Universiteit Brussel (VUB), Belgium

ABSTRACT

IT professionals who want to deploy foundation technologies such as groupware, CRM or decision support tools, but fail to justify them on the basis of their contribution to Knowledge Management, may find it difficult to get funding unless they can frame the benefits within a Knowledge Management context. Determining Knowledge Management's pervasiveness and impact is analogous to measuring the contribution of marketing, employee development, or any other management or organizational competency. This chapter addresses the problem of developing measurement models for Knowledge Management metrics and discusses what current Knowledge Management metrics are in use, and examines their sustainability and soundness in assessing knowledge utilization and retention of generating revenue. The chapter discusses the use of a Balanced Scorecard approach to determine a business-oriented relationship between strategic Knowledge Management usage and IT strategy and implementation.

INTRODUCTION

"Knowledge has become the key economic resource and the dominate — and perhaps even the only- source of competitive advantage." Peter Drucker, Managing in a Time of Great Change (1995)

Knowledge Management may be defined as a set of processes for transferring intellectual capital to value – processes such as innovation and knowledge creation and knowledge acquisition, organization, application, sharing, and replenishment (Knapp, 1998). Enterprises work in generating value from knowledge assets by sharing them among employees, departments and even with other companies in an effort to devise best practices.

From the point of view of information technology (IT) investment, it is important to note that the definition itself says nothing about technology; while Knowledge Management is often facilitated by IT technology, by itself it is not Knowledge Management.

"Now that knowledge is taking the place of capital as the driving force in organizations worldwide, it is all too easy to confuse data with knowledge and information with information technology." Peter Drucker, Managing in a Time of Great Change (1995)

The definition of knowledge is a complex and controversial one, and 'knowledge' can be interpreted in many different ways. Much of the Knowledge Management literature defines knowledge in broad terms, covering basically all the "software" of an organization. This involves the structured data, patents, programs and procedures, as well as the more intangible knowledge and capabilities of the people. It may also include the way that organizations function, communicate, analyze situations, come up with novel solutions to problems and develop new ways of doing business. Knowledge Management in an organization can also involve issues of culture, custom, values and skills as well as the enterprise's relationships with its suppliers and customers. Knowledge Management is a strategic, systematic program to capitalize on what an organization "knows" (Knapp, 1998).

Managerial interest in Knowledge Management stems from a number of economic facts about knowledge utilization in today's environment. These facts are shown in Figure 1.

- Long-run shifts in advanced industrial economies which have led to the increasingly widespread perception of knowledge as an important organizational asset.
- The rise of occupations based on the creation and use of knowledge.
- Theoretical developments - for example, the resource-based view of the firm — which emphasize the importance of unique and inimitable assets such as tacit knowledge.
- The convergence of information and communication technologies, and the advent of new tools such as Intranets and groupware systems.
- A new wave approach to packaging and promoting consultancy services in the wake of the rise and fall of Business Process Reengineering (BPR).

Figure 1. Economic Reasons for the Interest in Knowledge Management (KPMG Consulting, 2000)

- Long-run shifts in advanced industrial economies that have led to the increasingly widespread perception of knowledge as an important organizational asset.
- The rise of occupations based on the creation and use of knowledge.
- Theoretical developments -- for example, the resource-based view of the firm -- which emphasize the importance of unique and inimitable assets such as tacit knowledge.
- The convergence of information and communication technologies, and the advent of new tools such as Intranets and groupware systems.
- A new wave approach to packaging and promoting consultancy services in the wake of the rise and fall of Business Process Reengineering (BPR).

Organizational requirements for Knowledge Management involve *leveraging* intellectual capital, not just retaining it, which requires attention to what have been recognized as "knowledge enablers", i.e., structures and attributes that must be in place for a successful Knowledge Management program. Besides technology, these enablers include content, learning, culture and leadership (KPMG Consulting, 2000). In order to measure Knowledge Management, some attention needs to be paid to measurement of the enabling factors as well.

The chapter discusses two possible sets of metrics, containing both qualitative and quantitative components, to aid management in understanding Knowledge Management initiatives and the necessary IT investments in relation to the company's strategic direction. This chapter addresses the question: What set of industry standard metrics, containing both hard and soft elements, can be adapted for use in Knowledge Management initiatives?

The chapter first explores literature on current Knowledge Management metrics. It then examines, from published research, how Knowledge Management is viewed in organizations to examine the sustainability and soundness in metrics that assess knowledge utilization and retention of generating revenue. It then creates an extension of the Balanced Scorecard framework in terms of two possible perspectives, so as to address Knowledge Management metrics. The conclusion of the chapter ties the Knowledge Management perspectives suggested in this research to the original intent of a Balanced Scorecard so as to show the relationship to strategy.

BACKGROUND ON KNOWLEDGE MANAGEMENT

How is Knowledge Management Measured?

A number of Knowledge Management thought leaders, such as Larry Prusak and Thomas Davenport, have stated a belief that it is impossible to develop direct, meaningful measures of knowledge assets. They believe it is possible to measure only the outputs of knowledge, given that knowledge is, by definition, intangible and therefore unobservable. But by developing an understanding of what makes a "unit" of knowledge, one might be able to create the necessary relationship between knowledge and the value it creates for the organization.

Although the focus on corporate culture and organizational change may extend the timeframe for a Knowledge Management program, only measurable benefits justify increased duration and cost in the eyes of senior management. Those benefits include better preparation for implementation and the ability to take advantage of existing technology.

Knowledge Management investments are thus likely to include the extension of existing enterprise software to eliminate barriers between transactional applications and repositories of corporate knowledge. Increasingly, companies will exploit corporate knowledge and provide it to users within the context of business problems, a more effective alternative to simply storing this content in and accessing it from a centralized knowledge repository (Dyer & McDonough, 2001). Given that there is no clear single activity that is Knowledge Management, it is more how and when Knowledge Management is integrated into organizational activities that can be measured.

In general, however, intellectual and knowledge-based assets fall into one of two categories: explicit or tacit. Included among the former are assets such as patents, trademarks, business plans, marketing research and customer lists. Generally, explicit knowledge consists of anything that can be documented, archived and codified, often with the help of IT. Much harder to grasp is the concept of tacit knowledge, or the know-how contained in people's heads. The challenge inherent with tacit knowledge is figuring out how to recognize, generate, share and manage it. While IT in the form of e-mail, groupware, instant messaging and related technologies can help facilitate the dissemi-nation of tacit knowledge, identifying tacit knowledge in the first place is a major hurdle for most organizations (Surmacz & Santosus, 2001).

A recent Platinum Technologies study found that only 20% of Knowledge Manage-ment programs have used some form of metrics on how business performance is influenced. This may be that traditionalist management who use ROI for calculations have to find an appropriate equation for Knowledge Management (Shand, 1999). Platinum Technologies developed a method that was two parts hard, or quantifiable, measurement and one part soft, or more qualitative, measurement. By consolidating and better managing the different delivery channels for sales and marketing, Platinum significantly reduced the costs of maintaining and distributing collateral. This represents a hard and indisputable measure. The other third (the "soft" measure) resulted from increases in sales productivity, a measure with less clear impact on revenue. Senior management appreciated the difference between these hard and soft measures, espe-cially the lesser emphasis on a metric that could be easily contested. This appreciation needs to be considered when assessing harder metric methods like return on investment (ROI) (Shand, 1999).

When is Knowledge Management Measured?

Figure 2 emerged from observing the numerous organizations that participated in an American Productivity and Quality Center (APQC) project (2001) and how they measure the value of Knowledge Management. During its 2000 consortium learning forum entitled 'Successfully Implementing Knowledge Management', APQC focused on

Figure 2. APQC Project

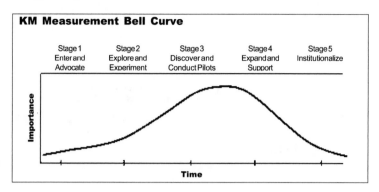

how some of the most advanced early Knowledge Management adopters implement a Knowledge Management initiative, mobilize resources, create a business case, and measure and evolve their Knowledge Management programs. This multi-client benchmarking project helped APQC and project participants identify measurement approaches, specific measures in use, and how measures impact and are impacted by the evolution of Knowledge Management.

In the earliest stages of Knowledge Management implementation, formal measurement rarely takes place, nor is it required (APQC, 2001). As Knowledge Management becomes more structured and widespread and companies move into Stages Two, Three, and Four, the need for measurement steadily increases. As Knowledge Management becomes institutionalized — a way of doing business — the importance of Knowledge Management-specific measures diminishes, and the need to measure the effectiveness of knowledge-intensive business processes replaces them.

According to the APQC, the key is to begin to ensure that direct business value is perceived by the organization as a result of the knowledge-enabling projects. During its 2000 consortium learning forum entitled 'Successfully Implementing Knowledge Management', APQC focused on how some of the most advanced early Knowledge Management adopters implement a Knowledge Management initiative, mobilize resources, create a business case, and measure and evolve their Knowledge Management programs. This multi-client benchmarking project helped APQC and project participants identify measurement approaches, specific measures in use, and how measures impact and are impacted by the evolution of Knowledge Management.

The APQC Knowledge Management Measurement Bell Curve can be seen to parallel the five stages of the IT Balanced Scorecard (BSC) Maturity Model developed by Van Grembergen and Saull (2000), in that as the use of Knowledge Management matures in the organization, defined and managed measurement processes develop and become linked to business process cycles. The Van Grembergen and Saull (2000) IT BSC Maturity Model highlights five maturity levels (Initial, Repeatable, Defined, Managed and Optimized) to classify to what extent the IT BSC Maturity Model is integrated into the strategic and operational planning and review systems of the business and IT.

As seen from the APQC data, it is important to establish a mechanism to capture the hard and soft lessons learned in the Knowledge Management pilots with their initial IT investments, as these will be the building blocks for the later Knowledge Management implementations (APQC, 2001).

CHALLENGES IN DEVELOPING KNOWLEDGE MANAGEMENT METRICS
Knowledge Management Diversity of Practice

The economic facts listed in the Introduction section may help to explain the breadth of interest in Knowledge Management ranging across many different industrial sectors. They also can help to explain the diversity in the actual practices which have been labeled as Knowledge Management (KPMG Consulting, 2000). Although such practices share a common interest in targeting knowledge rather than information or data, they tend to perform distinctively different functions depending on the business context. Therefore, measuring the impact of these different functions on the business potentially requires different approaches.

One may distinguish between at least four different types of Knowledge Management (Business Process Resource Centre, 2001; Sveiby, 2001a):

Valuing Knowledge. This approach is of interest in consulting firms and financial institutions — for example, the Skandia organization — and in management accounting areas. Knowledge is viewed as 'intellectual capital', and the focus is on quantifying and recognizing the value of the organization's knowledge-base. An example of this approach would be PLS-Consult in Denmark, who categorizes customers according to value of knowledge contribution to the firm and follows up in its management information system.

Exploiting Intellectual Property. This approach appeals to firms with a strong science and R&D base — typified by a number of pharmaceutical firms, the Buckman Labs organization, and so on — which are looking beyond the conventional approach based on patents, etc. to more effective ways of tapping into the commercial value of their existing knowledge-base.

An example of this approach would be the development of the Boeing 777, which was the first "paperless" development of aircraft. It included customers in design teams, with more than 200 teams with a wide range of skills who both designed and constructed sub parts, rather than the usual separated organization of design team and construction team. Suppliers world-wide used the same digital databases as Boeing.

Capturing Project-Based Learning. As firms increasingly move towards innovation and project-based organization, many are recognizing the need to capture the learning from individual projects and make it available throughout the organization. Consultancies, professional service firms, aerospace companies, etc., are in the vanguard of developing systems to codify and communicate such knowledge. The client who initiated this research effort also looks at Knowledge Management in this respect.

An example of this approach can be seen in the use of Knowledge Management by firms such as McKinsey and Bain & Co. These two management consulting firms have developed 'knowledge databases' that contain experiences from every assignment, including names of team members and client reactions. Each team must appoint a 'historian' to document the work.

Managing Knowledge Workers. The shift towards knowledge work in many sectors creates problems for traditional ways of managing and motivating employees. In many firms, Knowledge Management reflects managers' desire to increase the productivity of knowledge workers, breaking down some of the barriers to knowledge-sharing which are associated with 'professionalism' (BPRC, 2001).

An example of this approach can be seen at Analog Devices in the US. CEO Ray Stata initiated a breakdown of functional barriers and competitive atmosphere and created a collaborative knowledge sharing culture from the top down. The company encourages a 'community of inquirers' rather than a 'community of advocates'.

Why the different methods of viewing Knowledge Management is important is related to the knowledge, information and work flows associated with each type of Knowledge Management, as shown in Figure 3.

Knowledge Management and IT Investment

The impact of Knowledge Management on IT investment can be related to the effect that each Knowledge Management initiative will have on increased costs of deployed services and technology tools. Based on a 2001 survey of 566 respondents done together by KM magazine and market research firm IDC, these survey results estimate that an average Knowledge Management budget will increase from $632,000 in 2000 to more than $1 million in 2002. These figures fall lower than expected, according to IDC, because two-fifths of the respondents represented companies with 500 or fewer employees, which shows the pervasiveness of the Knowledge Management concept into the reaches of the small and medium size businesses. Past data that emphasized larger companies showed an average budget of $2.7 million in 2000.

Figure 3. Four Stages of Knowledge Transition (Wiig, 1997)

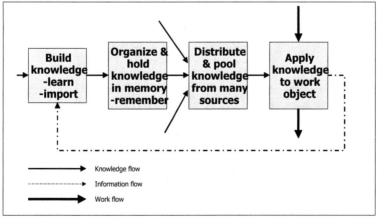

IDC believes from this survey that the budgets specifically designated for "Knowledge Management" initiatives decrease as these efforts become part of other technology or business process investments (Dyer & McDonough, 2001). For example, a company may perceive itself as investing in a customer service solution, though one with significant Knowledge Management capability, rather than categorize this investment as a Knowledge Management initiative. This again shows a need for measuring Knowledge Management and associated IT investment in a way to show its role in the organizational structure, benefiting business processes.

Measuring these roles via the Balanced Scorecard approach has already been established for evaluating IT and its investments, as Gold (1992) and Willcocks (1995) have already indicated in a conceptual manner and that has been further developed by Van Grembergen and Van Bruggen (1997), Van Grembergen and Timmerman (1998) and Van Grembergen (2000). Extending this Balanced Scorecard approach to the Knowledge Management environment would assist companies in understanding the use of Knowledge Management in relation to their knowledge capital resources, including IT implementation.

CONNECTION BETWEEN KNOWLEDGE MANAGEMENT METRICS DEVELOPMENT AND A BUSINESS PROCESS VIEW OF KNOWLEDGE MANAGEMENT

Irrespective of the terms used, the practical management objectives of measuring Knowledge Management are similar: to find out how well the organization has converted human capital (individual learning/team capabilities) to structural capital (organizational knowledge or 'what is left when people go home', such as documented processes and knowledge bases) and thereby moved from tacit to explicit knowledge, and reduced the risk of losing valuable knowledge if people leave the organization. Loss of 'corporate memory' as a result of downsizing is one of the prime reasons given for adopting formal Knowledge Management practices. Other factors often mentioned include global competition and the pace of change; organizations see Knowledge Management as a means of avoiding repetition of mistakes, reducing duplication of effort, saving time on problem-solving, stimulating innovation and creativity, and getting closer to their customers (Corrall, 1999).

For all the interest and money spent on Knowledge Management there seems to be relatively few attempts to actually quantify the impact and results in business terms. The rationale is that knowledge exists in the context of its use (Svoika, 2001). Superior Knowledge Management frees companies to operate on fewer assets, collect their cash faster and have less volatility. The challenge is to make sure that the scope and the goal of the process is clear and focused, by providing a method to display indicators that measure objectives and are focused on the mission as set by the management. Given that Knowledge Management requires a mix of technical, organizational and interpersonal skills, the mix and emphasis varies according to responsibilities, but everyone involved needs to be able to understand the business and communicate business needs effectively, one approach to do this for Knowledge Management could be the Balanced Scorecard.

Van Grembergen and Timmerman (1998) and Martinsons, Davison and Tse (1999) were some of the first to have suggested that the Balanced Scorecard can be the foundation for the strategic management of information systems in organizations. Martinsons et al. (1999) use the Balanced Scorecard metrics to guide attainment of efficiency and effectiveness not only of information systems development but also of the use of the resulting information systems products in the operation of the business. They propose adaptations of the Balanced Scorecard framework based on the premise that IT is essentially an internal support function within an organization in contrast to the original framework, which focused on the impact of the business on the external market. The same analogy might be used for Knowledge Management, given that knowledge supports the activities of the organization.

POSSIBLE SOLUTION OF UTILIZING BALANCED SCORECARD TO KNOWLEDGE MANAGEMENT METRICS

Rationale for Balanced Scorecard in Knowledge Management

In their book, *The Balanced Scorecard*, Kaplan and Norton (1996) set forth a hypothesis about the chain of cause and effect that leads to strategic success. Kaplan and Norton (1996) have introduced the Balanced Scorecard at an enterprise level. Their fundamental premise is that the evaluation of a firm should not be restricted to a traditional financial evaluation but should be supplemented with measures concerning customer satisfaction, internal processes and the ability to innovate. Kaplan and Norton (1996) distinguish Financial, Internal, Customer, and Learning and Growth perspectives on organizational processes essential to an overall strategy. In looking at the original Kaplan and Norton (1996) implementation, Knowledge Management clearly fits within, if it does not define, the Learning and Growth aspect of their framework. If this is true, Knowledge Management outputs will impact on other processes. This is one reason why the significance in measuring Knowledge Management benefits or costs to other processes in organizations is an important area for extending the present Kaplan and Norton work (Firestone, 1998). Managers can track measures as they work toward their objectives, and measurement metrics aid in showing how to build internal capacity, such as human capital, tacit knowledge, and a knowledge culture. And a metrics framework keeps measures from being ad hoc, providing a reference point for Knowledge Management measurement after an implementation.

Use of Perspectives in Knowledge Management

Perhaps the key to proper initiatives and drivers is selecting the right perspectives, to use the Kaplan and Norton phrase, to view the interaction between Knowledge Management and the organizational strategy. This paper suggests that there might be two unique approaches to perspectives that would enhance the measurement of Knowledge Management in the organization.

First Approach. The first approach would be to use the different types of capital available in an organization, as shown in Figure 4, to be the four different scorecard perspectives of how Knowledge Management is leveraged in the organization. This approach is already in use in areas such as the US government, who is empowered to these Knowledge Management initiatives by the Clinger-Cohen Act of 1996 (Public Law 104-106), formerly known as the IT Management Reform Act. This Act requires CIOs in government to focus on the core competencies which represent skills and knowledge needed for effective mission support using information technologies. The four capitals which make up a knowledge-centric organization (KCO) are (Neilson, 2000):

- *Human capital* is all individual capabilities, the knowledge, skill and experience of the employees and managers.
- *Intellectual capital* includes the intangibles such as information, knowledge and skills that can be leveraged by an organization to produce an asset of equal or greater importance than land, labor and capital.
- *Structural capital* includes the processes, structures and systems that a firm owns less its people.
- *Social capital* is the goodwill resulting from physical and virtual interchanges between people with like interests and who are willing to share ideas within groups who share their interests.

The dynamic mixing of human, intellectual, social and structural capital provides the fuel for creating and using knowledge. As retention and recruitment are major concerns in both public and private industry, an organization's success in leveraging its knowledge capital will ensure an organization remains competitive. The four capitals shown in the diagram can be directly related to the traditional Balanced Scorecard method in the following manner, shown in Table 1.

Figure 4. Leveraging Knowledge Management with Balanced Scorecard

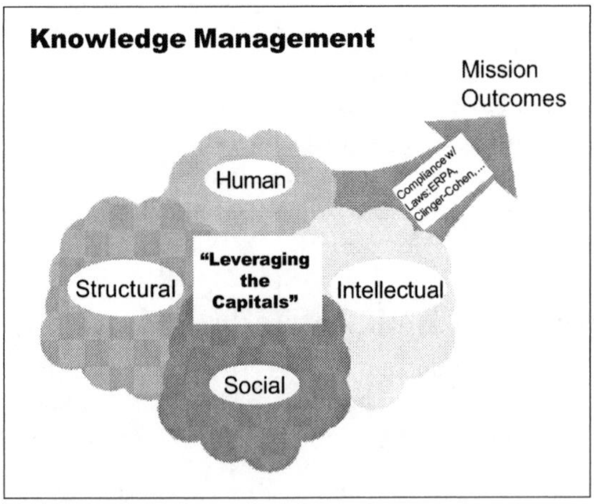

Table 1. KCO Method

Balanced Scorecard Perspective	Generic Measures	Intellectual Capital Perspective	Generic Measures
Financial	ROI, EVA	Intellectual	A wide variety of measures exist. See Table 2.
Customer	Satisfaction, retention, market and account share	Social	Social capital was originally defined and measured by the World Bank in terms that related entirely to density of horizontally organized social networks, subsequent investigations have resulted in complicating any such straightforward measurement. A variety of measures exist, one interesting one from Cap Gemini Ernst & Young and Henley College is called the KOPE survey. This covers the following categories: Knowledge Management strategy and link to business (K), organizational and cultural enablers (O), process enablers (P) and enabling technologies (E). For each of these categories there are between 10 and 13 dimensions. The survey has been designed to allow organizations to identify strengths and weaknesses in their Knowledge Management practices (Truch, 2001).
Internal	Quality, response time, cost and new product introductions	Structural	Andriessen and Tiessen's (2000) Value Explorer™ is an accounting methodology proposed by KPMG for calculating and allocating value to 5 types of intangibles: (1) Assets and endowments, (2) Skills & tacit knowledge, (3) Collective values and norms, (4) Technology and explicit knowledge, (5) Primary and management processes.
Learning and Growth	Employee satisfaction, IS availability	Human	From the work of Jac Fitz-Enz (1994), Human Capital Intelligence sets of human capital indicators are collected and benchmarked against a database. Similar to HRCA (Johansson, 1996), which calculates the hidden impact of HR related costs that reduce a firm's profits.

As seen in financial and other 'hard' measurements, when it comes to numbers, all sorts of measurements come out of the woodwork. Table 2 is a listing of various IC measurements compiled by Karl Erik Sveiby, one of the predominant academics in this field.

Edvinsson and Malone (1997) measure intellectual capital with an 'all-encompassing' reporting model. They assume that if enough aspects of intellectual capital can be captured, one can have a 'complete' understanding on the utilization. What is missing from their approach is an integrated framework to show how the indicators are related. This is where an approach such as Balanced Scorecard can be useful. However, Sveiby (1997) claims his Intangible Asset Monitor is a more suitable approach for Knowledge Management than Balanced Scorecard, based on its notion of a 'knowledge perspective' of a firm. Thus he believes his Intangible Asset Monitor becomes a more demanding option for a management team; to get the best value, one should start by redesigning the strategy to be more 'knowledge focused'. Both approaches are fine for firms such as consulting houses and other service-oriented industries that focus on utilizing human capital. It is difficult to assume that other firms will have been converted enough to

Table 2. Types of Intellectual Capital Measurement (Sveiby, 1997)

Authors	Name of Measurement	Description
Edvinsson and Malone (1997)	Skandia Navigator™	Intellectual capital is measured through the analysis of up to 164 metric measures (91 intellectually based and 73 traditional metrics) that cover five components: (1) financial, (2) customer, (3) process, (4) renewal and development, and (5) human.
Lev (2001)	Value Chain Scoreboard™	Matrix of non-financial indicators arranged into three categories according to the cycle of development: Discovery/Learning, Implementation, and Commercialization.
Roos, Roos, Dragonetti and Edvinsson (1997)	IC-Index™	Consolidates all individual indicators representing intellectual properties and components into a single index. Changes in the index are then related to changes in the firm's market valuation.
Pulic (2000)	Value Added Intellectual Coefficient (VAIC™)	Measures how much and how efficiently intellectual capital and capital employed create value based on the relationship to three major components: (1) capital employed, (2) human capital, and (3) structural capital.
Sveiby (1997)	Intangible Asset Monitor	Management selects indicators, based on the strategic objectives of the firm, to measure four major components of intangible assets: (1) growth, (2) renewal, (3) efficiency, and (4) stability.

Knowledge Management to be knowledge-centric in their strategy. Therefore, we turn to our other suggestion, using a resource-based management approach.

Second Approach. The other approach possible in viewing the role of Knowledge Management in organizational strategy via a Balanced Scorecard would be to use a resource management-based approach, focusing on intellectual capital resources combined with business processes of the organization. This is based on the work of Mouristen et al. (2001) at the Copenhagen Business School, using the work on intellectual capital balance statements of Sveiby (1997). Table 3 correlates this approach to the Balanced Scorecard perspectives of Kaplan and Norton (1992).

To tie the use of Knowledge Management into an organization, using either method, the relationship between Knowledge Management and organizational strategy must be understood and goals clearly defined. Figure 5 demonstrates the relationship between business processes in Balanced Scorecard terminology, and the use of Knowledge Management in an organization.

Correlating Knowledge Management Metrics to Strategy

Both of the suggested approaches above make use of such concepts as intellectual capital and the value of core processes in the organization. The rationale for this is that intellectual capital as a concept says more about the future earning capabilities of a

Table 3. Intellectual Capital Resources in a Balanced Scorecard Approach

Balanced Scorecard Perspective	Generic Measures	Intellectual Capital Statements (Mouristen et al., 2001)	Generic Measures
Financial	ROI, EVA	Employees	Measurement of intellectual capital, as discussed above in Table 3
Customer	Satisfaction, retention, market and account share	Customers	Customer satisfaction with 'quality of service' and product, related to Knowledge Management efforts
Internal	Quality, response time, cost and new product introductions	Processes	Internal hours on Knowledge Management process improvement, average response times for information gathered using Knowledge Management
Learning and Growth	Employee satisfaction, IS availability	Technology	Investment in Knowledge Management technology, number of hits on Knowledge Management project web sites, employee satisfaction with Knowledge Management project sites

company than any of the conventional performance measures we currently use (Roos, 1996). Kaplan and Norton (1996) discuss using Balanced Scorecard measures for assessing potential investment. In creating a mechanism that ties long-term objectives into measurable metrics, they claim that executives can see the relationship between investment and strategic plans. But should knowledge be considered a cost element, or a revenue generator? This is one challenge in using the Balanced Scorecard versus other mechanisms in industry. Until we view business processes in other ways than the traditional production function (input, output), approaches such as the Balanced Scorecard are still valid in the knowledge economy.

Figure 5. Knowledge Management Linkage: Cause and Activity

But there is also an element of debate of whether Balanced Scorecard mechanisms help promote the use of knowledge in organizations. Kaplan and Norton (1996) in their approach did not question the foundation of 'what constitutes a firm', but regard the notion of the firm as given by its strategy. They just want managers to take a more 'balanced view'. As they argue in their book (1996, p. 8): "The Balanced Scorecard complements financial measures of past performance with measures of the drivers of future performance. The objectives and the measures of the Scorecard are derived from an organization's vision and strategy." If Knowledge Management is not part of the firm's strategic view, then investment and management of intellectual assets will not take priority. Use of Knowledge Management requires firms to think about knowledge as a production element, but trends in Knowledge Management lead us to believe that knowledge will be viewed as a service economy, therefore with more intangible measurement. This is also true in the education field, where universities such as Manchester Metropolitan University's business school are examining how to measure their own use of Knowledge Management, and are exploring these two approaches discussed in the chapter as possible metric views.

Many corporations have not clearly articulated a need to manage knowledge. Of the 158 companies surveyed by the Conference Board (Beyond Knowledge Management: New Ways to Work and Learn, 2000), 80% had launched some kind of Knowledge Management activity, but only 15% had specific, stated Knowledge Management objectives and goals. Competitive necessity dictates that executives understand how Knowledge Management and knowledge sharing impact the bottom line, but many do not. KPMG Consulting reports that while most of those it surveyed understand that Knowledge Management can boost profits and reduce costs, less than 30% expected it to help increase their company's share price. The most important and useful metrics are those that directly inform the improvement of business performance and that can best be considered within the context of a learning process that embeds the metrics within the work process.

One Learning process, as an example, is that used by BP - Amoco (BP) as a central part of their Knowledge Management strategy — 'Learn Before, Learn During, Learn After'. Essentially BP - Amoco embeds Knowledge Management within the everyday work process by making it a normal part of doing business. At the beginning of any project they conduct a 'Peer Assist' (alternatively known as 'Prior Art'), where they get knowledgeable colleagues together to consider all that BP - Amoco knows about this particular subject. 'Learn During' involves a version of the US Army's well-known 'After Action Review' (AAR). BP - Amoco use the AAR after each 'identifiable event' rather than at the end of a project; thus it becomes a 'live' learning process that constantly informs the direction of the project. The third part is what BP - Amoco call a 'Retrospect', which is a team meeting designed to identify 'what went well', 'what could have gone better' and 'lessons for the future'.

By ensuring that time is made available within the actual project and that this learning process does not become extra work, BP - Amoco has managed to make it a normal part of doing business. The results have been real tangible business benefits visible in dollar terms that have turned around critics: "the Schiehallen oil field, a North Sea field considered too expensive to develop until a team spent six months pestering colleagues to share cost-saving tips. They were called wimps for not rushing out to 'make hole', but the learn-before-doing approach saved so much time on the platform (at $100,000 to $200,000 a day, not counting drilling costs) that they brought the field into production for $80 million less than anyone thought possible." Indeed, Tom Stewart recently stated about the CKO of BP: "Greenes is, as best I can figure, Knowledge Management's top moneymaker" (World Bank, 1999). This Learning Cycle then becomes the facilitating infrastructure for developing a process of Knowledge Management Metrics which allow the identification of real business value in each aspect of the necessary IT and other investments.

SUMMARY

Measurement has always been perceived as a science of precision; however, the measurement practice in most organizations today is anything but precise. Indeed, the issues looked for in scorecards, such as customer and employee satisfaction, and in intellectual capital, tend to require less precision and entertain more interest in trends than in exact figures. Kaplan and Norton's (1996) cause-and-effect hypothesis could be essential to understanding Knowledge Management metrics in a way that the Balanced Scorecard prescribes. Although the Balanced Scorecard can form the foundation for organizational strategic success, it is, however, not sufficient in itself. Along with strategies, there must be initiatives, such as business process improvement efforts, to steer the organization in the right direction and improve Knowledge Management implementation.

At the heart of an ideal definition of knowledge capital is the creation and provision of 'value'. Without linkage to strategic initiatives reflected through forms of measurement or recording of value, whether simply anecdotal or more quantifiable, Knowledge Management might degenerate into superficial business management hype. A conscious effort in conceptualizing, designing and putting to practice metrics like the ones described above may, however, assist in realizing the true worth of Knowledge Management.

As regards IT investment in particular, information technology may help growth and the retention of organizational knowledge if care is taken to continuously recall that IT is only a part; corporate culture and work practices being equally relevant to the whole. Information technologies best suited for this purpose should be expressly designed with Knowledge Management and organizational capital in view. Thus, while technology can support Knowledge Management, it is not the starting point of a Knowledge Management program.

REFERENCES

Andriessen & Tiessen (2000). *Weightless weight – Find your real value in a future of intangible assets*. Pearson Education London.

APQC. (2001). How to measure the value of Knowledge Management. *Knowledge Management Review, March/April.* Available online: http://www.apqc.org/free/articles/APQCKMR.pdf.

Business Process Resource Centre [BPRC]. (2001). *Defining Knowledge Management*. Warwick Business School, Business Process Resource Centre. Available online: http://bprc.warwick.ac.uk/Kmweb.html.

Corrall, S. (1999). Are we in the Knowledge Management business? *Knowledge Management, 18*. Available online: http://www.ariadne.ac.uk/issue18/knowledge-mgt/.

Drucker, P. F. (1995). *Managing in a time of great change*. New York: Truman Talley Books/Dutton.

Dyer & McDonough. (2001). The state of KM. *Knowledge Management*, (May). Available online: http://www.destinationkm.com/articles/default.asp?ArticleID=539.

Edvinsson, L., & Malone, M.S. (1997). *Intellectual capital: Realizing your company's true value by finding its hidden brainpower*. New York: Harper Business.

Firestone, J. (1998). Knowledge Management metrics development: A technical approach. *Executive Information Systems, Inc. White Paper No. Ten*, (June). Available online: http://www.dkms.com/dkmskmpapers.htm.

Fitz-Enz, J. (1994). *How to measure human resource management*. McGraw-Hill.

Gold, C. (1992). *Total quality management in information services – IS measures: A balancing act*. Boston, MA: Research Note Ernst & Young Center for Information Technology and Strategy.

GSA. (1996). Performance-based management: Eight steps to develop and use information technology performance measures effectively (p. 106). Washington: GSA.

Johansson, U. (1996). *Increasing the transparency of investments in intangibles*. Available online: http://sveiby.konverge.com/articles/oecdartulfjoh.htm.

Kaplan, R. S., & Norton, D. P. (1992). The balanced scorecard - measures that drive performance. *Harvard Business Review, 70*(1), January/February, 71-79.

Kaplan, R. S., & Norton, D. P. (1996). *The balanced scorecard - Translating strategy into action*, xi, 322. Boston, MA: Harvard Business School.

Knapp, E. (1998). Knowledge Management. *Business & Economic Review, 44*(4), July/September.

KPMG Consulting. (2000). *Knowledge Management Research Report 2000*, 1-13. Available for download on the KPMG main web site as 2000 Knowledge Management Survey.

Lev, B. (2001). Intangibles: Management, measurement and reporting. Brookings Institution. Washington. Cited in FASB Special report April 2001: *Business and Financial Reporting, Challenges from the New Economy*.

Martinsons, M., Davison, R., & Tse, D. (1999). The balanced scorecard: A foundation for the strategic management of information systems. *Decision Support Systems, 25*, 71-88.

Mouritsen, J., Larsen, H. T., Bukh, P. N., & Johansen, M. R. (2001). Reading an intellectual capital statement: Describing and prescribing Knowledge Management strategies. *Proceedings of the 4th Intangibles Conference, Stern School of Business, New York University* (May).

Neilson, R. (2000). Interview with Dr. Robert E. Neilson, Chief Knowledge Officer and Professor at the Information Resources Management College (IRMC) of the National Defense University, *CHIPS Magazine*, (Spring). Available online: http://www.chips.navy.mil/interview_with_dr.htm.

Pulic, A. (2000). An accounting tool for IC management. Available online: http://www.measuring-ip.at/Papers/ham99txt.htm.

Roos, J. (1996). Intellectual capital: What you can measure, you can manage. *IMD Perspectives for Managers, No. 10.* Available online: http://alexandrie.imd.ch/gotorec0302&id=&lang=?rec=031029057920189&act=0/.

Roos, J., Roos, G., Dragonetti, N. C., & Edvinsson, L. (1997). *Intellectual capital: Navigating in the new business landscape.* Macmillan, Houndsmills, Basingtoke.

Shand, D. (1999). Return on knowledge: Proving financial payoffs from Knowledge Management investments plagues experienced and novice practitioners. *Knowledge Management,* (April). Available online: http://www.destinationkm.com/articles/default.asp?ArticleID=725.

Strassmann, P. (1999). Measuring and managing knowledge capital. *Report on Knowledge, Technology and Performance.* Available online: http://www.strassmann.com/pubs/measuring-kc/.

Surmacz, J., & Santosus, M. (2001). The ABCs of Knowledge Management. *CIO.* Online edition, available: http://www.cio.com/research/knowledge/edit/kmabcs.html#what.

Sveiby, K. E. (1997). *The new organizational wealth: Managing and measuring knowledge based assets.* San Francisco, CA: Berrett Koehler. Chapter on measuring available online: http://203.147.220.66/IntangAss/MeasureIntangible Assets.html.

Sveiby, K. E. (2001a). Methods for measuring intangible assets. Available online: http://www.sveiby.com.au/articles/IntangibleMethods.htm.

Sveiby, K. E. (2001b). What is Knowledge Management? Available online: http://www.sveiby.com/articles/KnowledgeManagement.html.

Svioka, J. (2001). Knowledge pays. *CIO Magazine,* (February 15). Available online: http://www.cio.com/archive/021501/new.html.

Truch, E. (2001). Mapping advances KM journey. *Knowledge Management,* (April). Available online: http://www.kmmag.co.uk/APRIL01/FORUMapr.HTM.

Van Grembergen, W. (2000). The balanced scorecard and IT governance. *Information Systems Control Journal* (previously *IS Audit & Control Journal*), 2, 40-43.

Van Grembergen, W., & Saull, R. (2000). Aligning business and Information Technology through the balanced scorecard at a major Canadian financial group: Its status measured with an IT BSC Maturity Model. *Proceedings of the Hawaii International Conference on System Sciences (HICSS), Maui, Hawaii* (January).

Van Grembergen, W., & Timmerman, D. (1998). Monitoring the IT process through the balanced score card. *Proceedings of the 9th Information Resources Management (IRMA) International Conference, Boston, May* (pp. 105-116).

Van Grembergen, W., & Van Bruggen, R. (1997). Measuring and improving corporate information technology through the balanced scorecard technique. *Proceedings of the Fourth European Conference on the Evaluation of Information technology, Delft* (October, pp. 163-171).

Van Grembergen, W., & Vander Borght, D. (1997). Audit guidelines for IT outsourcing. *EDP Auditing, Auerbach 72-30-35* (June, pp. 1-8).

Wiig, K. M. (1997). Roles of knowledge-based systems in support of Knowledge Management. In J. Liebowitz & L. C. Wilcox (Eds.), *Knowledge Management and Its Integrative Elements* (pp. 69-87). Boca Raton: CRC Press.

Willcocks, L. (1995). *Information management. The evaluation of information systems investments.* London: Chapman & Hall.

World Bank. (1999). Knowledge Management metrics - A learning process. Comments from a speech from Laurence Smith, Knowledge & Learning Consultant, World Bank. Available online: http://www.zigonperf.com/resources/pmnews/knowledge_metrics.html.

Chapter VIII

Measuring ROI in E-Commerce Applications: Analysis to Action

Manuel Mogollon
Nortel Networks, USA

Mahesh S. Raisinghani
University of Dallas, USA

ABSTRACT

Measuring the Return on Investment (ROI) of a project is a key element of the IT Governance process. The research in this paper aims to provide an overview of how to calculate the ROI for e-commerce applications so that this information, and the attached ROI Calculator Tool Template, can be used by organizations to reduce the time in preparing the ROI for a project. Although there is much written about ROI, there is not that much said about how to prepare one, specifically for an e-commerce project. By reading this research paper and by using the ROI Calculator Tool, any IT group or organization that is going to deploy e-commerce applications will have a starting point for calculating ROIs.

SECTION I: INTRODUCTION

In an interview, Harvard Business School's Michael E. Porter had this to say about e-commerce: "The Internet as a family of technologies will have a very powerful effect on operational effectiveness. We'll see deeper integration among service, sales, logistics, manufacturing and suppliers. The first level of that will improve efficiencies, reduce transaction costs and reduce inventory" (Byrne, 2001, p. 114). Businesses are looking for those e-commerce applications, such as relationship-marketing software, integration software, customer management, E-Procurement, and collaborative software that can achieve the business integration discussed by Michael Porter to boost productivity and/ or create new sources of revenue.

In the late '90s, companies discovered the seemingly endless possibilities of the Internet, and they began to purchase e-commerce applications with little or no financial controls. There was a sense of urgency that precluded taking the time to measure the real economic value of these investments — companies were moving at Internet speed. Then the US economy slowed dramatically, Information Technology (IT) budgets dropped considerably, and projects were competing for the few available dollars (Sammer, 2001). In this new economy, IT managers must convince management of the value of a project, compare projects to determine which offers the best return, and justify the implementation of an e-commerce application.

A survey carried out by the Meta Group in November 2001 led to the following conclusion: "ROI has become one of the most important evaluation criteria for IT groups" (Ferengul, 2001, p. 1). The renewed focus on ROI as an evaluation tool came after years of IT departments using some internal measurements, such as the number of visitors to a web site and network uptime, that didn't have any meaning for business executives who were trying to link a project to the overall performance of a company. Today, Return-on-Investment is becoming an important tool for IT departments and corporate management to measure and quantify the return of e-commerce applications in relation to other investments.

In April and May of 2000, International Data Corporation (IDC) conducted an Internet Executive e-Panel survey about ROI (Rosenthal, 2000, pp. 3-8). These were some of IDC's findings:

- On the question of whether ROIs are calculated for e-business projects: 33% of the respondents said yes, 16% said they didn't know, and 50% said no.
- Reasons given as to why ROIs should not be calculated:

    • The time and energy aren't worth it because ROI is an imperfect calculation. Not enough firm data to base calculations on.

    • The ROI is usually positive in e-commerce applications so there is no need to calculate it.

    • Similar companies may already be doing it. Even though ROIs are not comparable from company to company, if another company has similar programs, tools, and features, then it may not be necessary to spend the time and resources to calculate the ROI for each new element in e-commerce.

- Reasons given for why ROIs should be calculated:

    • The ROI process will help determine the metrics that matter.

    • Once the ROI is calculated, it is easier to do it for similar projects.

    • The ROI process will provide a sense of which e-commerce applications produce the best and fastest results.

The results of IDC's survey summarize the disjunctive situation in which organizations find themselves today. On the one hand, they recognize that e-commerce applications need to be implemented quickly, but they do not want to spend too much time calculating the ROI for a project because they have problems finding the required data. On the other hand, they recognize that e-commerce investments are substantial and risky, and that they should have a thorough measure that projects the risk, yield, and benefit of a project. Organizations can resolve this situation by reducing the time to prepare ROIs, either by developing internal standardized processes and procedures, or by licensing tools already developed by companies specialized in preparing ROIs for e-commerce.

This chapter aims to provide an overview of how to calculate the ROI for e-commerce applications so that this information, and the attached ROI Calculator Tool Template, can be used by organizations to reduce the time in preparing the ROI for a project.

ROI

In any type of investment, investors want to find out what their return will be. This is a key element of the IT Governance process. The return is calculated by dividing the profit by the investment. It is a straight calculation when measuring a capital investment, a loan, for example, where the return on investment is the interest received. However, for a project, the investment is measured as the total cost in time, dollars, or any other unit required to plan, execute, and complete the project. The return is the savings, also in time, dollars or any other measurable unit generated by the project. The project's ROI is, then, the ratio of the total return divided by the total cost of the project.

The difficulty in calculating ROI is in determining what constitutes the total cost of the project and what constitutes the total return of the project. The problem results from the fact that even though most of the cost and return factors are easily measurable and have tangible values, the values of other factors involved are more difficult to determine. An example in point is the factor of customer satisfaction; it can be measured (surveys, focus groups), but its value may be hard to specify. ROI has been used as a measure of a firm's performance in various studies (Byrd & Marshall, 1997; Mahmood & Mann, 1993a, b; Woo & Willard, 1983). In this chapter, the term ROI will be used as the benefit generated by a project, regardless of whether it is measured by the Payback Period, Net Present Value, or Internal Rate of Return (IRR).

FINANCIAL ASPECT OF MEASURING THE ROI OF AN E-COMMERCE APPLICATION

In finance, the following standard steps are followed to evaluate the capital budgeted for a project (Brigham & Houston, 1999, p. 502):
1.   Determine the cost of the project.
2.   Estimate the expected cash flow from the project, including the salvage value of the assets at the end of the project's expected life.
3.   Estimate the project's risks.

4.   Determine the project cost of capital based on the project's risk.
5.   Calculate the present value of the expected cash flow.
6.   Compare the present value of the expected cash flow with the required outlay. If the present value of the cash flow exceeds the cost, the project should be implemented. Otherwise, it should be rejected.

The same steps used in the finance field to evaluate the capital budgeted for a project can be used to establish a process to measure the return for e-commerce applications. The steps are as follows:
1.   Determine the internal and external costs of implementing and maintaining current and new e-commerce application processes.
2.   Calculate the cost savings between the current process and the new process, and add the benefits in productivity and efficiency. All process improvements should be measurable; however, it may be easier in some cases than others to assign a dollar value.
3.   Calculate the e-commerce application's risk.
4.   Determine the company's cost of capital for that specific e-commerce application.
5.   Calculate the net present value.
6.   Compare the present value of the expected cash flow with the required outlay. If the present value of the e-commerce application cash flow exceeds the cost, the project should be implemented. Otherwise, it should be rejected.

The above steps use the Net Present Value (NPV) to measure the return for a project; however, other finance methods can be used to rank projects. Those methods include the following:
1.   Payback Period, defined as the expected number of years required to recover the original investment.
2.   Discounted Payback, similar to regular payback, except that the expected cash flows are discounted by the project's cost of capital.
3.   Internal Rate of Return, which is defined as the discount rate that equates the present value of a project's expected cash inflows with the present value of the project's costs.

A more complete definition of these terms (including NPV) is presented in Appendix A: ROI Terms.

The development of a comprehensive method for calculating the ROI for an e-commerce application is shown in the following steps. Initially, a project's ROI can be calculated as:

$$ROI = \frac{Project\ Return}{Cost\ to\ Implement\ the\ Project}$$

However, to further define the method, a project's return can be determined by calculating the expenses that a company will incur with the new process and subtract that from the expenses that a company incurs with the current process. Normally, when a new process is implemented, a value may be given to some other benefits. In this chapter "other benefits" will be included so as to learn how they could be measured and valued. Now, the ROI formula can be written as:

$$\text{ROI} = \frac{\text{Current Process Cost - New Process Cost + Other Benefits}}{\text{Cost to Implement the Project}}$$

Ideally, when implementing a new process, companies will incur a one-time only cost, the initial investment; however, this is not normally the case. New processes may involve some new costs for maintenance and operation. Therefore, it is not only necessary to determine the initial investment, but, also, how much it is going to cost a company to operate and maintain this new process for a certain period of years. The time factor, measured in cost per year, needs to be included in the ROI formula. In an e-commerce application, a period of three years is usually selected to calculate the ROI because IT and e-commerce technologies change very rapidly (Wu, 2000). The ROI formula now can be written as:

$$\text{ROI} = \frac{(\text{Current Process Cost - New Process Cost + Other Benefits}) \text{ per year } \times \text{ 3 Years}}{\text{Initial Investment } + (\text{Operation and Maintenance per year}) \times 3 \text{ Years}}$$

If the Net Present Value (NPV) method is used, then the formula for ROI can be written as:

$$\text{ROI} = \frac{\sum_{1}^{n} \text{PV} (\text{Current Process Cost - New Process Cost + Other Benefits}) \text{ per year})}{\sum_{1}^{n} \text{PV} [\text{Initial Investment } + (\text{Operation and Maintenance per year})]}$$

When calculating ROI, some companies have the policy that if something cannot be assigned a value in dollars, even if it is measurable, it should not be included in the ROI calculation. Other companies allow intangibles to be included in the ROI. In the ROI formula above, the intangibles, also called soft benefits, are included in "**Other Benefits**". The following list shows some tangible and intangible benefits used in calculating ROIs:

Tangible Benefits
(Measurable and a dollar value can be assigned)
- Sales increase
- Production increase
- Reduction in operating cost
- Reduced network downtime
- Increased mean time before failure
- Reduced time to configure a data network

Intangible Benefits
(Measurable, but difficult to assign a dollar value)
- Customer satisfaction
- Customer retention
- Managerial know-how
- Employee retention
- Stronger channel ties
- Increased customer base
- Improved employee morale

BUSINESS PROCESS ANALYSIS ASPECT OF MEASURING THE ROI FOR E-COMMERCE APPLICATIONS

The main reason a company implements an e-commerce application is because it seeks to improve a process. "It is about doing the same work, in a more effective and efficient way such that it will increase capabilities and reduce costs," states Dr. David Gordon, a professor at the University of Dallas and an expert in operations management and process improvement (D. Gordon, personal communications, February 9, 2002). In his opinion, customers and process meet in areas such as proposals, installations, purchasing/ordering, billing, inquires, and complaints. Dr. Gordon further states that organizations can improve some of these processes by using electronic applications. By improving the processes, Dr. Gordon means the following:

- Learning about the cause-and effect mechanisms that impact process performance
- Improving the effectiveness of their current processes
- Developing and implementing counter-measurements
- Confirming the results of the improvement

In its simple form, the ROI is given by:

$$\text{ROI} = \frac{(\text{Current Process Cost} - \text{New Process Cost} + \text{Other Benefits}) \text{ per year} \ \times \ 3 \text{ Years}}{\text{Initial Investment} + (\text{Operation and Maintenance per year}) \times 3 \text{ Years}}$$

As can be seen in the ROI formula, the different costs are related to a process, and this is where "Business Process Analysis" becomes important to help determine the different activities involved in a process.

Current and New Process Costs

When measuring a proposed project's ROI, the first step is to determine and identify the project's benefits. It should be established what the main reason for the project is, e.g., to increase sales, the number of sales leads, production, and customer satisfaction, or to reduce the sales cycle, network downtime, operating costs, the cost per sales lead, and the time to configure a data or voice network. An e-commerce application may replace or reduce the current sales force and distribution channels, or it may increase the performance of the current sales force and distribution channels.

It must also be determined what the expected results are. Statements such as, "The objective of this project is to improve..." help identify these results.

Once the proposed project's expected results are established, the following basic Business Process Analysis procedures should be used for collecting the information regarding the current process (Gordon, 2002):

- Understand the current process.
  - Go to the actual location of the process.
  - Observe the actual process in action — walk through the process.
  - Get the facts by talking to those directly involved.
- Map the actual process by developing a process flow chart.
- Select the right metrics and make sure they describe the business case.
- Collect performance data.
- Rate the process.

The same should be done for the new e-commerce application process; variations in performance, schedule, and cost should be noted.

Mapping the Process

After developing an understanding of the current process, the following steps should be used to map the current process and the new process:

1. Prepare a Task/Activity matrix with all the events that now take place in the current process. Highlight the activities that are going to be eliminated and the activities that are going to be improved with the new process. The matrix should include the tasks and the activities in each task. See Table 1, "ROI Step-by-Step".

2. Prepare two flow diagrams showing all the events that take place in the current and in the proposed new process indicating who the contributors are, i.e., the departments in which the events take place, for each. Figures 1 and 2, "ROI Step-by-Step", show the customer and company events that take place. By putting the events for the current and the new processes in the same time format, it is easy to visualize the efficiency of the new process.

3. Determine, in hours, how long it takes for the vendor and for the customer to carry out a transaction at a specific point in the flow chart. Then, multiply that number of hours by the labor rate for the vendor and for the customer. The result in dollars is the labor cost of each transaction at a specific event. Figures 2 and 3 show an example of the labor cost for the actual and for the new proposed process. When the labor costs per transaction are added in, the total cost per transaction of the current and the new process is given.

Initial and Recurring Costs

"Underestimating implementation costs is not as big a problem as underestimating maintenance and operation costs," states Barbara Gomolski (2001, p. 12) in her article, "The Cost of E-Business". She points out that, typically, ongoing costs of new systems are 40% to 60% of the first year's expenses. If the first year's expenses of hardware, software, and staffing are $1 million, then it will cost a company approximately $500,000 per year to maintain and integrate that system.

Many consultants and IT managers share Barbara Gomolski's opinion about the cost of maintaining and integrating an e-commerce solution. Therefore, when trying to determine the cost of a new process, not only the initial investment should be considered, but also, the recurring cost that will continue after the project has been implemented. A good example of this point is that of DoveBid, which, in November 1999, rolled out an online auction service for surplus capital assets. IDC prepared a bulletin describing DoveBid's ROI for adding e-commerce to an existing bricks and mortar company. In that bulletin, IDC showed that most of the costs were not related to acquiring information technology, but to other expenses. For its B2B auction site, DoveBid spent 1% on IT equipment and software purchases, 10% in consulting fees, 34% in staffing, and 55% in advertising (Rosenthal, 2000). The advertising costs were relatively high since this was a new service, but it was important to let the potential customers know about its B2B auction web site.

Initial and recurring costs can fall into any of the seven broad categories (Oleson, 2000): Consulting, Personnel, Software, Hardware, Implementation & Integration, Training, and Facilities (office space expense). Initial costs are also called implementation or start-up costs. Recurring costs are also called operation or maintenance costs. The following table shows some of the costs that need to taken into account when calculating the initial and the recurring costs.

Calculating the Intangibles — Other Benefits

Measuring the ROI for an e-commerce application is no different than measuring it for any other project in the sense that all projects have tangible and intangible benefits. The initial investment and the operational and maintenance expenses for implementing an e-commerce project are tangibles, also referred to as "hard dollars", which can be translated directly into a value such as money or time.

The intangibles, also called "soft dollars", are benefits that sometimes cannot be tied directly to cash. This is mainly due to a lack of information that can be used to establish a level of value. For example, it is difficult to give a value to the intangibles on the following detailed list (Schlegel, 2001; Meta Group, 2001):

• Increased customer satisfaction
• Increased customer retention
• Greater customer reach
• Increased customer base
• Reduced customer contact/support requirements
• Reduced fulfillment and customer response errors
• Increased employee retention
• Improved employee morale
• Improved customer or employee knowledge and learning
• New business processes
• Decreased time-to-market

Table 1. Initial and Recurring Costs

Consulting	• Create an overall strategy and a road map for project implementation.

Personnel	• Project Management
	• Assign a Project Manager.
	• Hold customer meetings to determine the concept, project plan/requirements, training needs, and to evaluate progress.
	• IT Staff
	• Create an entire IT department for the e-commerce application. People to hire may include:
	  o Department Director
	  o A Webmaster in charge of developing and maintaining the site
	  o A database engineer
	  o Site programmers
	  o Customer service representatives

IT Purchases	• Purchase as needed: • E-Commerce application platforms • Additional hardware • Security software, firewalls, and VPNs • Database software • Payment authorization software
Implementation and Integration	• Build and verify configurations. • Develop and maintain new software. • Make web site enhancements. • Do test configuration.
Training and Documentation	• Provide internal and/or external training. • Provide workforce transformation/change management training.
Facilities	• Lease more office space. • Modify existing physical facilities.
Advertising	• Do Internal and/or external promotion.

- Reduced ordering process complexity
- Better or improved managerial know-how
- Stronger channel ties
- Better delivery date information
- More productive time on web site
- Increased brand awareness
- Better business alignment

- More product information availability
- Improved communications
- Organizational flexibility
- Streamlining of knowledge
- Ability to work from a remote location
- Reduced days sales outstanding (average collection period)
- Increased efficiency
- Improved information access, search, and retrieval time

In any ROI calculation, it will take significant time and effort to determine the value for these intangibles. Product Managers, Sales Teams, and personnel in IT and the e-commerce departments might be of assistance in placing values on intangibles (the Finance department generally seems to generate hard dollar values for tangibles). Some examples of methods and tools used to identify intangibles include workshops, roundtable discussions, and desk research. Once the values have been established and agreed upon, it makes sense to use them in all ROIs calculations, not only to save time, but also to make all ROI calculations comparable.

THE DECISION TO IMPLEMENT AN E-COMMERCE APPLICATION

After calculating a project's ROI, how can it be determined whether the project should be implemented or not? If a project has an ROI of 15%, for example, is that enough return for a company to implement the project?

After calculating the ROI for an e-commerce application, it is necessary to compare it to a point of reference; in this case, it should be compared to the company's cost of capital. The cost of capital of a company is a complex interaction of several factors such as interest rate depreciation, taxes, and capital gains. Different projects have different risk levels, so companies have different costs of capital for different types of projects. A company's chief financial officer can determine the company's cost of capital for e-commerce applications. This cost of capital may be different for other company investments because of the different risks involved in e-commerce projects vs. other expenditures. In a simple ROI calculation, not using Present Value, the ROI of a project should be higher than the cost of capital to be considered for approval. In an ROI calculation using the Present Value, the Cost of Capital is the rate used to discount the cash inflows and outflows.

CONCLUSION

When companies include e-commerce initiatives in their visions and strategies, they should consider e-commerce applications that are in accord with those visions and strategies and that have appropriate ROIs.

When calculating the ROI for an e-commerce project, it is necessary to take into account the following:

- ROIs are not perfect calculations and are not comparable between different companies, between similar e-commerce projects from two different companies, or from strategic business units within the same company.

- ROIs should not only be used as an approval tool, but, also, to continue measuring the project's benefits during the project's life. Once approved, the same data used to evaluate the ROI can be used to determine if final project costs are on target or not. A higher or lower actual project cost will result in a different ROI than projected.

- If two companies are implementing the same e-commerce application, having a higher ROI for one only means that a greater improvement was achieved in that company and not necessarily that the implementation was better.

- Measuring ROI is a way to present the value of an e-commerce application in a language understood by corporate finance and management.

It is generally assumed that measuring the ROI for an e-commerce application is associated more with finance rather than with the business process analysis, but, in reality, it is the other way around. Finance rules are standard and the same formulas can be used to calculate the ROI for any project; for example, this article includes a template with all the standard finance formulas. On the other hand, the activities are different from project to project and to uncover and quantify those activities requires much more effort. Calculating the intangible benefits is another area that will require a lot of time because standardized information is not available. In addition, how one company calculates intangibles might substantially differ from how other companies might do it. This chapter includes some e-commerce intangibles collected during the research phase that might be used as a starting point.

SECTION II: ROI CASE STUDY

In this section, a case study, the **ROI of an Online Tool for Configuring and Pricing a Data Product,** will be used to show, step-by-step, how to conduct an ROI calculation. The numbers used in this case study are hypothetical.

Situation Snapshot. In data communications, a high level of expertise is required to configure a data product. Engineering rules to configure a product are established by the engineers who developed the product or by other engineers who learned to operate the product. Knowledge and access to information is, therefore, in the hands of a relative few, and it would not be cost effective to train the number of engineers needed to be available to answer sales support technical questions.

Situation Verification. ABC, Inc., a global telecommunications company that sells data products to its distributors, has received complaints from the distributors that it takes too much time to configure and price a data product. Distributors have mentioned that it takes four days to configure ABC's product, while it only takes them four hours to configure data products from DEF, Inc., a competitor of ABC.

Severity Assessment. Distributors' dissatisfaction with the time it takes to configure data products is causing ABC to lose revenue.

Improvement Target. The Internet allows companies to post information on their web sites that users can access to carry out electronic commerce activities, such as finding prices and ordering products. ABC, Inc. would like to develop or purchase an online tool to allow its distributors to configure data products for complex networks and to determine the price of those configurations. An online e-commerce application to configure and price data products would have a beneficial impact on the overall cost associated with doing business, but ABC, Inc. would like to calculate the ROI for such an e-commerce application before deciding whether to make the financial investment involved in developing or purchasing the application.

The following are the steps to calculate the ROI for this business case:

ROI Depth and Level of Complexity

If several persons are asked to individually prepare an ROI for a specific project, chances are that each ROI would be calculated in a different way. Everyone's approach to the ROI is different, but even though the ROIs are prepared using different approaches, the results may be the same. However, it is recommended that before starting work on an ROI, several interviews should be conducted to gather information about how ABC, Inc. measures ROIs, so that the ROI is prepared according to the company's or manager's guidelines. ABC's e-commerce group can help determine the issues involved in measuring the ROI of e-commerce applications. Questions such as the following could be put to five executives from e-commerce:

- Is it worth it to measure the ROI of an e-commerce project?
- Do you feel that ABC, Inc. should measure the ROI for all e-commerce projects?
- Does ABC, Inc. have some guidelines for calculating the ROI for e-commerce projects?
- What is your opinion of using intangibles in the ROI calculation?
- Do you track the ROI of an e-commerce project to determine whether the results meet the expectations?
- Do you track the actual implementation costs against the ROI analysis to determine how actual costs compare to forecasted costs?
- What is the typical time frame, or amortization schedule, that ABC, Inc. uses when calculating an ROI for an e-commerce project?

The answers to these questions provide input as to the type and depth of the ROI measurement. The Center of Public Technology of the University of North Carolina recommends using a three-tiered set of business case methodologies that are scaled to the size and complexity of an IT or e-commerce initiative. The following are the three possible methodologies (Rivenbark, FitzGerald, Schelin, Yates, & Runkle, 2001, p. 23):

- Tier 1 (least complex/cost): Total Cost of Ownership;
- Tier 2 (middle/cost): Total Cost of Ownership, Payback Period, Benefit/Cost Ratio, Internal Rate of Return, and Return on Investment;
- Tier 3 (most complex/cost): Applied Information Economics. [1]

ABC's executives agreed that the two-tiered model should be used because the ROI measurement is not very complex and the investments are not very high. The executives also decided not to include intangible benefits in the ROI calculation.

Identifying the Processes

The second step in preparing the ROI analysis for ABC, Inc. is to determine the activities that take place for the current process, as well as the activities that would be involved with the new process.

Again, interviews could be used to gather information; five persons who are very familiar with the current process could be asked the following questions:

- Is the company satisfied by the way the current process is carried out, or has someone requested changing the process?
- Can the process be improved? How? What advantages would these changes bring to the company's process?
- How do employees or customers carry out the current process?
- How much time do employees spend on the current process, broken down by activity?
- How does this time compare to the time another company spends executing a similar process?
- What would be the impact of changing the current process to the new process?
- How would changes made to the current process affect other departments or customers?
- Are there any existing corporate obligations that would make the adoption of the new process more costly?

The answers to these questions allows a company to determine whether or not the new process includes all suggested changes, and if there are other costs not taken previously into account.

If ABC would choose to purchase a new e-commerce application, instead of developing one internally, then a sales representative from the company that sells the application could be interviewed and asked the following questions:

- What activities are involved in using the new application and which ones from the current process would not be necessary? Show them in a flow chart.
- What is the total cost of ownership including the initial investment and the recurring cost-maintenance and operational costs?
- What information and resources does a company need to provide to your company to maintain the accuracy of the new application tool?
- What new hardware or additional software would it be necessary to have once the e-commerce application is implemented?
- Are there any existing features in the current process environment that will be lost by going to the new e-commerce application?
- How much time does it take a company to implement your application?
- Which other companies use your application?

Table 1. Activity Matrix for Configuring Data Products at ABC, Inc.

Tasks	Activity	Activity	Activity	Activity
Maintain	Train customers' sales engineers	Update engineering configuration rules		
	Send pricing catalogs to customer	Load pricing catalogs		
Product Configuration	Prepare configuration draft	Send configuration for review		
	Verify configuration	Resolve differences	Send final configuration to customer	Review and approve configuration
	Key-in configuration	Price configuration		
Purchasing & Shipping	Place order	Match or approve	Check equipment for shipping	
Installation	Check received items	Check for missing parts	Order missing parts	

Note: The activities highlighted in the lighter shade can be eliminated. The ones highlighted in the darker shade can be improved.

Current and New Process Costs

With the information collected during the interviews, a Task-Activity matrix should be prepared; it should list all the activities, grouped by tasks, as shown in Table 1. This table shows the activities that take place at ABC, Inc. when configuring a data product. The activities are grouped by their main four tasks: Maintain, Product Configuration, Purchasing and Shipping, and Installation. The next step is to determine which activities are going to be eliminated and which ones are going to be improved with the new process. Based on this information, a flow diagram for both processes, the current and the new improved process, should be drawn as shown in Figures 1 and 2.

The flow-matrix diagram for the current and new processes shown in Figures 1 and 2 includes the activities, contributors, the time spent in each activity, the labor rate for each activity, and the total cost of the process on one worksheet.

Combining all this information on one worksheet helps not only to visualize the process, but also to enter the information and calculate the cost for each process (Shiavi, Bourne, Brodersen, & Dawant, 1995; Brassard & Ritter, 1994).

This flow-matrix diagram model is used in the ROI Calculator Tool that is part of this chapter.

Calculate the Initial and Recurring Costs

The cost for the initial investment and the recurring costs are entered on the worksheets. In ROI Calculator Tool, the highlighted part of the ROI formula at the top of the worksheet shows which part of the ROI formula the information is for.

Figure 1. Current Process Flow-Matrix Diagram

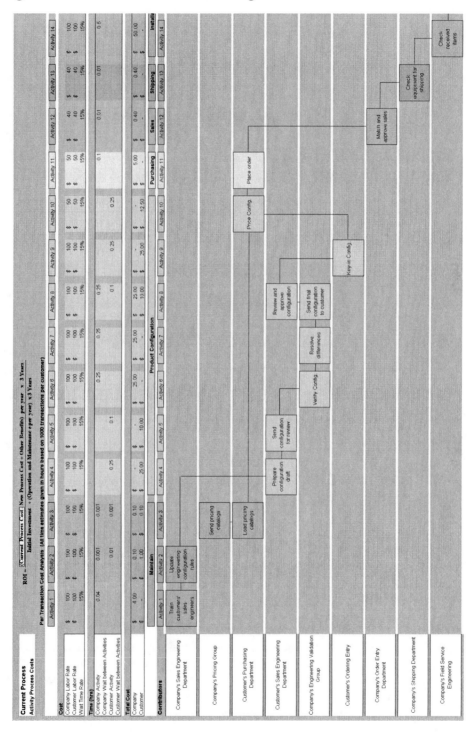

Figure 2. Proposed Process Flow- Matrix Diagram

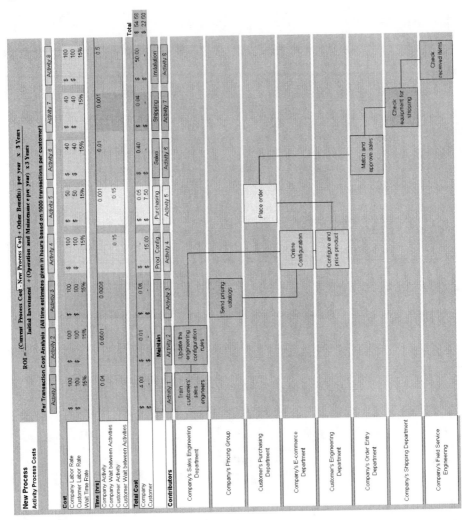

For each of the worksheets, the ROI Excel tool provides additional formulas to calculate the cells not highlighted in yellow.

Preliminary ROI Results

Before trying to calculate the intangible benefits, it must be determined whether the ROI results satisfy the company's requirements. The ROI tool provides two calculations, the Simple ROI and the Net Present Value (NPV) ROI (Oleson, 2000; Brigham & Houston, 1999). It is necessary to enter the company's Federal and State Taxes Bracket, the Company Discount Rate, and the Straight-line Depreciation (Yrs) on the worksheet to calculate the NPV.

Figures 5 and 6 show the ROI results without taking into account the intangible benefits.

Figure 3. Initial Investment

$$ROI = \frac{(\text{Current Process Cost} - \text{New Process Cost} + \text{Other Benefits}) \text{ per year } \times \text{ 3 Years}}{\boxed{\text{Initial Investment}} + (\text{Operation and Maintenance per year}) \times 3 \text{ Years}}$$

Project Initial Investment - Company & Customer

		One time Project Costs	
		Company	Customer
Consulting			
Create an overall strategy and a road map for project implementation.			
Consultant one-time fee		$ 25,000	$ 15,000
Travel		$ 800	$ 1,200
Project Management			
Project Management (Personnel)		$ 12,704	$ 1,851
Travel		$ 7,500	$ 2,400
Implementation and Integration			
- Building and verification of new configurations.		$ 6,000	$ 3,000
- Development and maintenance of new software.		$ 3,000	$ 2,000
- Web site enhancements.		$ 3,000	$ 2,000
- Test configuration		$ 4,000	$ 3,000
Training During Initial Installation			
- Internal and/or external training		$ 1,500	$ 1,200
- Documentation and other training expenses		$ 3,000	$ 1,000
Total Initial Application Costs		$ 66,504	$ 32,651
IT Capital Equipment Purchases			
Network Infrastructure		$ 41,000	$ 12,000
Software		$ 260,000	$ 50,000
Total Capital Equipment		$ 301,000	$ 62,000

The Simple and NPV ROI calculations for this example, Figures 5 and 6, show that this project has a very high return, and that it is not necessary to measure and calculate the intangible benefits (which would be shown above as "Other Benefits").

Calculating the Intangible Benefits

Intangibles, by definition, are assets that are not physical in nature; for example, goodwill, patents, trademarks, customer satisfaction, and copyrights. For those intangibles that can be measured, a dollar value can be assigned or calculated.

ABC's executives agreed not to include the intangible benefits in the ROI calculation for their e-commerce tool. However, because this is a very important aspect of an ROI calculation, some examples on how to calculate intangibles such as customer satisfaction, customer and employee retention, decreased time to market, reduced sales outstanding, and web improvement are discussed in this section.

The ROI Calculator Tool calculates several intangibles such as customer satisfaction, customer and employee retention, decreased time to market, reduced sales outstanding, and web improvement. These measurements are discussed in this section.

Figure 4. Recurring Costs (Operation and Maintenance)

$$\text{ROI} = \frac{\text{(Current Process Cost - New Process Cost + Other Benefits) per year } \times \text{ 3 Years}}{\text{Initial Investment } + \boxed{\text{(Operation and Maintenance per year)}} \times \text{ 3 Years}}$$

Project Recurring Costs - Operation & Maintenance
Company & Customer - Cost per Month

	Company	Customer
Operation & Maintenance		
Communications		
- Additional communications expenses - per month	$ -	$ -
Software Licensing and Support Fees (Per Month)		
Licensing Fees	$ 1,200.00	$ 600.00
Software Maintenance Charges	$ 1,200.00	$ 600.00
Service and Usage Charges	$ 1,200.00	$ 600.00
IT Staffing Labor		
IT department for the new E-Commerce application. People to hire may include:		
- Department Director	$ 5,250	$ -
- Webmaster in charge of developing and maintaining the site	$ 3,281	$ -
- Database engineer	$ -	$ 1,167
- Site Developers to maintain new software	$ 7,875	$ 2,042
- Customer representatives	$ 6,563	$ -
Ongoing Training		
- Internal and/or external training	$ 800	$ 1,200
- Workforce transformation / Change management		
Facilities		
- More office space (negative if less office space)*	$ -	$ -
- Modifications to physical facilities.	$ -	$ -
Custom Solution Costs	$ -	$ -
Total Operation & Maintenance per Month	$ 27,369	$ 6,208

Customer Satisfaction

Customer satisfaction reduces cost or increases profits in the following areas:

- Increase in customer retention
- Increase in sales
- Increase in customers' Life Time Value (LTV)
- Increase in market share
- Decrease in customer support service
- Reduction in the purchasing decision lead-time, prospect, sale
- Creation of new customer value propositions and profit models
- Enhancement and strengthening of customer relationships
- Leveraging of talent to drive higher productivity improvements

Figure 5. Preliminary Simple ROI Results Without Intangible Benefits

Project's Simple ROI					

$$ROI = \frac{(\text{Current ProcessCost} - \text{New ProcessCost} + \text{Other Benefits}) \text{ per year } \times \text{ 3 Years}}{\text{Initial Investment } + (\text{Operation and Maintenance per year}) \times 3 \text{ Years}}$$

The current and new process costs are per transaction., therefore, it is necessary to determine the number of transactions per year.

Business Assumptions

Increase in number of transactions per year		0%	10%	10%	
Number of transactions per year	10000	10000	11000	12100	

Company

Benefits	Start-Up	Year 1	Year 2	Year 3	Total
- Current Process Cost		$ 1,650,000	$ 2,035,000	$ 2,238,500	$ 6,123,500
- New Process Cost		$ (545,800)	$ (600,380)	$ (660,418)	$ (1,806,598)
- Other Benefits		$ -	$ -	$ -	$ -
Total Benefits per Year		$ 1,304,200	$ 1,434,620	$ 1,578,082	$ 4,316,902
Initial and Recurrent Costs					
- Initial Investment	$ 367,504				$ 367,504
- Operation & Maintenance Costs		$ 328,425	$ 328,425	$ 328,425	$ 985,275
Total Operation & Maintenance Costs	$ 367,504	$ 328,425	$ 328,425	$ 328,425	$ 1,352,779
Total Savings per Year	$ (367,504)	$ 975,775	$ 1,106,195	$ 1,249,657	
Company's 3-year ROI	319%				
Payback in Years	0.38				

Customer Business Assumptions

Increase in number of transactions per year		0%	10%	10%	
Number of transactions per year	5000	5000	5500	6050	

Customer

Benefits	Start-Up	Year 1	Year 2	Year 3	Total
- Current Process Cost		$ 418,000	$ 459,800	$ 505,780	$ 1,383,580
- New Process Cost		$ (112,500)	$ (123,750)	$ (136,125)	$ (372,375)
- Other Benefits		$ -	$ -	$ -	$ -
Total Savings per Year		$ 305,500	$ 336,050	$ 369,655	$ 1,011,205
Initial and Recurrent Costs					
- Initial Investment	$ 94,651				$ 94,651
- Operation & Maintenance Costs		$ 74,500	$ 74,500	$ 74,500	$ 223,500
Total	$ 94,651	$ 74,500	$ 74,500	$ 74,500	$ 318,151
Total Savings per Year	$ (94,651)	$ 231,000	$ 261,550	$ 295,155	
Customer's 3-year ROI	319%				
Payback in Years	0.41				

Increasing Customer Retention

Companies are always losing customers and, normally, it costs more to get a new customer than to keep one. That is why there is so much emphasis on retaining customers. The value of a customer to a company is the profit that (s)he brings to the company. Suppose, for example, that ABC, Inc. is losing 10% of its customers a year and would like to implement a Customer Relationship Management (CRM) application with the idea of losing only 4% of its customers. Furthermore, ABC, Inc. is more interested in retaining its more valuable customers rather than those that bring less profitability. Figure 7 shows a customer retention calculation (Synergex, 2001; Novo, 2000).

The following methodology shows how to calculate the different values. First, the customers should be grouped into profitability segments; then it should be determined,

Figure 6. NPV ROI without Intangible Benefits

Company's ROI Using Net Present Value (NPV)			

Business Assumptions

		0%	10%	10%
Increase in Number of Transactions per Year		0%	10%	10%
Number of Transactions per Year	10000	10000	11000	12100

Financial Analysis Summary Without Intangible Benefits

Company (10000 Transactions)		**Customer (5000 Transactions)**	
Project's Net Savings Present Value	$ 1,542,261	Project's Net Savings Present Value $	358,927
3-year ROI	461%	3-year ROI	458%
Internal Rate of Return	157.6%	Internal Rate of Return	157.6%
Payback in Years	0.65	Payback in Years	0.64

Annual Benefits	Start-Up	Year 1	Year 2	Year 3
- Current Process Cost		$ 1,850,000	$ 2,035,000	$ 2,238,500
- New Process Cost		$ (545,800)	$ (600,380)	$ (660,418)
- Other Benefits (Intangibles)		$ -	$ -	$ -
Total Savings / Year		$ 1,304,200	$ 1,434,620	$ 1,578,082

Expensed Costs	Initial Investment	Year 1 Expenses	Year 2 Expenses	Year 3 Expenses
- Initial Investment	$ 66,504			
- Operation & Maintenance Costs		$ 328,425	$ 328,425	$ 328,425
Total per Period	$ 66,504	$ 328,425	$ 328,425	$ 328,425

Capital Cost and Depreciation Schedule	Initial Investment Expenses	Year 1 Depreciation	Year 2 Depreciation	Year 3 Depreciation
Initial Investment/Depreciation	301000	60200	60200	60200
Total Depreciation per Period	301000	60200	60200	60200
Salvage Value at End of Project				120400

Basic Financial Assumptions

Federal and State Taxes	50%
Discount Rate (Cost of Capital)	17.0%
Straight-line Depreciation (Yrs)	5

Net Cash Flow	Investment Cost	Year 1	Year 2	Year 3
Total Benefits	N/A	$ 1,304,200	$ 1,434,620	$ 1,578,082
Less Expended Cost	$ (66,504)	$ (328,425)	$ (328,425)	$ (328,425)
Savings Cash Flow Before Taxes	$ (66,504)	$ 975,775	$ 1,106,195	$ 1,249,657
Depreciation Adjustment		$ (60,200)	$ (60,200)	$ (60,200)
Oper. Income Before Taxes (EBIT)	$ (66,504)	$ 915,575	$ 1,045,995	$ 1,189,457
Taxes on Operating Income (50 %)	$ 33,252	$ (457,788)	$ (522,998)	$ (594,729)
Net Operating Profit After Taxes (NOPAT)	$ (33,252)	$ 457,788	$ 522,998	$ 594,729
Add Back Depreciation		$ 60,200	$ 60,200	$ 60,200
Operating Cash Flow	$ (33,252)	$ 517,988	$ 583,198	$ 654,929
Capital Equipment Initial Investment	$ (301,000)			
Adjust for Salvage Value				$ 120,400
Net Cash Flow	$ (334,252)	$ 517,988	$ 583,198	$ 775,329

Financial Analysis

Project's Net Savings Present Value	$ 1,542,261
3-year ROI	461%
Internal Rate of Return (IRR)	157.6%
Payback in Years	0.65

of the 10% that ABC is losing, what percentage is in each profitability segment. This calculation is not necessary, but it helps to visualize which segment of the customers ABC, Inc. is losing. This is given by:

Customers Lost = No. of Cust. x Total % of lost cust. x % per segment
Customers Lost = 100,000 x 10% x 18% = 1800

Second, the number of retained customers in each of the profitability segments that ABC, Inc. would like to have should be determined. This is given by:

Retained Customers = No. of Cust. x % goal for retined cust. x % per segment
Retained Customers = 100,000 x 6% x 25% = 1500

Figure 7. Measuring Customer Retention

Increasing Customer Retention							
Percentage of customers lost per year				10%			
Percentage of customers retained goal				6%			
Number of customers				100,000			
Profitability Segment	Average Profits	% of Lost Customers. *	Lost Cust.	% of Retain Cust. **	Retained Cust.	Cust. Retent. Gain Profits	
1	$600	18%	1,800	25%	1,500	$	900,000
2	$550	15%	1,500	20%	1,200	$	660,000
3	$500	10%	1,000	20%	1,200	$	600,000
4	$450	35%	3,500	20%	1,200	$	540,000
5	$400	5%	500	5%	300	$	120,000
6	$350	5%	500	5%	300	$	105,000
7	$300	5%	500	5%	300	$	90,000
8	$250	3%	300	0%	-	$	-
9	($15)	2%	200	0%	-	$	-
10	($25)	2%	200	0%	-	$	-
	Totals	100%	10,000	100%	6,000	$	3,015,000

* Percentage of customers lost per year in each profitability segment
** Percentage of retained customers in each profitability segment

Third, the profits that ABC, Inc. will gain by retaining some of the customers that it is currently losing should be calculated. This is given by:

Gained Profits = No. of Retined Cust. x Profit per segment
Gained Profits = 1500 x 600 = 900,000

This procedure shows that if ABC, Inc. implements a CRM application that results in retaining some of the customers that it is now losing, ABC, Inc. will have a profit of $3,015,000. The same approach can be used for a company that is gaining more customers than it is losing. The percentage of gained customers can be used instead of the percentage of retained customers.

Customer Life Time Value (CLV)
Many online activities are aimed at retaining customers and the Customer Life Time Value (CLV) can be used to justify the investment by showing projected increase sales or margins (Hansen, 2000, pp. 179-180).

CLV is a measure of the Net Present Value of a customer's profitability for the period, in years, that a customer will remain loyal and keep buying a company's product. It includes the initial cost of getting a new customer.

Figure 8. Customer Life Time Value

Customer Life Time Value (LTV)		
Cost of getting a new customer	$	245
Number of years customer keeps buying from the company		3
Customer profitability per year	$	1,000
Discount Rate (Cost of Capital)		17%
Net Present Value of Customer Life Time Value (LTV)	$	1,965

Figure 9. Increasing Employee Retention

Increasing Employee Retention			
Total number of employees who left the previous year			1236
Total number of employees who left the current year			437
		Retained Employees	799

Retained Employees	Cost to Hire and Train New Empl.	Total Savings per Year
799	$ 546	$ 436,254

Increasing Employee Retention

To determine the savings to a company for increasing employee retention, the number of retained employees should be multiplied by how much it costs a company to hire and train a new employee. In the example below, after implementing an e-business application, a company was able to reduce the number of employees leaving the company in a particular year from 1,236 to 437. If the cost to hire and train a new employee was $546 per employee, then the total savings to the company was $436,254.

Decreasing Time-to-Market

Engineers exchanging information online can reduce the time spent in product development considerably. The main effect of introducing a product earlier to the market is that it brings future sales to the present. The profit to the company is the Net Present Value of those sales.

Figure 10. Decreasing Time-to-Market

Decreasing Time-to-Market			
Date when product will be available for sale			8/15/2002
Forecasted product general availability			2/15/2002
		Decreasing Time-to-Market in Days	181

Reduction in Time to Market	Forecasted Prod. Sales per Year	Cost of Money	Total Savings
181	$ 2,568,747	5.25%	$ 66,875

Decreasing time-to-market may allow a company to increase market share, thus, increasing sales.

Qty of Units Sold per Year	Increase in Market Share	Profit per Unit	Total Revenue Increase
5,685	6.00%	$ 354.00	$ 120,749

Decreasing time-to-market may allow a company to charge more for a product, thus, increasing profits

Qty of Units Sold per Year	Units sold in 181 days	Increase in Price/Profits	Total Revenue Increase
5,685	2820	$ 75.00	$ 211,500
Total Decreasing Time-to-Market Savings			$ 399,125

Figure 11. Web Updates

Web Updates				
Average Time to Update a Web page	Number of Web Pages Updated Daily	Web Designer Salary	Company Benefits Overhead	Total Cost
Old process				
0.25	560	$ 75,000	1.75	$ 9,188
New process				
0.05	560	$ 75,000	1.75	$ 1,838
			Total Savings per Day $	7,350
Assuming 2000 working hours per year.				

An early product introduction may also allow a company to increase market share or to charge more for the product.

Web Updates

The number of web pages that a company posts on the Internet or on their intranet increases continuously. Updating web pages is costly, and the solution is to update the web pages by using a software application that updates the pages dynamically.

Reduced Days Sales Outstanding (Average Collection Period) and Float

Companies using ABC's Online Tool for Configuring and Pricing Data products will be able to place the order and pay online with a wire transfer by using Electronic Data Interchange (EDI) in their ordering management process. Wire transfers expedite receivables collection by several days.

The benefit to ABC, Inc. for collecting accounts receivable earlier is what a company can earn on those early cash flows; that is, the investment produced when the savings are reinvested or by having to borrow less money to maintain the required cash balance to operate.

Increasing Online Sales

When customers purchase products online, they do not require any assistance, thus reducing the cost of sales. The resulting cost savings can be calculated by

Figure 12. Reducing Collection Period

Reducing Days Sales Outstanding (Average Collection Period) and Float
Electronic Payment reduces the average collection period and float (time the check is in the mail)

Last year number of days sales outstanding (DSO)	45
Forecasted days sales outstanding with new process implemented	22
Float (time the check is in the mail, time to process the check, and time to clear)	6
Forecasted Reduction In Days Sales Outstanding	29

Reduction in Acc. Receivable DSO	Average Acc. Receivables	Cost of Money	Total Savings
29	$ 24,256,385	5.25%	$ 101,179

Figure 13. Increasing Online Sales

Increasing Online Sales
Cost savings of online sales (consumers doing much of the sales work themselves)
Time a salesperson expends per sale   10   minutes

Aver. Salary Salesperson	Benefits Overhead	Cost per Sale	No. of Online Transactions per Month	Total Savings per Month
$ 40,000.00	1.75	$ 5.83	15,000	$ 87,500.00

Assuming 2000 working hours per year.

determining how long it takes a salesperson to do a sale and multiplying that by his/her salary.

Providing Online Support

When customers find the information that they need at a company's web site, it saves the company the cost of technical support. A way to calculate the savings of online support is to determine offline and online support costs per unit and then to multiply the difference by the percentage of customers looking for support online.

Finding Information on the Web Site

The calculation of this savings is similar to the savings calculation of Providing Support Online. Determine the number of visitors who will not call the company and multiply it by the cost of each phone call.

Figure 14. Increasing Online Sales

Providing Online Support

Cost of product support offline per unit	$ 4.50
Cost of product support online	$ 1.50
Total savings per product sold	$ 3.00
Number of products sold per month	25,000
Percentage of cust. looking for support online	40%
Total Savings per Month $	30,000

Figure 15. Finding Information on the Web Site

Finding Information on the Web Site

Number of visitor sessions per month	100,000	Visitors looking for information
Percentage online successful sessions	50%	Percentage of visitors who find the information
Total successful sessions	50,000	Visitors who will not call for information
Cost of answering a customer's call	$ 4.50	Cost to the company of customer's calls
Total Savings per Month $	225,000	

As mentioned previously, there is no specific information available about calculating intangibles in e-commerce. The issue is complex because companies might see the value of an intangible from different perspectives and take different approaches to arriving at a dollar value. Therefore, as part of the IT Governance, each company should create its own library of intangibles values and use it for all ROI calculations.

CONCLUSION

The case study, the ROI of an Online Tool for Configuring and Pricing a Data Product, was used to show, step-by-step, how to conduct an ROI calculation. The ROI results were taken directly from the ROI Calculator Tool that includes all the financial formulae. As part of the IT Governance, organizations should develop or purchase a particular ROI calculator and use it for all e-commerce ROI calculations. The use of the same tool for all projects will provide ROI results based on the same standards, in addition to reducing the time needed to prepare a ROI calculation for an e-commerce project.

REFERENCES

Brassard, M., & Ritter, D. (1994). *The memory jogger II, A pocket guide tools for continuous improvement and effective planning* (pp. 13-16, 56-62). Methuen, MA: GOAL/QPC.

Brigham, E. F., & Houston, J. F. (1999). *Fundamentals of financial management* (pp. 502, 548, 496-528). Forth Worth: Harcourt College Publishers.

Byrd, T. A., & Marshall, T. E. (1997). Relating Information Technology investment to organizational performance: A causal model analysis. *Omega, 25*(1), 43-56.

Byrne, J. (2001). Caught in the Net. *Business Week,* (August 20), 114.

CIOView. (2001). *Financial primer: How to calculate ROI, NPV, Payback and IRR.* Available online: http://www.cioview.com/white/finl_primer.htm.

Ferengul, C. (2001). *Service management strategies.* (November 14), 1. Meta Group.

Friedlob, G. T., & Plewa, F. J. (1996). *Understanding return on investment* (pp. 184-191, 207-212). New York: John Wiley & Sons.

Gomolski, B. (2001). The cost of e-business. *Infoworld,* (December 12), 12.

Gordon, D. (2002). *Business process analysis: Enhancing organizational effectiveness* (p. 3). Dallas, TX: University of Dallas.

Hanson, W. (2000). *Principles of Internet marketing* (pp. 179-180). Cincinnati, OH: South-Western College Publishing.

Mahmood, M. A., & Mann, G. J. (1993a). Measuring the organizational impact of IT investment: An exploratory study. *Journal of Management Information Systems, 10*(1), 97-123.

Mahmood, M. A., & Mann, G. J. (1993b). *Impact of IT investment: An empirical assessment. Accounting Management and Information Technology, 3*(1), 23-32.

Meta Group. (2001). *How do I get CRM right in a slowing economy? A third-quarter ROI update.* Meta Group.

Novo, J. (2000). *Drilling down, turning customer data into profits with a spreadsheet* (pp. 17-29). Saint Petersburg: Deep South Publishing Company.

Oleson, T. (2000). *The Return on Investment associated with eBusiness and eCommerce.* IDC #21462, (February), 35-36.

Rivenbark, W. C., FitzGerald, K. M., Schelin, S. H., Yates, W. H., & Runkle, T. (2001). *Information technology investments — Metrics for business decisions* (pp. 23, 18-20). Center for Public Technology, Institute of Government, The University of North Carolina at Chapel Hill. Available online: http://www.cpt.unc.edu/pdfs/Final%20Report.pdf.

Rosenthal, R. (2000). *ROI, eCommerce, and the metrics that matters: Results from IDC's Internet executive ePanel.* IDC #22743, (July), 7-8.

Rosenthal, R. (2000). *Using ROI to help a bricks-and-mortar enter eCommerce: DoveBid.* IDC #21506, (January), 9.

Sammer, J. (2001). Project ROI. *Business Finance Magazine.* Available online: http://www.businessfinancemag.com/archives/appfiles/Article.cfm? IssueID=357& ArticleID=13824.

Schlegel, K. (2001). *Web & collaboration strategies.* (March 27). Meta Group, File: 990.

Shiavi, R., Bourne, J., Brodersen, A., & Dawant, M. (1995). *Review of Japanese and other quality tools. Vanderbilt University, Introduction to Quality Improvement.* Available online: http://www.vanderbilt.edu/Engineering/CIS/Sloan/web/es130/quality/qtooltoc.htm.

Synergex. (2001). *Where is the ROI in CRM?* Available online: http://www.synergex.com/whitepapers/WhereisROIinCRM.htm.

Woo, C. Y., & Willard, G. (1983). Performance representation in Business Policy Research: Discussion and recommendation. *The Twenty-third Annual National Meetings of the Academy of Management, Dallas, Texas, USA.*

Wu, J. (2000). *Calculating ROI for business intelligence projects.* (December 12). Base Consulting Group. Available online: http://www.baseconsulting.com/html/articles_and_white_papers.html.

ENDNOTES

[1]   The Applied Information Economics was developed by Douglas Hubbard and uses concepts in Economics, Statistics, Decision Analysis, Information Theory, and other disciplines to measure intangibles in an ROI calculation. See Appendix A for more information.

APPENDIX A: ROI TERMINOLOGY

A definition of some terms associated with ROI terminology and the pros and cons of using alternative measurements will assist in understanding the measured benefits of a project. The ROI term definitions were compiled from different sources (Rivenbark et al., 2001; Friedlob & Plewa, 1996; CIOView, 2001; Brigham & Houston, 1999).

ROI

The ROI is the total quantitative ***net savings*** or return, in hours, dollars, or other measurable units, generated by an initial investment, ***divided by the total cost of the*** *initial investment*. It is the most popular metric. The following are some of the pros and cons:

- Evaluates the earnings derived from a project, an improving effort, or e-commerce application investment as compared to the expended initial investment required to complete the venture.
- Used widely as a key metric to sell IT and e-commerce application investment decisions.
- Indicates the percentage return over a specific period of time without taking into account the magnitude of the project. Would it be better to have 80% ROI in a $20,000 project or 40% ROI in a $200,000 project?

Net Present Value

The Net Present Value is the calculation of the project's cash inflows and outflows discounted to the present time using a pre-selected project Discount Rate or Cost of Capital. The difference between the discounted inflows and outflows is the Net Present Value. A positive NPV means that the project will earn a return greater than an investment would make using the discount rate. A negative NPV means that the company would be better served investing in another project because the project's return is less than the discount rate. The following are some of the pros and cons:

- Helps determine if the project, the improving effort, or the e-commerce investment should be made.
- Gives a direct measure of the project's dollar benefits, but does not explicitly produce a project's rate of return.
- Recognizes the time value of money. Negative NPV: implies that the initiative should be avoided if it is directed at cost reduction. Positive NPV: implies that the assessment should include further risk analysis.
- Does not provide information about when the project savings occur.
- Used to evaluate mutually exclusive projects, especially those that differ in scale and/or timing.
- Used to compare independent projects with approximately the same initial investment. In independent projects, NPV and IRR measurements always lead to the same results; for example, if NPV indicates accept project, IRR also indicates accept project.

Internal Rate of Return

The Internal Rate of Return is the investment yield rate produced by a project when the Net Present Value calculation is used; it is the rate at which the present value of the inflows is equal to the outflows. The IRR is compared to the Project Discount Rate. If the IRR is higher than the Project Discount Rate, the project is selected; if it is lower, the project is rejected. The pros and cons of using IRR for implementing decisions include the following:

- Is the project's expected rate of return.
- Does not give any indication of the magnitude of the project involved.

- Determines the interest rate and then compares this rate to a minimum required rate of return, project discount rate, or cost of capital.
- Is useful in the ranking of IT investment initiatives because they can easily be compared.
- Is very useful for comparing dissimilar initiatives because it is a rate or ratio.
- Is best used in conjunction with Net Present Value.

Payback Period

The Payback Period calculation determines the amount of time, either estimated or measured, that is required for a project, improvement effort, or e-commerce application savings to recoup the initial investment, to show a profit, or to break even. Typically, projects with a short payback period are preferred because they are assumed to indicate a lesser degree of risk. The following are some of the pros and cons:

- Does not provide information about the magnitude of the savings or how the investment performs beyond the payback period.
- Ignores the timing of cash flows.
- Emphasizes liquidity and allows managers greater flexibility in planning for the availability of funds (shorter payback projects).
- Should not be used as sole indicator for a project, improvement effort, or e-commerce application investment.

The Discounted Payback is similar to regular payback, except that the expected cash flows are discounted by the project's cost of capital.

Applied Information Economics (AIE)

AIE, developed by Douglas Hubbard, a principal of Hubbard Ross, applies proprietary scientific and mathematical methods to compute the return on the IT investment — including all intangibles. The following are some of the pros and cons:

- Synthesizes a variety of techniques from scientific and mathematical fields (decision theory, financial theory, and statistics).
- Determines the value of information, methods for modeling uncertainty in estimates, and treating the IT investment as a type of investment portfolio.

Investment

The estimated or measured total cost in hours, dollars, or other units that an improvement effort requires to be planned, executed, and completed.

Cost of Capital

The Cost of Capital is the interest rate that a company has to pay to providers of capital when borrowing capital to implement a particular business or project. Also called Project Discount Rate.

APPENDIX B: COMPANIES OFFERING ROI MEASURING SERVICES

Sometimes organizations are considering e-commerce applications that would require a large financial investment to implement, and they would prefer the help of an external company to prepare the ROI. In doing the research for this paper, the authors have come across names of some companies that offer those services. The authors have neither had any contact with these companies, nor recommend any of them.

The following is a list of companies that provide those services.

1.   CIOview: http://www.cioview.com/
2.   Glomark Corp.: http://www.glomark.com/html/index.htm
3.   Hubbard Ross, LLC.: http://www.hubbardross.com/

Chapter IX

Technical Issues Related to IT Governance Tactics: Product Metrics, Measurements and Process Control

Michalis Xenos
Hellenic Open University, Greece

ABSTRACT

This chapter deals with some technical aspects of the strategies for IT Governance described in most of the other chapters of this book. The practical application of IT Governance techniques and tactics requires the collection and analysis of measurable data that guide estimation, decision-making and assessment, since it is common sense that one can control and manage better what one is able to measure. This chapter aims at introducing the reader to product metrics and measurements, proposing methods relating to the implementation of a measurements program and analyzing how product metrics can be used to control IT development.

Focus is placed on product metrics, measurements and process control techniques. Such instruments can aid significantly in monitoring the development process and making IT-related tasks more transparent to IT managers. They also aid in design, prediction and assessment of the IT product quality. They provide data used for decision-making, cost and effort estimation, fault prevention, and testing time reduction. Moreover, the use of product metrics and process control techniques can direct the standardization of IT products and IT development process, as well as the assessment of the process maturity of outsourcing partners. The most commonly used metrics are presented and the reader is introduced to measurement techniques and automation tools. This chapter also discusses the application of metrics along with statistical process control tools. Finally, it provides suggestions on how product metrics can aid towards IT Governance not only in large IT companies, but also in the smaller ones.

INTRODUCTION

IT Governance is *"the organizational capacity exercised by the board, executive management and IT management to control the formulation and implementation of IT strategy and in this way ensure the fusion of business and IT"* (Van Grembergen, 2002). Previous chapters of this book have successfully introduced the reader to several IT Governance techniques and tactics that aid the board, executive and IT managers in monitoring, standardization, quality assessment, cost estimation and cost cut down; in short, they aid in controlling the IT process. The practical application of such techniques and tactics requires the collection and analysis of measurable data that guide estimation, decision-making and assessment. It is common sense that one can control and manage better what one is able to measure, as pointed out by DeMarco (DeMarco, 1982).

This is what product metrics do. They provide measurements of factors related to IT development that can be utilized as input to process control techniques. Product metrics, along with the measurement techniques used for their collection, constitute the means to measure desirable product characteristics and in this way allow control of the product development process. This chapter focuses on product metrics, measurements and process control techniques. Such instruments can aid significantly in monitoring the development process and making IT-related tasks more transparent to IT managers. They also aid in design, prediction and assessment of the IT product quality. They provide data used for decision-making, cost and effort estimation, fault prevention, and testing time reduction. Moreover, the use of product metrics and process control techniques can direct the standardization of IT products and IT development process, and the assessment of the process maturity of outsourcing partners.

Measurements and Process Control in the IT Industry

IT management and development has suffered from many disasters. Examples of such crises can be found in the classic article of Gibbs (Gibbs, 1994) describing software crises in both the private and government sectors. In other cases, the delivered IT quality suffers from such defects that can lead in significant losses. For instance, one of the recent US National Institute of Standards and Technology reports states that insufficient software costs the US as much as US $59 billion a year and that up to US$22 billion of that could be saved if licensed software had just 50% fewer defects (Miller & Ebert, 2002). These facts lead to a reasonable question: How can IT management be aided in reducing errors and improving IT quality? Two of the means significantly aiding towards this direction are software metrics and process control techniques. Such metrics and techniques are extensively used in the current IT development industry and form an inseparable part of the development process of companies that have achieved CMM fourth and fifth levels, as documented in recent reports (McGarry & Decker, 2002).

Metrics are used in IT development process to measure various factors related to software quality and can be classified as product metrics, process metrics and recourse metrics. Product metrics are also called software metrics. These are metrics that are directly related to the product itself, such as code statements, delivered executables, and manuals, and strive to measure product quality, or attributes of the product that can be related to product quality. Process metrics focus on the process of IT development and measure process characteristics, aiming to detect problems or to push forward successful practices. Resource metrics are related to the resources required for IT development and their performance. This chapter focuses on product metrics and how such metrics can aid towards IT development process control and respectively towards IT Governance.

This Chapter

It could be said that this chapter deals with some technical aspects of the strategies for IT Governance described in most of the other chapters of this book. Having read about different strategies and tactics for IT Governance (see the relevant chapter) the reader is now introduced to metrics and how these can aid IT managers. Issues such as cost estimation, monitoring the IT development process, diagnosing the IT development effectiveness, decision making on outsourcing and integration, and quality assessment can be achieved more easily with the use of metrics and process control.

In order to present examples of the techniques covered, this chapter goes into some technical details that require knowledge of basic software engineering principles and statistical control. Such technical details are presented using smaller fonts. If the reader is not interested in them, he may skip areas marked with smaller fonts and still be able to follow the main objectives of the chapter.

This chapter aims at introducing the reader to product metrics and measurements, proposing methods relating to the implementation of a measurements program and analyzing how product metrics can be used to control IT development.

What this chapter presents is an overview of software metrics and measurement techniques. The most commonly used metrics are presented and the reader is introduced to measurement techniques and automation tools. Furthermore, emphasis is placed on the correlation of internal metrics with external product quality characteristics by combining internal and external measurements. Finally, this chapter discusses the application of metrics along with statistical process control tools. Such tools aim at keeping the IT development process under control. Emphasis has also been placed on the argument that product metrics can be utilized not only by large IT developing enterprises, but also by smaller or medium-size ones, since in many cases the misinterpretation of measurements in small and medium enterprises has led to the belief that product metrics are only meant for large companies. This chapter provides suggestions on how product metrics can aid towards better IT Governance not only in major IT companies, but also in the smaller ones.

The chapter is organized as follows: the chapter begins with a literature review of product metrics, measurements and statistical process control. Next, the chapter discusses issues related to measurements, use of product metrics, correlation between internal and external measurements and the application of Statistical Process Control during IT development and offers solutions and recommendations. This is followed by a discusses on future trends and a summary of conclusions of this chapter.

BACKGROUND

Metrics and Measurements

Measurement is the process by which numbers or symbols are assigned to attributes of entities in the real world in such a way as to describe them according to clearly defined rules (Fenton & Pfleeger, 1997). In IT development, measurements are conducted by using metrics. A *metric* is an empirical assignment of a value in an entity aiming to describe a specific characteristic of this entity. Measurements have been part of our everyday life for a long time and were introduced to IT Governance in order to satisfy the need to control IT development, since according to DeMarco (1982), IT development cannot be controlled if it cannot be measured.

The first software metrics were proposed in the mid '70s by McCabe (1976) and Halstead (1977), while the first practical use of metrics — in terms of correlation with development data — was presented by Funami and Halstead (1976). In the following years, a large number of software metrics were proposed, some of which still have practical use and can lead to improvements in IT development, while others proved to be quite impractical in certain occasions. The proliferation of metrics was followed by more practical proposals on how to interpret results from metrics, such as the one by Fitzsimmons and Love (1978) or the one by Shepperd and Ince (1990). A number of methods combining metrics into measurement methodologies were also presented, such as the ones proposed by Hansen (1978) and by Xenos and Christodoulakis (1994).

A company involved in IT development can select from a variety of applied metrics those that are more suitable to be included in its IT Governance techniques. This selection can be made from a large volume of proposed product metrics that include code metrics such as lines of code, software science metrics, complexity metrics, readability metrics such as the one proposed by Joergensen (1980), data metrics, nesting metrics, and structure metrics such as the one proposed by Henry and Kafura (1981). There are also process metrics, estimation metrics, design metrics, testing metrics, reliability metrics, maintainability metrics, reusability metrics, as well as metrics applied on specific programming languages, such as object-oriented metrics, or metrics for visual programming. Therefore, taking into account the volume of literature that exists about software metrics, it is no more a question of finding metrics for an IT project, rather than selecting the appropriate ones for better IT Governance. Given the great number of metrics (measuring almost everything), any attempt to select a metric without basing the selection on a detailed breakdown of the IT developing company needs and an extensive investigation of the proposed metric's applicability would result in minor benefits from its use or no benefits at all. To benefit from the use of metrics, apart from fully understanding the various existing metrics, one should also define well why one wants to measure, what to measure and when is the right time to measure it.

So the first question rising is: '*Why use metrics?*' The answer to this question is that metrics are needed to provide understanding of different elements of IT projects. Since it is not always clear what causes an IT project to fail, it is essential for IT Governance to measure and record characteristics of good projects as well as bad ones. Metrics provide indicators for the developed software. As Ragland (1995) states, indicators are metrics or combinations of metrics that provide insights of the IT process, the software project, or the product itself. Measurements aim at the assessment of the status of the development process and the developed product. Therefore, metrics can be used for performance evaluation, cost estimation as Stamelos and Angelis (2001) have proposed, effort estimation, improving IT productivity, selecting best practices and — in general — for improving IT Governance.

This discussion leads to the next question: 'What to measure?' As previously mentioned, process and product are what we need to measure. One may argue that since the result of IT projects is software, what we need to measure is software and only software. This is not true. According to Deming (1986), if the product you have developed is erroneous, do not just fix the errors, but also fix the process that allowed the errors into the product. This way you will not have to keep fixing the error in subsequent productions. Therefore, both process and product metrics and measurements are important in IT Governance. You have already read about techniques for governing IT process in the other chapters of this book, including Statistical Process Control shortly described in this chapter. For this reason, the remaining part of this chapter focuses on the IT product itself; it focuses on product metrics.

The use of product metrics is highly related to the maturity of the IT development company. The adoption of standards such as the ISO 9001 (1991) or the CMM (Paulk, Curtis, Chrissis, & Weber, 1993) enforces a disciplined IT development process that has to be in control, therefore measured. Methods like Goal Question Metric (Solingen & Berghout, 1999) enable developers to implement realistic product metrics in the development process tailored to their needs. The more mature the company is with regard to the employment of IT Governance techniques, the more improved its measurement plan is and the more complex and detailed can be the metrics introduced into the IT development process.

It must be noted that before selecting the appropriate metrics, it is very important to define the desired product quality characteristics. The selection of these quality characteristics aids in defining what needs to be measured and what needs not, depending on the particular needs of the IT developing company. In the early 70's, McCall, Richards and Walters (1977) defined a framework for measuring such characteristics and proposed the Factors Criteria Metrics model — also known as FCM model — a model for defining what is software quality in terms of sub-characteristics. Incorporating FCM and experience from similar proposals, years later, the ISO standard ISO 9126 (1996) standardized what product quality is in terms of sub-characteristics. Therefore product metrics can be used within the IT development procedure in order to measure such product characteristics related to product quality. The amount and the identity of the characteristics to be measured are related to the IT Governance maturity and are further discussed in this chapter.

Having defined the goals and reasons for measuring, the next question that arises is: 'When to measure?' Although measurements should be conducted throughout the entire IT development life cycle, their scope varies depending on the IT development phase. Different measurement goals are defined at different development phases and therefore different kinds of metrics should be used. In the early phases of IT development, metrics are used mainly for estimation purposes. It is useful to collect metrics relating to different projects, so that they can serve as historical data for future projects, aiding in better IT Governance. As far as current projects are concerned, such data from past projects are used to assist in estimation (Putnam & Myers, 1992). E.g., size metrics are frequently used to predict the size of a new project. This is how COCOMO (Boehm, 1989), a well-known estimation model operates. Function points (Albrecht, 1979) are also commonly used for estimation.

In the intermediate phases of IT development process, metrics are used for project monitoring purposes while, in the meantime, code metrics are used to prevent errors. Furthermore, defect reports during testing are used for evaluating product quality and calibrating the measurement methods of the early phases. This purpose is also served by collecting external measurement data following project delivery, namely during the beta testing or maintenance phases of a project. So the time to measure is determined by the requirements and the aims of the measurement program and can vary from one IT project to another.

Summarizing, using an oversimplifying statement it could be said that metrics are an important instrument of IT Governance, aiding in making estimations in the early phases of an IT development project, preventing problems in intermediate phases and evaluating quality in the late project phases.

Statistical Process Control

Product metrics are essential instruments for the application of Statistical Process Control techniques into IT Governance. *Statistical Process Control* — briefly called SPC — is a collection of problem-solving tools useful in achieving process stability and improving capability (Montgomery, 1991). SPC has been successfully used by industry for the production of material goods aiming at the detection of non-conformity and the elimination of waste during production of such goods. The following paragraphs, in smaller fonts, discuss further some SPC technical details.

One of the concepts of SPC techniques is to monitor specific product quality characteristics over time. This is a way of monitoring whether the development process is 'in control', or whether it has become 'out of control' and consequently should be stopped and adjusted. Sample measurements relating to specific product characteristics are compared to each other using a control chart illustrating when problems occur in the development process and human intervention is required. Figure 1 is an example of a control chart. The vertical axis illustrates the values of the measured characteristic, while the horizontal axis represents periodical sample measurements. The measurement period may cover a few minutes or days or even weeks, although in industry it usually covers minutes rather than days or weeks. The center horizontal line represents the desired value of the measured characteristic — also called mean value, since in a process that is in control the mean value measured must be the desired one — while the other two horizontal lines represent the upper and lower acceptable limits of the characteristic's values. The dots depict sample measurements, while the lines connecting the dots are drawn simply to make monitoring easier by visualizing the dots sequence.

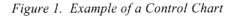

Figure 1. Example of a Control Chart

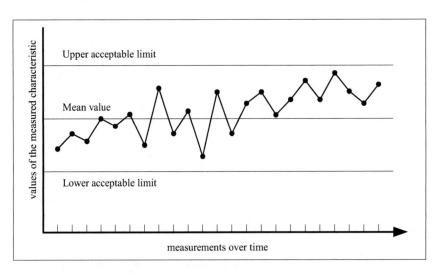

As long as the dots appear within the area between the upper and lower limit lines, the monitored process is considered to be in control. Such a process is considered as a stable one, and more than 99% of its variation falls within three sigma of the mean value of the measured quantity, i.e., within the upper and lower limit. On the other hand, a dot falling out of the upper or lower limit is an indication that the process might be out of control and further investigation or even adjustment is required. A process, however, may still be considered as out of control even if all measurements fall within the acceptable limit lines. This is the case when the dots seem to behave in a systematic, therefore non-random, manner. For example, in the control chart of Figure 1, ten consequent measurements fall above the desired value. This is strong indication that the development process may be out of control and in need of adjustments. Therefore, a process that stays within the upper and lower limit and does not show other indications of non-random behavior is a controlled process; in such cases the past performance of a process can be used to predict its future performance.

The aforementioned example makes it clear that SPC can only be applied on characteristics that can be objectively measured. This is very common in the production of material goods, where SPC has been applied with success, i.e., the values of the characteristics are most often dimensions or weights, such as the size of an instrument in millimeters or its weight in kilograms. This is rarely the case, however, in software development, where objective measurements are not always easy to obtain. For this reason, product metrics are essential in the IT development process since they are the instruments used to provide objective values that make the application of SPC techniques feasible. Summarizing, the exploitation of product metrics for obtaining values of the desired quality characteristics has made SPC a valuable tool aiding towards the improvement of the IT development process (Burr & Owen, 1996).

Sample cases of SPC use into IT development have been presented by Weller (2000), who used SPC for source code inspections to assess product quality during testing and to predict post-ship product quality, and by Florac, Carleton and Barnard (2000), who used SPC for analyzing the IT development process of a NASA space shuttle project that was related to 450,000 total delivered source LOC. To conclude, SPC can be used in IT development in conjunction with product metrics so as to exploit measurements of past performance and aiming to determine the IT development process capability to meet user specifications.

PRODUCT MEASUREMENTS AND PROCESS CONTROL IN IT

Issues, Controversies, Problems

This section classifies product metrics into two categories, internal and external, and provides a short definition and examples of each category, descriptions of how they can be used in IT development and what kind of problems one might have to face during their application. Emphasis is placed on the correlation between internal and external metrics, since this issue is highly related to the use of metrics. The discussion also focuses on the way that the company size affects the application of metrics, since in certain cases the conduction of measurements appears to be problematic in small and medium size IT companies. For this reason, issues related to the application of product metrics in small or medium size companies are further analyzed in this section.

Product metrics can be categorized (Fenton & Pfleeger, 1997) as internal product metrics and external product metrics. *Internal product metrics* are those used to measure attributes of the product that can be measured directly by examining the product on its own, irrespectively of its behavior. *External product metrics* are those used to measure attributes of the product that can be measured only with respect to how the product relates to its environment.

Internal Metrics

Internal metrics can be classified into three categories based on the product attributes they measure. These categories are: size, complexity and data metrics.

- *Size metrics* are metrics that measure attributes related to the product size. Such attributes are the lines of code, the percentage of comments within the code, the volume of the basic entity used for design (for example the number of fifth level bubbles on a DFD-based design), the volume of the documentation, etc.
- *Complexity metrics* are metrics that measure attributes related to program complexity. Such attributes are the complexity of the program flow graph, the nesting levels, the object-oriented methods complexity, etc.
- *Data metrics* are metrics that measure attributes related to the data types and data structures used by the product. Such attributes are the volume of the data, the complexity of data structures, the volume of recursive data structures, the volume of data used per class in object-oriented programming, etc.

A more technical discussion enriched with typical examples of metrics from each category is presented in the following paragraphs using italics.

Among the internal size metrics, *perhaps the most simple and most extensively used is the* LOC *metric, where* LOC *stands for Lines Of Code. However, despite its simplicity, defining* LOC *is not always an easy task. Namely, before applying the* LOC *metric, one must define whether* LOC *includes comment lines or not, empty lines or not and how code statements that have been broken into many lines will be counted. Of course, a measurements program based on a well-defined quality manual helps a lot towards solving such issues, as discussed in the following subsection of this chapter.*

Although LOC *is a commonly used size metric, it does not always provide accurate estimations of product size, due to differentiation in the definition of the 'line' according to each programmer's style. A way to face this problem (Halstead, 1977) is to break down code statements into tokens, which can either be operators or operands. In this way, measurements are not based on counting ambiguous lines anymore, but on counting well-defined tokens. Following the above concept, a set of metrics is defined by considering that* n_1 *is the number of distinct operators,* n_2 *the number of distinct operands,* N_1 *the total occurrences of operators in the measured module and* N_2 *the total occurrences of operands. Examples of such metrics are program size given by equation (1), effort estimator given by equation (2) and language level given by equation (3).*

$$N = N1 + N_2 \tag{1}$$

$$E = \left(N_1 + N_2 \right) \cdot \log_2 \left(n_1 + n_2 \right) \cdot \frac{n_1 \cdot N_2}{2 \cdot n_2} \tag{2}$$

$$\lambda = \left(N_1 + N_2 \right) \cdot \log_2 \left(n_1 + n_2 \right) \cdot \left(\frac{2 \cdot n_2}{n_1 \cdot N_2} \right)^2 \tag{3}$$

As regards metrics (1), (2) and (3), N provides a sufficient metric of the module's size, which has proven to be more accurate than LOC, *while* E *is a good indicator of the effort spent on a module. Moreover, the* E *metric has proven (Fitzsimmons & Love, 1978) to be highly correlated to the number of defect reports for many programming languages. The* λ *metric is an indicator of how well the programming language has been used and variations of a standard measured* λ *of the programming language or languages the IT company is using for a particular IT project can aid in the early detection of problematic modules.*

There are many more internal metrics, such as comments ratio, number of methods per class, number of public instance methods per class (Lorenz & Kidd, 1994), ratio of methods per class, etc. For example, function points, which fall into this category, are a very widespread internal metric used for estimation. Function points are based on an empirical relation and the assignment of weighting factors used to make countable measures of information domain to contribute to the total count of function points (Arthur, 1985).

As regards the internal complexity metrics, *perhaps the most well-known and commonly used one is cyclomatic complexity (McCabe, 1976). Simplifying the formula so as to be applicable for each module, the cyclomatic complexity metric* V(g) *could be calculated as in (4), where* e *is the number of edges and* n *is the number of nodes of the measured module's flow graph* g.

$$V(g) = e - n + 2 \qquad\qquad (4)$$

Table 1 shows a small sample of code in Object Pascal, produced using Borland Delphi for an image-processing tool. The numbers in the left of the code are used simply to aid in corresponding code statements with the graph points. The simplified flow graph g *for this small module of sample code is illustrated in Figure 2. Consequent statement lines have been merged to a single node for simplicity purposes.*

Using the formula (4) and the flow graph g *of the module shown in Table 1 it is easy to calculate the cyclomatic complexity* V(g) *of* g, *as shown in formula (5). High values of* V(g) *normally correspond to complex modules, while small values correspond to less complex ones. A module having only sequential statements and no control or decision commands would have* V(g)=1.

$$V(g) = 16 - 11 + 2 = 7 \qquad\qquad (5)$$

Other well-known complexity metrics are the depth of inheritance tree (Bansiya & Davis, 1997) used in object-oriented programming, the method complexity metric (Lorenz & Kidd, 1994), etc. Complexity metrics have been used with success to predict error-prone parts of code, but as discussed in this chapter their use can be problematic

Table 1. Sample code in Object Pascal

```
01    for i:=0 to m*n-(signlength*8) do
          begin
          if i mod step = 0 then
02            begin
              pgrs1.position := trunc (i/step);
              label5.caption := inttostr(info)+'%';
03            end;
          for j:=i to i+(signlength*8)-1 do
04            begin
              temp1[j-i]:=levels[level,j];
              end;
          sim:=0;
05            for j:=0 to (signlength*8)-1 do
              begin
06                if temp1[j]=signature[j] then
                  begin
                  sim:=sim+1;
07                end;
              end;
08            similarity[i]:=sim;
          if similarity[i]>=signlength*8 then
09            sign_found:=sign_found+1;
10            end;
11    similarity[i+1]:=0;
```

Figure 2. The G Flow Graph

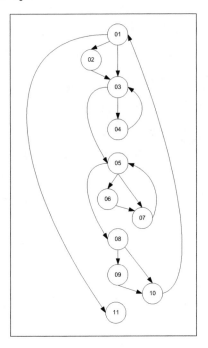

or lead to false estimations if they are not conducted in the framework of a complete method.

Finally, as regards the internal data metrics, *such as the Data Structure Complexity Metric (Tsai, Lopez, Rodriguez, & Volovik, 1986), or the Number of Abstract Data types, or the Data Access Metric (Bansiya & Davis, 1997), it is important to note that one of their major benefits is that they can provide predictions before the completion of a large portion of code, since in most cases the data structures are known in advance. Data metrics are used to measure the volume or the complexity of data and data structures used in a module, or modules of code. The use of data metrics, as well as complexity and size metrics is further discussed in this chapter.*

As far as internal product metrics in general are concerned, it is important to mention that one of their major *advantages* is that they are easy to automate and therefore data collection can be made in an easy, automated and cost-effective way. Furthermore, the measurement results can also be analyzed with an automated way using statistical techniques, and thus conclusions can be drawn rapidly. Tools such as Athena (Tsalidis, Christodoulakis, & Maritsas, 1991), QSUP (Xenos, Thanos, & Christodoulakis, 1996), Emerald (Hudepohl, Aud, Khoshgoftaar, Allen, & Maykand, 1996), GQM automation (Lavazza, 2000), etc. have made internal measurements very easy to conduct. The screenshot from the metrics form of QSUP shown in Figure 3 is an example of the simple and automated way in which such measurements can be made.

Figure 3. A Screen Shot from QSUP

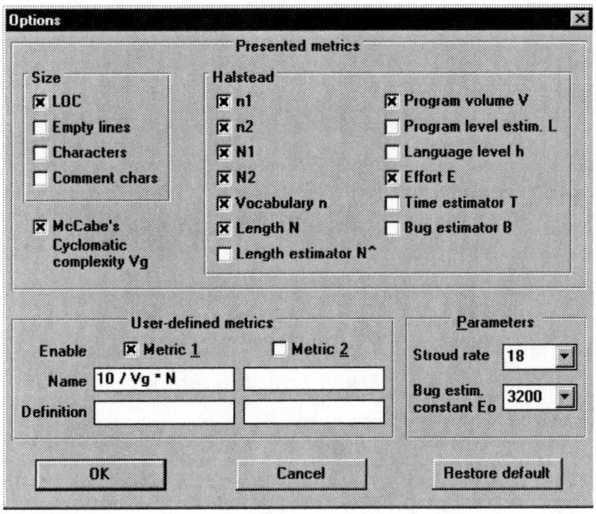

On the other hand, it should be mentioned that among the *disadvantages* of internal product measurements is the fact that they are, in many cases, difficult to interpret. In other cases, the internal quantities measured are not clearly related to the external quality characteristics that one wants to assess. Such problems can only be solved in the framework of a well-defined measurement method that combines internal and external metrics.

External Metrics

Based on ISO 9126 (1996) standard, the *external factors* affecting IT product quality are Functionality, Usability, Efficiency and Reliability. As far as the definition of these factors is concerned (Kitchenham & Pfleeger, 1996), functionality refers to a set of functions and specified properties that satisfy stated or implied needs. Usability is defined as a set of attributes that bear on the effort needed for the use and on the individual assessment of such use by a stated or implied set of users. Efficiency refers to a set of attributes that bear on the relationship between the software's performance and the amount of resources used under stated conditions, while reliability refers to a set of attributes that bear on the capability of software to maintain its performance level, under stated conditions, for a stated period of time.

External metrics are used to measure directly these four factors or the characteristics of which these factors are composed. Unlike internal metrics (that measure internal characteristics of IT products and aim to relate measurements of such characteristics to these factors), external metrics measure directly these factors or their characteristics. Such metrics in many cases can be based on subjective estimates. Among the means employed by external metrics are surveys on user opinion providing valuable measurements for IT functionality or usability. Measures like defect reports, or mean time between failures (Fenton & Pfleeger, 1997) are used to determine IT product reliability, while measures like memory usage are used to determine efficiency.

As already mentioned, the application of external metrics implies that a certain extent of subjectivity is involved; even metrics that appear to be objective are often characterized by some degree of subjectivity. For example, defect reports seem to be a solid metric that can be used to objectively measure reliability. But the number of defect reports submitted by a user is influenced by issues such as the time and the extent of IT product usage, the user experience and even the user's motivation to edit and submit a defect report. Therefore, such metrics must be analyzed very carefully and under a framework that will take under consideration such issues. Among the external metrics that have to seriously consider the problem of subjectivity are survey-based external metrics. The following paragraphs, in italics, are a technical discussion on the use of questionnaire-based external metrics.

Perhaps the most direct way to measure users' opinion for an IT product is simply to ask them. This in most cases is done either by user interviews, or by questionnaire-based surveys. Interviews may cost significantly more than surveys — especially in cases that users are geographically dispread — and in many cases the interviewees' opinions might affect users' judgment. On the other hand, as argued by Kaplan, Clark and Tang (1995), surveys allow focusing on just the issues of interest, since they offer complete control of the questions to be asked. Furthermore, surveys are quantifiable and therefore are not only indicators in themselves, but also allow the application of more sophisticated analysis techniques to the measurements, techniques that are appropriate to organizations with higher levels of quality maturity.

However, the less costly, questionnaire-based surveys have to face problems too. The most common problem is the low response rate (especially when the means used for the conduction of the survey are mail, or e-mail); therefore a large number of users is required to guarantee an adequate volume of responses and thus effectiveness of the survey. Furthermore, surveys have to deal with problems such as the subjectivity of measurements and the frequency of erroneous responses. Questionnaire-based surveys sent by mail to a number of recipients should follow guidelines (Lahlou, Van der Maijden, Messu, Poquet, & Prakke, 1992) on how the questionnaire should be structured in order to minimize subjectivity due to misleading interpretations of questions or choice levels. For example, questionnaires should be well structured and the questions should follow a logical order, while references to previous questions should be avoided. Moreover, it is recommended that questions with pre-defined answers are used instead of open questions, where possible, while concepts such as probability, which may confuse the user, should be avoided, etc.

Still, due to the nature of surveys, errors are likely to occur. Actually, what is meant by errors, in this case, are responses not actually representing user opinion. Such errors could occur because, for example, the user answered the questionnaire very carelessly and made random choices when confused, or because the user was enthused in the first questions, but lost interest somewhere in the middle of the questionnaire and thus made some random choices in order to finish it, or simply because he misunderstood some of the questions and unintentionally made errors.

A technique for error minimization involving the use of safeguards to detect errors has been proposed by Xenos and Christodoulakis (1997). A safeguard is defined as a question placed inside the questionnaire so as to measure the correctness of responses provided by users. Therefore, safeguards are not questions aiming to measure users' opinions about the quality of an IT product, but questions aiming to detect errors. Specifically, safeguards are either questions that can be answered by only one particular answer and any other answer is considered as an error, or repeated questions (phrased differently in their second appearance) placed into different sections of the questionnaire to which exactly the same choices are offered as candidate answers — or repeated questions with exactly the same phrasing but to which completely different types of answers are offered. One major advantage of this type of surveys, which will be further discussed, is the fact that they place focus on IT user requirements and may aid in calibrating metrics, controlling measurement results, and providing confidence to both the company and the users.

$$QWCO_S = \frac{\sum_{i=1}^{n}\left(O_i \cdot E_i \cdot \frac{S_i}{S_T}\right)}{\sum_{i=1}^{n}\left(E_i \cdot \frac{S_i}{S_T}\right)} \tag{6}$$

Apart from error detection, this technique also weights users' opinions based on their qualifications and a metric $QWCO_S$ *is calculated using formula (6). In this formula,* O_i, *measures the normalised score of user i opinion,* E_i *measures the qualifications of user i, n is the number of users who responded to the survey,* S_i *is the number of safeguards that the user i has replied correctly to, and* S_T *is the total number of safeguards included. Since the use of the* $QWCU_S$ *technique implies the existence of at least one safeguard in the questionnaire, the division by* S_T *is always valid.*

Summarizing, one of the major *advantages* of external metrics is that they measure directly the desired external IT product quality characteristics; thus no further analysis or interpretation is needed. Additionally, external metrics contribute to a great extent to what is considered to be one of the main goals of IT product quality: user satisfaction. On the other hand, *disadvantages* and problems should be taken seriously under consideration when deciding to use external metrics, the most important of which being that such metrics are not objective and additional effort is required to ensure their objectivity. Furthermore, they are not as cost effective as internal measurements and in many cases it is difficult to conduct measurements due to high error rates, especially in cases that error detection techniques have not been used during measurements.

Correlating Metrics to Software Quality Characteristics

In this subsection the correlation between internal and external metrics is discussed. Being well aware of the correlation between internal and external measurements is essential for defining the goals and the techniques of an overall measurement method

Table 2. External IT Projects Measurements

External Measurement Results for 46 IT Projects											
0.06	0.12	0.12	0.14	0.16	0.19	0.19	0.21	0.23	0.26	0.28	0.30
0.30	0.30	0.31	0.33	0.33	0.34	0.34	0.39	0.40	0.41	0.42	0.43
0.43	0.44	0.45	0.46	0.47	0.48	0.48	0.48	0.51	0.54	0.56	0.60
0.60	0.60	0.67	0.75	0.76	0.78	0.80	0.86	0.88	0.94		

appropriate for the specific needs of an IT developing company and therefore being able to aid towards better IT Governance.

Some interesting results regarding the correlation of internal and external measurements were derived from a study (Xenos, Stavrinoudis, & Christodoulakis, 1996) in the framework of which measurements were conducted for 46 IT projects. Table 2 shows the normalized measurement results of the 46 software products measured using the aforementioned QWCO$_S$ external metric in assenting order — worst measurements first. The measurements of user perception of IT product quality were derived from a questionnaire-based survey.

The internal measurement results relating to the 46 projects of the research – in the same order as in Table 2 — are shown in Table 3. The internal measurement results are normalized and were calculated using a formula combining a number of internal metrics. This combination metrics formula — called *CMF* — does not measure a physical product quantity, but combines all internal metric results. Its solid purpose is to provide a collective mechanism for comparison as shown in equation (7).

$$CMF = 0.2 \cdot \lambda_{wa} + 0.2 \cdot R + 0.4 \cdot V_{wa} + 0.2 \cdot T \qquad (7)$$

The metrics that form part of the *CMF* are the weighed average language level λ_{wa}, the essential size ratio R, the weighed average cyclomatic complexity V_{wa} and the data structure complexity metric T.

As expected, the external measurements results for the 46 projects followed an almost normal distribution with mean value 0.437 and standard deviation 0.219, as shown in Figure 4 illustrating the grouped frequency distribution — using a class range of 0.13 — around the marked midpoints.

In a similar manner, the combined internal measurement results are shown in Figure 5. Internal measurements also followed a normal distribution with mean value 1.149 and standard deviation 0.363, as can be seen in Figure 5.

Table 3. Combined Internal IT Projects Measurements

Combined Internal Measurement Results for same IT Projects											
0.581	0.965	1.545	0.742	0.444	0.727	1.223	0.694	0.780	0.660	1.043	0.693
1.488	0.806	1.022	1.162	0.980	0.916	1.000	1.599	0.779	1.400	1.316	0.946
1.276	1.494	0.952	0.930	0.970	1.050	1.100	0.934	1.143	1.200	1.220	1.031
1.434	1.234	1.996	1.428	1.358	1.826	1.493	1.750	1.730	1.785		

Figure 4. Distribution of External Measurements

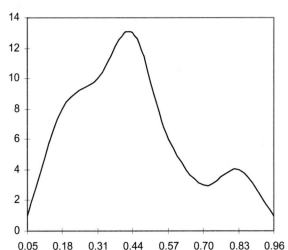

The preparatory work described above provided a uniform manner to represent external and internal IT product quality measurements, and consequently the examination of the correlation was feasible. The scatter plot in Figure 6 illustrates the correlation of internal and external measurements for these 46 IT projects. The horizontal line represents the measurements conducted using the external metrics while the vertical line represents the measurements conducted using the internal metrics. Projects are marked as points in the coordinates for each measurements line. The diagonal line — also called correlation line — represents the way in which correlation would be illustrated if the two measurement methods were 100% correlated.

Figure 5. Distribution of Internal Combined Measurements

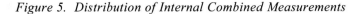
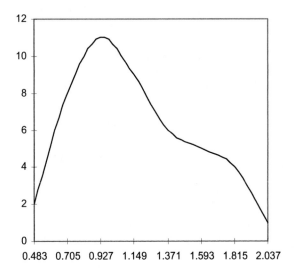

Figure 6. Correlation Among Internal and External Measurements

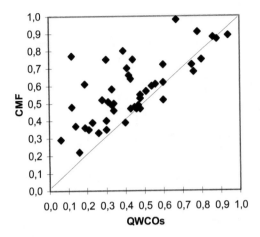

Points marked below the correlation line represent measurements that produced a higher score when external metrics of IT product quality were conducted, compared to their low internal measurements score. Namely, points below the correlation line represent IT projects that, despite their low internal measurements score, produced higher external metrics scores, i.e., the end-user perception of quality was not as low as expected based on the internal metrics results. On the other hand, points marked above the correlation line are measurements that produced a lower score when external metrics were conducted, compared to their high internal measurements score. Namely, these points represent IT projects that, despite their high score in internal measurements, users' perception for quality was not equally high.

As made evident by the scatter plot of Figure 6, only a very small number of projects fail the internal quality measurements and still achieve high external measurements scores; very few points are marked below the correlation line and no points are marked in great distance below this line. On the contrary, many points are marked just over the correlation line and even further above. This easily leads to the conclusion that satisfaction of internal quality standards — as measured using internal metrics — does not guarantee a priori success in meeting customers' perception of quality as measured using external metrics. IT products that receive high scores in an internal measurements program are not always likely to received high user acceptance.

Problems and Issues Using Metrics in Small and Medium Size IT Companies

Although in major IT developing companies metrics are most often an integral part of IT development process aiming at better IT Governance, this is rarely the case in small or medium size software development companies where none or very few software measurements are included in the development process. This is either due to lack of awareness of metrics and measurement methods, or due to the belief that measurements are impractical for small or medium size software development companies, cost too much and should always be conducted within a quality assurance framework available only in larger companies. This is partially true, since "measuring in the small" — a common expression for measurements in small and medium size companies — had to overcome several problems. However, it is just as true — as discussed further in this subsection — that measuring in the small is feasible and practical and all the problems can be overcome. Moreover, the use of metrics entails a great number of benefits for the small and medium IT developing companies that will implement them.

As already mentioned, small software developing companies willing to incorporate the application of metrics in their IT development process will often have to confront a number of *problems and limitations*. One of them lies in the fact that most small companies do not base software development on a standard. The lack of a standard framework that will constitute a guide for the use of metrics often leads to the conclusion that measurements are a luxury that they cannot afford. High cost is also an important reason why such companies often resist the application of new methods, tools, technologies and developing standards.

Furthermore, it should be considered that the majority of small companies operate under tremendous pressure. Given the limitations of their resources, they cannot afford a failure. Time is very critical and, since reaching a maturity level is a long-term commitment, this commitment is overlooked due to constant 'fire fighting'. The developing methods are in most cases ad hoc and chaotic and even in the cases of successfully completed projects, a major problem is the inability to repeat successful practices. Moreover, roles in small software development companies are sometimes unclear. There is no clear distinction between process and product engineers, or project management and development. The lack of formal development methods often results in unavailability of project records — historical data of any kind, not necessarily measurement data — and process documentation. The processes are kept in the minds of the development engineers and consequently are not documented in order to become common practice for every company employee.

Summarizing, the lack of a standard framework for measurements, the lack of formal development methods, the necessity for everyday 'fire fighting', as well as the lack of time to keep records are the basic limitations relating to the use of metrics in small IT developing companies.

On the other hand, small IT developing companies enjoy a number of *benefits* that are very helpful for the application of a measurements program. A fact that has not been stressed enough, but is true in most cases, is that the human resources of small software companies comprise skilled, well-educated and usually experienced engineers. This statement is based on the fact that in small companies the engineers are usually the same people who decided to take the risk of founding the company, and these people are, in most cases, experienced in the field of software engineering, or else they would not risk in a field for which they have little knowledge. Such engineers — although in many cases are obliged to use ad hoc methods for developing a project — already have the knowledge and the experience to acknowledge why metrics will improve their development process and what the benefits of a measurements program are. As a result, they are usually capable of selecting themselves the metrics appropriate for their IT projects and stay committed to them, not feeling that metrics are simply part of a process that has been enforced on them, which is often the case with employees in larger companies.

Another benefit of small software developing companies is the enhanced internal communication. While large companies spend a lot of management time and effort to select proper communication approaches and organize and reinforce internal communication processes, in small companies good internal communication is an everyday reality. Thus, internal distribution of data — among which is measurement data — is facilitated, while there is no need for complex communication procedures. Even more important than internal communication, however, is enhanced external communication. While in large companies there is no direct communication between product engineers and the customer, in small companies communication with the customer is far more direct. There are no intermediaries between the customer and the developers and in most cases, this improves significantly the level of communication, despite the informality and lack of a communication method. This fact facilitates the collection of external measurements and their direct exploitation.

Another benefit regarding the use of metrics in small companies is that the difference of belief-systems of product engineers, i.e., software developers and developer managers, and process engineers, i.e., software quality engineers, quality assurance engineers, process improvement specialists and change agents reported (Mackey, 2000) for large IT development companies is not a problem in small and medium ones. Specifically, one of the major problems when implementing metrics is that processes and actions initiated by process engineers tend to be underestimated by product engineers and considered as bureaucratic slowdown of their work. In small and medium IT developing companies, where engineers play multiple roles — since they are mainly product engineers but also have to consider process improvement — it is easier to use metrics and appreciate their value.

Finally, it should be stressed that the collection and use of historical data required for measurements is facilitated thanks to the small size of the company, the few years of operation — in most cases — and the employment of the same individuals for a long period of time, which is also common in most cases. It is easier for personnel to note improvement owed to the use of metrics in practice and be convinced to use them. Realizing day after day in practice the real benefits of using metrics and achieving product and process improvement is much more convincing that just reading about it in impersonal reports — which might be the case in large companies. Therefore, despite what is currently believed, small companies have many reasons motivating them to incorporate metrics in their IT developing process. Recommendations regarding the use of metrics in such small and medium size IT developing companies are discussed in the following subsection.

Solutions and Recommendations

This subsection discusses solutions and recommendations for dealing with the issues, controversies and problems presented in the preceding section. Firstly, practical guidelines are presented relating to how both internal and external measurements can aid towards better IT Governance within a rigorous framework combining them into a complete measurement method applicable for IT developing companies. Then, the application of this method in small and medium IT companies is discussed, while suggestions are made on how the problems analyzed in the preceding section can be overcome. Finally, this subsection closes with a discussion on how metrics can be used in accordance with SPC so as to aid towards better IT Governance.

Combining Internal and External Metrics into a Measurement Method

As already mentioned in the previous sections, internal and external measurements must be conducted under a well-defined framework with precise goals. Before selecting the appropriate metrics for any IT project, it is necessary that all metrics available for use in the IT developing company have been collected and documented in detail in the company's *quality manual*. This manual is a basic component of the metrics application process and includes the metrics, the measurement techniques as well as guidelines for the application of metrics, the data analysis and the corrective actions required for improving the IT developing process. It should also be mentioned that the quality manual includes all metrics that are available regardless of how many times they have been used, or the availability of measurements data from past IT development projects.

Then, for each IT project, a set of metrics appropriate for this particular project is selected from the quality manual. The criteria on which the selection of metrics is based are the particular quality factors that the IT project places emphasis on. This set of metrics is documented — using the guidelines available in the quality manual — and consists of the *quality plan* of the IT project. Thus, a project quality plan should include all the metrics, measurement guidelines and goals applicable for the IT project. It is self-evident that the project plan of a specific IT project may be entirely different from another project's plan and may use a completely different set of metrics. The selection of the appropriate metrics and measurement techniques is performed by the IT developing company quality manager in cooperation with the project manager and — if necessary — with the members of the development team. In most cases, the analysis of historical data and measurements collected from similar projects can prove to be very helpful for the definition of the project quality plan. Figure 7 presents an illustration of the above procedure.

The quality plan of each project includes internal metrics so as to provide an easy and inexpensive way to detect possible causes for low IT product quality, as this might be perceived by the end-users, and take early corrective action. Including internal metrics in the quality plan will help in preventing failures and non-conformities. For achieving

Figure 7. Selection of Metrics for Each IT Project

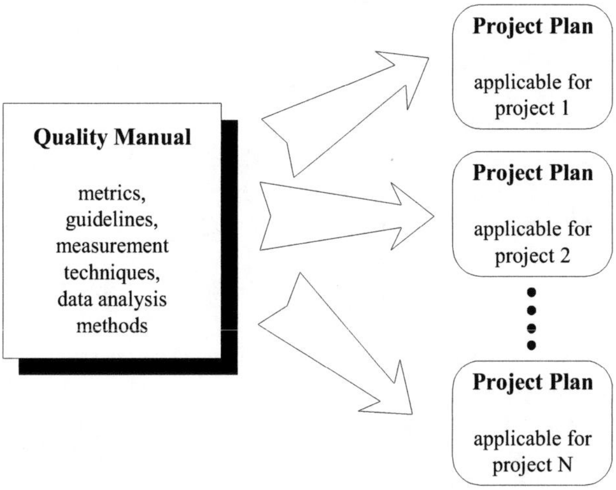

better results, it is recommended that every quality plan includes a combination of internal metrics. For example, high complexity measurements relating to a module are not necessarily an indication of bad practice and low quality, but if combined with additional measurements indicating a large volume of code and low data complexity, then this could be considered as an indication that problems might occur during testing or even after project delivery and that early corrective action should be taken.

A project quality plan should also include external metrics — applied during alpha or beta testing and post shipment — so as to measure external quality factors. Occasionally external metrics could be used in order to test the soundness of the internal metrics and measurements results and even to calibrate internal metrics. Such calibrations must be reflected with the appropriate changes in the quality manual.

It must be noted that the successful selection of metrics and measurement techniques to be included in the IT developing company quality manual is very dependent on company maturity. The adoption of sophisticated techniques and complex metrics by a company might prove to be ineffective, if it is not supported by years of experience with metrics and measurements and large volumes of data from past project measurements. IT developing companies should always keep this fact in mind and set feasible measurement goals, not aiming too high at the early stages of metrics application.

Application of Metrics in Small and Medium Size IT Companies

As regards *small and medium size IT developing companies*, it should be stressed that metrics *can* be used without causing a major increase in the development cost. For example, communication costs, which should be included in any cost estimation relating to quality, are high in larger companies but minimal in small ones. Moreover, a large number of metrics and tools are freely available, thus minimizing the cost of using metrics. Of course, it is important to note that metrics can help small IT developing companies to improve process efficiency, product quality and reach a higher level of maturity. It is suggested that metrics are used with a standard in mind. This does not imply that a small company should attempt to obtain an ISO9001 certification or a high level CMM assessment by using metrics. However, it is suggested that the IT company should attempt to build a quality manual based on a standard and introduce metrics into specific project quality plans following the guidelines of this standard. The primary goal should be the improvement of product quality and process efficiency by the use of metrics and a future goal might be the certification. In the case that the certification is achieved, it is not advisable to use it for marketing reasons only, such as promotional purposes, as often is the case, but mainly for improvement. Furthermore, it is recommended that the standard's guidelines are adopted step by step while the use of metrics, in the early stages of introduction of measurements in the company, should be limited to a selected set of well-known metrics, metrics that as Kautz (1999) said "must make sense for small IT companies."

The use of metrics in small and medium size companies should be based on each company's specific needs. A good approach would be to start with a method such as the GQM (Solingen & Berghout, 1999), by selecting very few metrics at first, based on specific and well-defined goals. Existing data that require little effort to collect (such as size measurements, testing time, data from the customers) should be collected and exploited. Experience reports (Grable, Jernigan, Pogue, & Divis, 1999) have shown that similar metrics have been used with success in small IT projects aiding towards better IT Governance. However, it should always be kept in mind that the main idea of metrics is to measure only what actually helps towards well-defined goals and is achievable considering time and personnel limitations. In the same way, the use of free methods and tools is advisable only provided that their selection is based on real company needs and not on the fact that they are free. Perhaps, the most important suggestion regarding the use of metrics is to rely on small companies' benefits — as presented in the preceding section — and take advantage of them as much as possible. For example, good external communication can significantly aid in better and easier collection of external product measurements. Engineers playing both roles of product and process engineer have the advantage of being in the position to interpret various internal and external metric results and at the same time benefit from process improvement to achieve more effective product development process.

Finally, especially for small and medium size IT developing companies, it must be noted that *commitment* is the most critical factor that determines success or failure of a measurement program. It is better to set minor goals and use very few metrics at first, increasing them progressively, rather than define an ambitious measurement program that will be abandoned later on, when the first need for 'fire fighting' will occur and preoccupy everyone. If commitment to metrics weakens, it is usually difficult to revive. Small IT development companies making limited use of metrics, based on their particular needs and aiming towards process improvement according to a chosen standard, have a lot to gain from this effort.

Using Metrics within a SPC Framework

As mentioned in previous sections, the use of statistical process control techniques to monitor if a process is in control — so as to implement corrective measures, if the process has become out of control, or simply to use data relating to past performance to predict future behavior — is feasible only if internal metrics are available to be used as basis for the application of SPC. These internal metrics are used to measure essential characteristics of the IT developing process.

It should be stressed that the selection of internal metrics is very important for the proper application of SPC techniques. On the other hand, examples of models that have failed to predict process behavior have been reported (Adams, 1984) and based on these examples the inclusion of internal metrics in such models was criticized (Fenton & Neil, 1999); specifically, it was argued that external metrics are the only true indicators of process performance. It is indisputable that external metrics are the final indicators of process performance, but such metrics cannot be available in the early phases of IT process development, where predictions need to be made. This luck of external metrics leads to the use of internal ones as well as the use of SPC techniques to monitor the IT development process. Of course, it goes without saying that the selection of the proper metrics to measure statistical relationships is a very important task and thus should be made very carefully and should always be verified by using external metrics. This is the proper way to apply SPC techniques, as discussed in the following examples.

As Florac and Carleton (1999) mention, the control chart, already presented earlier in this chapter, is an ideal tool for analyzing process behavior, because it measures process performance over time and provides an operational definition of stability and capability. It is a fact that IT developing companies are beginning to appreciate the value that control charts add to IT Governance by providing quantitative insights of the behavior of their IT development processes. Using control charts, it is easy to monitor when a metric exceeds the preset limits, which is interpreted that the IT developing process has become out of control and therefore immediate corrective action should be taken to eliminate abnormalities. A more technical discussion of the use of such charts follows.

Based on the main idea of the control chart illustrated in Figure 2, a number of control chart variations exist that can be used to monitor IT development process depending on the type of data measured. For individuals or attributes of data – such as counts related to occurrences of events or sets of characteristics — the most appropriate are XmR control charts. In XmR charts the middle bar, called Xbar, represents the mean value of the measured item. The mRbar is the mean value of the absolute differences of successive pairs of data. The upper control limit UCL and the lower control limit LCL in XmR control charts are fixed straight lines and the values of these lines can be calculated using formulas (8) and (9). Further details on how to develop and use XmR control charts can be found in Florac and Carleton (1999). Other than the formulas used to calculate the acceptable limits, namely the limits within which measurements should fall in order for the process to be considered in control, the XmR control charts are similar to the control chart example presented in Figure 2.

$$UCL = Xbar + 2.66 \cdot mRbar \qquad\qquad (8)$$
$$LCL = Xbar - 2.66 \cdot mRbar \qquad\qquad (9)$$

Figure 8 is an XmR chart illustrating the results of 20 inspecting sessions using the metric LOC per inspection person hour. An inspecting session is a formal review process conducted by a number of experienced code reviewers. The 'inspection person hour' stands for the number of reviewers multiplied by the number of the hours that the session lasted. Assuming that the 20 measurements collected are: 42, 55, 41, 50, 37, 17, 22, 44, 63, 68, 30, 42, 51, 62, 38, 34, 55, 39, 27, 49, the Xbar represents the average of all measurements, therefore, Xbar = 43.30. In a similar manner, mRbar is calculated as the average of the absolute differences of the 19 successive pairs of data, therefore mRbar = 15.21. Using formulas (8) and (9), UCL and LCL are calculated respectively: UCL = 83.76 and LCL = 2.84. Using the aforementioned calculations and a typical spreadsheet it is easy to monitor this IT process and to create the XmR control chart of Figure 8. In this example the monitored process is in control, since all measurements fall within UCL and LCL and there are no indications of non-random behavior.

When the measured item is a variable — as in the case of observations of continuous phenomena or counts that describe size or status — then the appropriate type of control charts for monitoring purposes are the u-charts. In this case, the upper and lower control limits (UCL and LCL) vary depending on each measurement, since the last measurement affects the limits and leads to the calculation of new ones. In the example illustrated in Figure 8, a u-chart is used to monitor the testing process and measures

Figure 8.   An XmR Chart of LOC Per Person Hours Spent During Formal Inspection Sessions

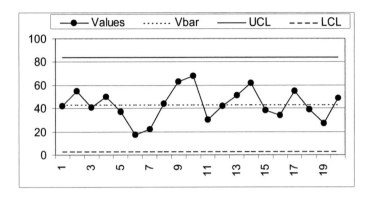

the number of defects per LOC *reported during a typical testing session. The session is typical in the sense that it is well planned and took place regularly and in a manner similar to the previous sessions. The center line is called* Ubar *and represents the number of defects reported during the testing session divided by the number of* LOC *examined. The upper and lower control limits are calculated using formulas (10) and (11).*

$$UCL = Ubar + 3 \cdot \sqrt{\frac{Ubar}{LOC}} \qquad (10)$$

$$LCL = Ubar - 3 \cdot \sqrt{\frac{Ubar}{LOC}} \qquad (11)$$

To create a u-chart *is very simple and can be done using a typical spreadsheet such as the one used to create the* u-chart *of Figure 9. It should be noted that for using the* u-chart *it is necessary to receive measurement data on a regular basis. SPC techniques cannot be applied with few sets of data received sporadically over time in a non-periodical manner. More important than regularity in data receipt is the issue of data homogeneity. This applies to all variations of control charts, since in order to use control charts data must be homogeneous. For example, the* u-chart *of a* LOC *metric based on the data received from the testing sessions of two different programming languages will not provide any valuable results; in this case, the IT development process can only be monitored using two different control charts. Similarly, when the testing sessions include retesting of some modules, again, two different control charts are required, since it is expected that retesting sessions will result in lower rates than the previous testing sessions, resulting in abnormal variation of the* Ubar *measurements. Consequently, the use of two control charts (one for testing sessions and one for retesting sessions) is necessary to ensure measurements validity.*

Figure 9. A U-Chart of Defects Per LOC Reported During Regular Testing Sessions

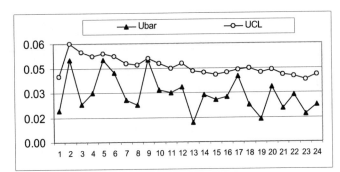

The example of the u-chart of Figure 9 is based on data from a past IT project. The LCL is not calculated or illustrated, since it is not applicable for the particular case. The testing sessions do not include any retest of revised code. Of course, in another case in the formulas (10) and (11), the upper and lower control limits would have been calculated using another metric instead of the number of defects divided by LOC.

The main purpose of control charts is to help in the detection of variations in the measured IT process. Variations can be common-cause variations or assignable-cause variations (Florac, Carleton, & Barnard, 2000). Common-cause variations are normal and rather expected variations caused by regular interaction among process components of the IT development process such as people, machines, material, tools, environment and methods. Such variations usually cause measurements to fluctuate within acceptable limits. Assignable-cause variations, on the other hand, are caused by events that are not part of the normal IT development process and indicate abnormal or sudden changes to one or more basic process components. In the majority of cases, assignable-cause variations cause measurements to fall out of the acceptable limits. Removing all assignable causes and preventing their reoccurrence will lead to a stable and predictable IT development process.

In order to statistically monitor the IT development process, one can choose from a number of internal metrics to be included in the model. This section presented a few examples of the application of SPC techniques contributing to the exploitation of past process performance to predict future behavior and detect abnormalities in the IT process. Using such techniques allows better control of the IT development process and therefore helps towards better IT Governance. For better results, as already discussed, one must always combine such techniques with external measurements under a measurements framework.

FUTURE TRENDS

For about three decades now, metrics are used for the estimation of product-related issues (such as product size, required effort, time required for testing, etc.) for early detection and prevention of problems during development and for product assessment after product release. More recently, during the past decade, the use of metrics for statistically controlling IT development process was proposed.

Although both these practices were proven to be successful in practice and aided significantly towards better IT Governance, the benefits from the use of metrics are not commonly recognized. This is partly due to the lack of awareness of metrics in small and medium size IT developing companies. Although in major IT companies metrics are extensively used, in many cases, small and medium ones are not even aware of the prospect and benefits of using metrics. However, this is constantly changing. More and more small and medium size IT developing companies become aware of product metrics and process control. Besides, the adoption of standards such as ISO9001, or assessment in CMM higher levels, has contributed to this change since both standards are encouraging the use of IT metrics.

Another issue that is expected to change in the near future is the availability of more sophisticated tools. Although many measurements tools are available, using a number of metrics, there are not many tools available yet that use past projects' measurement data in combination with current project data in order to aid in decision making. Combining metrics with decision support techniques, or methods for resolving uncertainty will lead to the development of valuable tools that will aid towards IT Governance. A recent approach towards this direction (Fenton, Krause, & Neil, 2002) is using IT metrics and Bayesian networks for effective IT Governance, especially for risk management, by automatically predicting defects in the released IT product. One of the benefits of such models is that they allow reasoning in both forward (used for prediction of future behavior) and backward direction (used for assessment of best practices or analysis of problematic practices and definition of revisions and corrective actions that need to be taken).

Finally, another future goal is related to the expansion of open source systems. An ambitious plan would be to use metrics for defining measurable goals related to source code modules available for inclusion in open source IT projects and therefore to define a set of standards that will determine the quality of such code. Selecting the appropriate set of metrics for open source modules and setting up the standards for such a measurements program is a difficult and ambitious future goal due to the particularities of open source development and the lack of formal management structure.

CONCLUSION

This chapter introduced the reader to software metrics that are used to provide knowledge about different elements of IT projects. It presented internal metrics that can be applied prior to the release of the IT product to provide indications relating to quality characteristics, and external metrics applied after IT product delivery to give information about user perception of product quality. It also analyzed the correlation among internal and external metrics and discussed the way in which such metrics can be combined into a measurements program. Emphasis was placed on small and medium size IT developing companies and how such companies can overcome problems relating to the application of metrics. Finally, the use of metrics within a statistically controlled IT development process was discussed.

Software metrics can be used to measure various factors related to IT product development. These factors include estimation, early detection and prevention of problems, product assessment, etc. Their utilization within a measurements framework in combination with the use of automated tools can aid towards IT development process control and better IT Governance, regardless of the IT developing company size.

REFERENCES

Adams, E. (1984). Optimizing Preventive Service of Software Products. *IBM Research Journal, 28*(1), 2-14.

Albrecht, A. J. (1979). Measuring application development productivity. *Proceedings of IBM Applications Development Symposium, Monterey, California,* (pp. 83-92).

Arthur, L. (1985). *Measuring programmer productivity and software quality.* Wiley-Interscience.

Bansiya, J., & Davis, C. (1997). Using QMOOD++ for object-oriented metrics. Automated metrics and object-oriented development. *Online Dr. Dobb's Journal.*

Boehm, B.W. (1989). *Software risk management.* Los Alamos: IEEE Computer Society Press.

Burr, A., & Owen, M. (1996). *Statistical methods for software quality.* Thompson Computer Press.

DeMarco, T. (1982). *Controlling software projects.* New York: Yourdon Press.

Deming, W. (1986). *Out of the crisis.* Cambridge: MIT Center for Advanced Engineering Study.

Fenton, N., Krause, P., & Neil, M. (2002). Software measurement: Uncertainty and causal modeling. *IEEE Software, 19*(4), 116-122.

Fenton, N. E., & Neil, M. (1999). A critique of software defect prediction research. *IEEE Transactions on Software Engineering, 25*(5), 675-689.

Fenton, N. E., & Pfleeger, S. L. (1997). *Software metrics: A rigorous & practical approach (2nd ed.).* London: International Thomson Computer Press.

Fitzsimmons, A., & Love, T. (1978). A review and evaluation of software science. *Computing Surveys,* 10.

Florac, W. A., & Carleton, A. D. (1999). *Measuring the software process: Statistical process control for software process improvement.* MA: Addison-Wesley.

Florac, W. A., Carleton, A. D., & Barnard, J. R. (2000). Statistical process control: Analyzing a space shuttle onboard software process. *IEEE Software, 17*(4), 97-105.

Funami, Y., & Halstead, M. (1976). A software physics analysis of Akiyama's debugging data. *Symposium on Computer Software Engineering, Polytechnic Institute of New York,* (pp. 133-138).

Gibbs, W. (1994). Software's chronic crisis. *Scientific American, 271*(3), 86-95.

Grable, R., Jernigan, J., Pogue, C., & Divis, D. (1999). Metrics for small projects: Experiences at the SED. *IEEE Software, 16*(2), 21-28.

Halstead, M. (1977). *Elements of software science.* North Holland: Elsevier Publications.

Hansen, W. (1978). Measurement of the program complexity by the pair (cyclomatic number, operator count). *ACM SIGPLAN,* 13.

Henry, S., & Kafura, D. (1981). Software structure metrics based on information flow. *IEEE Transactions on Software Engineering, SE-7* (5), 510-518.

Hudepohl, J. P., Aud, S. J., Khoshgoftaar, T. M., Allen, E .B., & Maykand, J. (1996). Emerald: Software metrics and models on the desktop. *IEEE Software, 13*(5), 56-60.

ISO9001: Quality management and quality assurance standards. (1991). International Standard. ISO/IEC 9001.

ISO9126: Information technology. Evaluation of software. Quality characteristics and guides for their use. (1996). International Standard. ISO/IEC 9126.

Joergensen, A. (1980). A methodology for measuring the readability and modifiability of computer programs. *BIT Journal, 20*, 394-405.

Kaplan, C., Clark, R., & Tang, V. (1995). *Secrets of software quality*. New York: McGraw Hill.

Kautz, K. (1999). Making sense of measurement for small organizations. *IEEE Software, 16*(2), 14-20.

Kitchenham, B., & Pfleeger, S. (1996). Software quality: The elusive target. *IEEE Software, 13*(1), 12-21.

Lahlou, S., Van der Maijden, R., Messu, M., Poquet, G., & Prakke, F. (1992). *A guideline for survey - techniques in evaluation of research*. Blussels: ESSC-EEC-EAEC.

Lavazza, L. (2000). Providing automated support for the GQM measurement process. *IEEE Software, 17*(3), 56-62.

Lorenz, M., & Kidd, J. (1994). *Object–oriented software metrics*. New York: Prentice Hall.

Mackey, K. (2000). Mars versus Venus. *IEEE Software, 17*(3), 14-16.

McCabe, T. (1976). A software complexity measure. *IEEE Transactions in Software Engineering, SE-2* (4), 308-320.

McCall, J. A., Richards, P. K., & Walters, G. F. (1977). Factors in software quality. (Vols I, II, III). *U.S. Rome Air Development Center Reports NTIS AD/A-049*, 14-55.

McGarry, F., & Decker, B. (2002). Attaining Level 5 in CMM process maturity. *IEEE Software, 19*(6), 87-96.

Miller, A., & Ebert, C. (2002). Software engineering as a business. *IEEE Software, 19*(6), 18-22.

Montgomery, D.C. (1991). *Introduction to statistical quality control* (2nd ed.). New York: John Wiley & Sons.

Paulk, M. C., Curtis, B., Chrissis, M. B., & Weber, C. V. (1993). *Capability maturity model for software (Version 1.1)*. Pittsburgh: Carnegie Mellon University – Software Engineering Institute: Technical Report. CMU/SEI-93-TR-024.

Putnam, L., & Myers, W. (1992). *Measures for excellence*. Cambridge: Yourdon Press.

Ragland, B. (1995). Measure, metric or indicator: What's the difference? *Crosstalk, 8.*

Shepperd, M., & Ince, D. (1990). The use of metrics in the early detection of design errors. *Proceedings of Software Engineering.*

Solingen, R., & Berghout, E. (1999). *The goal question metric method*. McGraw Hill.

Stamelos, I., & Angelis, L. (2001). Managing uncertainty in project portfolio cost estimation. *Information & Software Technology*. Elsevier Publications, *43*(13), 759-768.

Tsai, W. T., Lopez, M. A., Rodriguez, V., & Volovik, D. (1986). *An approach to measuring data structure complexity*. Compsac86. 240-246.

Tsalidis, C., Christodoulakis, D., & Maritsas, D. (1991). Athena: A software measurement and metrics environment. *Software Maintenance Research and Practice.*

Van Grembergen, W. (2002). Introduction to the Minitrack: IT governance and its mechanisms. *Proceedings of the 35th Hawaii International Conference on System Sciences (HICSS), IEEE.*

Weller, E. F. (2000). Practical applications of statistical process control. *IEEE Software, 17*(3), 48-55.

Xenos, M., & Christodoulakis, D. (1994). An applicable methodology to automate software quality measurements. *IEEE Software Testing and Quality Assurance International Conference, New Delhi* (pp. 121-125).

Xenos, M., & Christodoulakis, D. (1997). Measuring perceived software quality. *Information and Software Technology Journal.* Butterworth Publications, *39*(6), 417-424.

Xenos, M., Stavrinoudis, D., & Christodoulakis, D. (1996). The correlation between developer-oriented and user-oriented software quality measurements (a case study). *Proceedings of the 5th European Conference on Software Quality, EOQ-SC, Dublin: Ireland* (pp. 267-275).

Xenos, M., Thanos, P., & Christodoulakis, D. (1996). QSUP: A supporting environment for preserving quality standards. *Proceedings of the 6th International Conference on Software Quality, Dundee Scotland* (pp. 146-154).

SECTION III:

OTHER IT GOVERNANCE MECHANISMS

Chapter X

Managing IT Functions

Petter Gottschalk
Norwegian School of Management, Norway

ABSTRACT

This chapter discusses imperatives for IT functions, organization of IT functions, roles of IT functions, roles of chief information officers (CIOs) and key issues in IT management. CIOs in Norway find the role of entrepreneur most important and the role of liaison least important. Improving links between information systems strategy and business strategy is the highest ranking key issue in managing IT functions in Norway.

INTRODUCTION

The formal organizational unit or function responsible for technology services is often called the information technology department. The IT department is responsible for maintaining the hardware, software, data storage and networks.

The IT department consists of information technology professionals such as computer programmers, systems analysts, information architects, and project leaders. An IT manager, called the chief information officer (CIO), is heading the information technology department in many companies. The CIO is a senior management position to oversee the use of information technology in the firm. Often, the CIO is the most important person in IS/IT strategy work.

In larger organizations, we may find that the IT function consists of several IT departments in various business segments or at various business locations. While each department may have an IT manager, the CIO will coordinate all IT departments at the corporate strategic level.

IMPERATIVES FOR IT FUNCTIONS

For IT functions to be successful, Rockart et al. (1996) have listed eight imperatives:

1. *Achieve Two-Way Strategic Alignment.* The first imperative is to align IT strategy with the organization's business strategy. With more than 50 percent of capital equipment investment in the United States now being devoted to information technology, IT has clearly become a major resource for management in carrying out its strategic initiatives.

2. *Develop Effective Relationships with Line Management.* The key people using information technology in any organization are its functional, product, and geographical line managers. They provide the strategic and tactical direction and the commitment to implementation that converts visions of new systems into improved organizational processes. Thus IT personnel at all levels must develop strong, ongoing partnerships with line managers.

3. *Deliver and Implement New Systems.* Although the primary function of the IT department has been the development and operation of systems, today's approach to system development is radically different from the past. The task has changed from developing mainframe-based transaction-processing systems that support a single function to delivering desktop systems that address the integrated data needs of knowledge workers.

4. *Build and Manage Infrastructure.* IT is currently charged with creating an "IT infrastructure" of telecommunications, computers, software, and data that is integrated and interconnected so that all types of information can be expeditiously — and effortlessly, from the users' viewpoint — routed through the network and redesigned processes.

5. *Reskill the IT Organization.* For almost two decades, the basic approach to systems development did not change. COBOL was the major language, and the mainframe was the major platform on which systems were developed. Today, by far the largest number of systems are being built for client/server use. Developers in this environment must regularly learn new programming languages, operating systems, and communications protocols.

6. *Manage Vendor Partnerships.* Outsourcing some IT responsibilities to computing services firms can compensate for skill shortages in IT units and relieve management of the need to oversee tasks that are not competitive strengths or core competencies. IT management must be informed buyers and prime negotiators.

7. *Build High Performance.* In the future, IT units, like all other functions in the firm, must strive to meet increasingly demanding performance goals and improve their economic and operational track record.

8. *Redesign and Manage the Federal IT Organization.* For the past three decades, IT organizations have struggled with the "centralization-decentralization" issue.

The exact locus of all or part of IT decision-making power is critical, and getting the right distribution of managerial responsibilities is thus the eighth imperative.

ORGANIZATION OF IT FUNCTIONS

IT functions can be organized in different ways in organizations. Robson (1997, p. 309) makes distinctions between centralized, decentralized and devolved as follows.

Centralized: One single-access function: IT department provides one single service, with single-access provision. A centrally located IT department provision may be a continuation of always having been centralized or may be a regrouping in response to pressures for cost savings. The centralized approach to locating IT function is effective at gaining, or regaining, control over IT. The technology and systems infrastructure can be efficiently and effectively provided and there should be few problems of data format and security or software compatibility. With one agency in control there should be no confusion over responsibilities and it will have the power to impose standards that ensure that all related parts of the business are able to interface successfully. Despite frequently being used to reduce costs, the bureaucracy and inflexibility often associated with centralized IT function can cause costs to escalate uncontrollably. The early proponents of centralization stated that computing power was proportional to the square of the cost of the processor and indicated that there is economies of scale inherent in centralized IT department.

Radical changes in cost/performance ratios have challenged these proponents' views and so there may be other, more effective, routes to a cost-effective IT infrastructure. Centralized location can lead to confusing the issues of coordination (for instance in building the infrastructure) with those of control and ownership. A centrally located IT function may also be correlated with IT making a low contribution to the business since it may be preoccupied with the complexities of its own internal concerns and so be out of touch with business priorities and so not able to respond to them.

Decentralized: Lots of single-access functions: IT function being a number of smaller single-site, single-access centers, a collection of IT departments. The proliferation of multiple IT departments brings IT geographically closer to the user community but perhaps no nearer in culture or understanding. Decentralization has some powerful advantages. Since it can be much closer to the grass roots of the business, IT has a better chance of motivating and involving users and, by distributing the involvement, the logic is that users will act in a responsible way because they are responsible (and in control, and accountable).

Decentralization focuses less on IS costs and more on user effectiveness. Local IT staff is part of the business. More business-relevant systems should be created since, with fewer, more generalist, IT staff who have less chance of being distracted, business needs are the systems drivers rather than technical interest. In addition, simpler systems may result and 'small is beautiful' and 'simple engineering is good engineering'. Whilst these points give some benefits over the centralized location of IT function, there are drawbacks; primarily what is achieved is many groups all having the same problems so that the main disadvantage is one of duplication-driven higher costs plus staff isolation in the mini-IT sections.

Since IT departments deliver their services in much the same way there is little difficulty in changing from centralized to decentralized provision. The cyclical swing between prioritizing control of centralized location and the flexibility of this decentralized location happens perhaps every five to eight years. The ease with which the change can be made makes it clear that nothing is very radically different between them. The decentralization of IS resources may be one side of this continually flipping coin or be a stage in a progression towards devolved locations. Currently there are strong pressures to lower IT resource costs; there is a growing IS literacy within the entire user community, and there is phenomenal growth in end-user computing.

All of this suggests that decentralization cannot effectively provide the balanced complement to the high degree of standardization associated with centralized IS. Highly centralized IS tends to discourage creativity since IT function's fear of chaos if standards are relaxed is a major inhibitor to the high risk, high payoff application. It would seem that the necessary complement to centralized IT must be devolved IT that will transfer authority and responsibility to where IT and the business interface, so that business-relevant innovations can emerge and be delivered from the combination.

Devolved: Geographically and managerially dispersed: IT function is a web of lateral linkages plus a significant degree of end-user control over processing and applications systems development and environment. The distinction between decentralized IT function and devolved IT function is of the degree of dispersion of control and authority. This is perhaps the structural name for the collected set of activities that include departmental computing and all forms of user self-managed computing. The advantages and disadvantages of a devolved IT location flow from this dispersal of control. Devolution adds to the technical dispersion inherent in distributed computing but replaces the central IS control with organization-wide cooperation and coordination in order to gain integration. There is still a need for automated support of activities in a devolved environment. Rather than striving for the 'lights out' operation of centralized data centers, the thrust of automation should be a systems safety.

Devolved IT function leads to a dangerous potential confusion over who will be responsible for the, perhaps unglamorous, housekeeping aspects of IS; devolution must be about who is accountable for the system in all respects. Since the devolved location leads to a risk that the business of system protection falls as no one's responsibility the answer is to automate protection as far as possible. The other area to automate as much as possible is the management of the network backbone itself. Software updates, capacity loading adjustments, etc. can be added to basic system and data hygiene housekeeping.

The costs incurred in such housekeeping may be lower since the devolution means users have a direct, vested interest in cost effectiveness. Devolution would seem to be an option favored by organizations that have a good claim to understanding the appropriate role of IS to a competitive business.

Decentralization and distributed computing tend to create islands of technology whereas devolution puts the resources where they are needed by the business, and the main driver for devolution has been the need to get the IT function closer to the business and its customers. The central IT department disappears and is replaced by a utility service that provides the organization-level needs such as network facilities, corporate planning systems, and support for the process of establishing standards and principles for IT procedures. Some central coordination and planning will remain.

Robson (1997, p. 326) finds that devolution has been highly correlated with IS significantly contributing to the business and has been supported by five thrusts:

- Downsizing trends in processing power: Powerful desktop computers make local access to any nature of system a technical reality.
- Growth of standards: Particularly in the area of networking, these allow 'plug and go' capabilities that therefore demand far fewer IT specialist skills.
- Greater IT awareness: Amongst all managers there is greater interest in using and managing IT to the business' advantage.
- The need to match organizational unit autonomy: Including supporting business decoupling to enable divestment programs.
- The drive to manage costs: In enlightened organizations this is not only to cut them (often only with the result of weakening the organization and IS), but also to make them appropriate (that is, lower than the long-term gains). Devolution places costs where gains can be judged against business productivity.

Organization is concerned with the location of the IT department as well as the internal organization of the IT department. Location is where the department is sited. Alternatives for location include centralized, decentralized, dispersed, and outsourced. The centralized alternative is illustrated in Figure 1.

The internal organization of the IT function will vary, but a typical example is illustrated in Figure 2. To assist the CIO in managing IS/IT strategy work, there is often a strategic planner. The information architect works on information models to manage data resources. The operations manager is concerned with hardware, networks and operating systems, while the development manager is concerned with application systems. Help desk helps users help themselves with end user software and hardware.

As Knowledge Management becomes more important, we see organizational change. In Figure 3, a possible organization in a law firm is illustrated. Law firm partners typically own their law firm, and some partners are elected to the board. One partner serves as managing partner, managing the production of legal services. A chief executive officer (CEO) is in charge of all support functions in the firm. Both a chief knowledge officer (CKO) and a chief information officer (CIO) may report to the CEO. Responsibilities

Figure 1. Location of the IT Function in the Organization

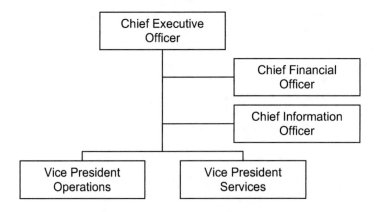

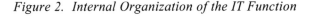
Figure 2. Internal Organization of the IT Function

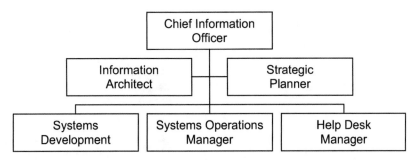

of the CKO include knowledge architecture, strategic planning, training and library. Responsibilities of the CIO include information architecture, systems development and operations, and help desk. A close cooperation between the CKO and CIO is needed to secure success of Knowledge Management initiatives in the firm.

ROLES OF IT FUNCTIONS

The IT function in a company may be defined as the collection of all dedicated IT responsibilities and applied IT competence within the company, involved in tasks like deploying IT, supporting the use of it, operating and maintaining, etc. or even doing tasks concerning exploitation of IT as a business critical component. In this book we have seen that the role of the IT function may vary; it could be merely technically oriented and nonstrategic, it could be a resource to support the business strategy, it could enable new business opportunities, and it could as well play a critical role to the survival of the company in the long run.

Criteria for evaluating the IT function in a company are strongly dependent on its role, and may vary from operations efficiency and cost minimalization, through ability to contribute to the achievement of business goals and some specified results to come from IT initiatives defined in the business strategy, to some long-term effects to the organization.

The primary task of the IT function at Stage 1 can be office automation, systems operation and user support. It is often required that the services are delivered according to predefined criteria, preferably defined in an SLA (Service Level Agreement). Here the IT function is evaluated for operations efficiency, cost efficiency or customer satisfaction. IT functions on this stage normally have a customer-supplier relation to the users of their services. At Stage 2 the primary task may be to utilize the technology in the best possible way to achieve the goals the business management has defined. At Stage 3 the enabling of new business goals by using IT may be the focus, while at Stage 4 the IT function is committed to exploit IT as one basic input factor to enabling the creation of value in the company.

This is illustrated in the two first columns of Figure 4. The matrix in addition indicates some typical deliveries or efforts of the IT function at the various stages, as well as organizational issues.

Figure 3. A Law Firm Organization Chart

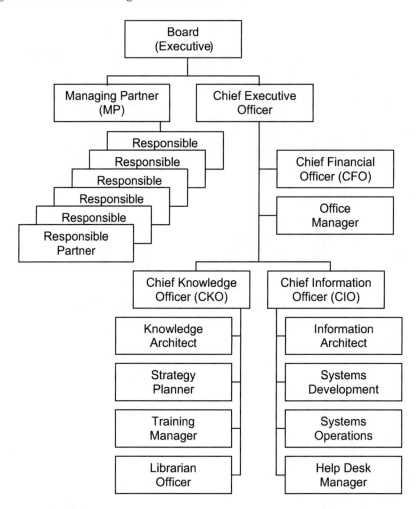

An empirical study in Norway showed that of 41 IS/IT plans, 15 documented that the firms were at Stage 1, with an IT function focused on technical matters. Thirteen firms had an IT function on Stage 2, while 14 firms gave the IT function influence on the business strategy and hence may be classified as being on Stage 3. Only two firms recognized their IT function as critical for the business ability to survive in the long run (Stage 4).

This model can be utilized in processes of organizational change or business improvement, to help identify the IT function's opinion of its own role today and where it wishes to be heading. The result can be valuable to defining the future needs of competence at the various stages and where this should be placed in the organization. Relations to external vendors and partners should also be examined as part of such a process.

A specific company may have IT functions on several stages. The model could be viewed as stairs where Stage 1 is the base step; to be able to have an IT function on, let's say, Stage 3 requires that the Stages 1 and 2 are well functioning. Concerning Stage 1, the tasks could be outsourced, but the stage still is part of the IT function because the responsibility of the impact on the organization prevails internally. An internal IT function has to control the external supplier.

The requirements to the IT function depend on the nature of the business. It is not a goal for every company to have IT functions on Stage 4. Stage 1 may prove sufficient to a firm with mainly manual processes. A dot.com, a web-based bank or a telecom service provider may require a Stage 4 type of IT function. This also applies to a lawyer firm in case online legal services are to be offered to customers on a self-service basis.

During recent years it has been a trend to make products of all that the IT function offers, believing that only by trading measurable items could the quality be optimized. Correctly, this applies to Stages 1 and 2, but it would rather be an obstacle to good quality of IT functions on Stages 3 or 4. Sharing of business responsibility is a main driver for quality at these two stages, although "business consciousness" and business under-standing is important at Stages 1 and 2 as well.

THE CHIEF INFORMATION OFFICER

The CIO can be defined as the highest-ranking IT executive who typically exhibits managerial roles requiring effective communication with top management, a broad corporate perspective in managing information resources, influence on organizational strategy, and responsibility for the planning of IT to cope with a firm's competitive environment. This definition is in line with research, which applied the following criteria when selecting CIOs for empirical observation: (1) highest-ranking information technol-ogy executive, (2) reports no more than two levels from the CEO, i.e., either reports to the CEO or reports to one of the CEO's direct reports, (3) areas of responsibility include information systems, computer operations, telecommunications and networks, office

Figure 4. The Role of IT Functions in Relation to Business (Source: Noralf Husby, IT-ledelse AS, Norway)

	Primary Task	Evaluation Criteria	Deliveries	Organizing
stage 1	• Cater for basic infrastructure • Office automation • Operate according to SLA	• Operations efficiency, reliability, security • Cost efficiency • Customer satisfaction	• Daily operations and user support • All deliveries could be defined as IT services	• Separate department with a Customer-Supplier interface to the business; most of the services could be bought externally
stage 2	• Support business strategy • Support business processes • Use of technology to achieve business goals	• Influence that IT has on the business bottom line • Achievement of goals (set by the business management)	• Support, like consultancy and systems development • Man-hours or turn-key; deliveries might be packaged as services	• Rather a partner than a Customer-Supplier relation • Center of excellence type of unit
stage 3	• Business enabler • Use of technology to set forth new business goals • Influencing business strategy	• Share of contributions to the creation of competitive advantages	• Cooperation - not 'deliveries' • Be a driving force uncovering possible options and benefits • Utilize the technology to give business value	• Strategic unit; balanced IT and business competence • Business participant and co-responsible
stage 4	• Integrated part of business • Manage IT as a critical success factor of the business • Conduct IT as one of the management diciplines	• Long term effect on the organization • Contributions to the ability to survive in the long run	• Integrated - not 'deliveries' • Developing IT as part of the business development	• Integrated in management and competence/discipline development, locally and centrally • Business responsible

automation, end-user computing, help desks, computer software and applications, and (4) responsibility for strategic IS/IT planning.

The CIO position emerged in the 1970s as a result of increased importance placed on IT. In the early 1980s, the CIO was often portrayed as the corporate savior who was to align the worlds of business and technology. CIOs were described as the new breed of information managers who were businessmen first, managers second, and technologists third (Grover et al., 1993). It was even postulated that in the 1990s, as information became a firm's critical resource, the CIO would become the logical choice for the chief executive officer (CEO) position.

As a manager of people, the CIO faces the usual human resource roles of recruiting, staff training, and retention, and the financial roles of budget determination, forecasting and authorization. As the provider of technological services to user departments, there remains a significant amount of work in publicity, promotion, and internal relations with user management. As a manager of an often virtual information organization, the CIO has to coordinate sources of information services spread throughout and beyond the boundaries of the firm. The CIO is thus concerned with a wider group of issues than are most managers.

While information systems executives share several similarities with the general manager, notable differences are apparent. The CIO is not only concerned with a wider group of issues than most managers, but also, as the chief information systems strategist, has a set of responsibilities that must constantly evolve with the corporate information needs and with information technology itself. It has been suggested that the IT director's ability to add value is the biggest single factor in determining whether the organization views information technology as an asset or a liability.

According to Earl and Feeny (1994, p. 11), chief information officers have the difficult job of running a function that uses a lot of resources but that offers little measurable evidence of its value:

"Chief information officers have the difficult job of running a function that uses a lot of resources but that offers little measurable evidence of its value. To make the information systems department an asset to their companies — and to keep their jobs – CIOs should think of their work as adding value in certain key areas."

Creation of the CIO role was driven in part by two organizational needs. First, accountability is increased when a single executive is responsible for the organization's processing needs. Second, creation of the CIO position facilitates the closing of the gap between organizational and IT strategies, which has long been cited as a primary business concern.

Alignment of business and IT objectives is not only a matter of achieving competitive advantage, but is essential for the firm's very survival. Though the importance of IT in creating competitive advantage has been widely noted, achieving these gains has proven elusive. Sustained competitive advantage requires not only the development of a single system, but the ability to consistently deploy IT faster, cheaper, and more strategically than one's competitors. IT departments play a critical role in realizing the potential of IT. The performance of IT functions, in turn, often centers on the quality of leadership, i.e., the CIO.

As early as 1984, some surveys suggested that one-third of US corporations had a CIO function, if not in title. While exact percentages differ, ranging from 40% to 70%, Grover et al. (1993) found that the number of senior-level information systems executive positions created over the past ten years had grown tremendously. The earliest scientifically conducted research on the CIO position examined 43 of 50 top-ranked Fortune 500 service organizations in the US, and noted that 23 (58%) of these organizations had the CIO position. In 1990, the 200 largest Fortune 500 industrial and service organizations were examined, and it was found that 77% of the industrials had a CIO position as compared with 64% of the service organizations. It is very likely that these numbers have increased in recent years.

Few studies have examined the reasons behind the creation of the CIO position in firms. Creation of the position effectively increases accountability by making a single executive responsible for corporate information processing needs. In a sample of Fortune 500 firms, i.e., appearing on the list for four consecutive years, 287 firms with CIOs were compared in 1995 to firms without CIOs on a number of variables hypothesized to predict creation of the position. It was observed that a number of characteristics of the corporate board, including the number of outside directors and equity ownership of the directors, predicted the existence of the CIO position. A firm's information intensity was also found to be positively related to the creation of the CIO position. Furthermore, the CIO position was more likely to exist when the CEO appreciated the strategic value and importance of IT.

The CIO title itself has become a source of confusion. The term CIO has been somewhat loosely defined and is often used interchangeably with various titles such as IT director, vice president of IS, director of information resources, director of information services, and director of MIS, to describe a senior executive responsible for establishing policy and controlling information resources. Sometimes, the CIO label denotes a function rather than a title. Studies relating to the CIO have focused on the evolution of the position and the similarities between the CIO and other senior-level executives.

The CIO label itself has been met with resistance, and some firms have replaced the title with alternative labels such as knowledge manager, chief knowledge officer (CKO) or chief technology officer (CTO). It has been found that the CKO has to discover and develop the CEO's implicit vision of how Knowledge Management would make a difference, and how IT can support this difference.

There are significant differences between the tasks of a CTO, CIO and CKO. While the CTO is focused on technology, the CIO focuses on information, and the CKO focuses on knowledge. When companies replace a CIO with a CKO, it should not only be a change of title. Rather, it should be a change of focus.

Applegate et al. (1996) indicate that the CIO is becoming a member of the top management team and participates in organizational strategy development. Similarly, it has been stated that CIOs see themselves as corporate officers and general business managers. This suggests that CIOs must be politically savvy and that their high profile places them in contention for top line management jobs. The results of these studies indicate that today's CIO is more a managerially oriented executive than a technical manager. Some provide a profile of the ideal CIO as an open communicator with a business perspective, capable of leading and motivating staff, and as an innovative corporate team player. Karimi et al. (2001) found that successful CIOs characterized themselves in the following way:

- I see myself to be a corporate officer.
- In my organization I am seen by others as a corporate officer.
- I am a general business manager, not an IT specialist.
- I am a candidate for top line management positions.
- I have a high profile image in the organization.
- I have political as well as rational perspectives of my firm.
- I spend most of my time outside the IT department focusing on the strategic and organizational aspects of IT.

Business strategist is likely to be among the most significant roles that CIOs will fulfill in the digital era, according to Sambamurthy et al. (2000). As a business strategist, the CIO must understand and visualize the economic, competitive, and industry forces impacting the business and the factors that sustain competitive advantage. Further, the CIO must be capable of plotting strategy with executive peers, including the chief executive officer (CEO), chief operating officer (COO), and other senior business executives (Sambamurthy, 2001, p. 285):

"Business strategist is likely to be among the most significant roles that CIOs will fulfill in the digital era. As a business strategist, the CIO must understand and visualize the economic, competitive, and industry forces impacting the business and the factors that sustain competitive advantage. Further, the CIO must be capable of plotting strategy with executive peers, including the chief executive officer (CEO), chief operating officer (COO), and other senior business executives. Not only are CIOs drawn into the mainstream of business strategy, but also their compensation is being linked with the effectiveness of competitive Internet actions in many firms. With an understanding of current and emergent information technologies and an ability to foresee breakthrough strategic opportunities as well as disruptive threats, CIOs must play a lead role in educating their business peers about how IT can raise the competitive agility of the firm. Obviously, to be effective business strategists, the CIOs must be members of an executive leadership team and part of the dominant coalition that manages the firm."

With an understanding of current and emergent information technologies and an ability to foresee breakthrough strategic opportunities as well as disruptive threats, CIOs must play a lead role in educating their business peers about how IT can raise the competitive agility of the firm. To be effective business strategists, the CIOs must be members of an executive leadership team and part of the dominant coalition that manages the firm.

Robson (1997) has suggested that CIOs have to be hybrid managers to be successful. Hybrid managers require business literacy and technical competency plus a third dimension. This third item is the organizational astuteness that allows a manager to make business-appropriate IS use and management decisions that enhance or set business directions as well as follow them. It is fairly well recognized that hybrid managers are problematic, perhaps requiring inbuilt talent and personal qualities, but can be encouraged or discouraged. For this reason undergraduate study can generally produce only hybrid users whilst postgraduate and post-experience study can support the development of hybrid managers.

Hybrid users are the people involved in user-controlled computing; they combine a degree of technical competence with business literacy required to fulfill their primary role (Robson, 1997, p. 367):

"This management description is not just another term to describe the users engaged in the user-controlled computing. A clear distinction exists between hybrid users and hybrid managers and this distinction is one of emphasis and purposes. Hybrid users are the people involved in user-controlled computing, they combine a degree of technical competence (perhaps defined by notions such as the end-user continuum) with, of course, the business literacy required to fulfil their primary role. Hybrid managers, as opposed to managers who are hybrid users, require this business literacy and technical competency plus a third dimension. This third item is the organisational astuteness that allows a manager to make business-appropriate IS use and management decisions that enhance or set business directions as well as follow them. It is fairly well recognised that hybrid users can be trained whereas the more sophisticated development of hybrid managers is problematic, perhaps requiring inbuilt talent and personal qualities, but can be encouraged or discouraged. For this reason undergraduate study can generally produce only hybrid users whilst postgraduate and post-experience study can support the development of hybrid managers.

The notion of hybrid management is an essentially British one and a significant amount of work on the concept of hybrid management has been done. Earl provided the initial working definition of hybrid managers that has been subsequently adopted by other works. Hybrid managers are a high risk, high cost, people infrastructure that enables the organisational integration of IS and business. This integration ensures both business-consistent IS and IS-exploitative business and so hybrid managers straddle two, previously disparate, disciplines. No amount of communication, or translation bridges, between the two separate disciplines can achieve the same degree of integration. Whilst there is no theoretical reason why hybrid managers cannot be drawn from any discipline, experience seems to show that it proves easier to add IS 'technical' knowledge to a base of business awareness than to inculcate IS technicians with a broader, organisational vision. The primary benefits of hybrid managers are that they create 'islands' of true business/IS understanding; these islands then provide the catalyst that leads to an organisational hybridisation. Even from the earliest stages of hybridisation programmes, organisational gains in flexibility and effectiveness are reported.

Since developing hybrid managers (or any form of management development) is a costly and uncertain exercise it can be a problem notion in recessionary times. The long-term benefits, rather than short-term gains, of such a development programme can look easy to 'trim' out of recession-hit budgets. And yet, paradoxically, it is precisely this type of people infrastructure that supports the cross-boundary, radical re-works typically associated with business process redesign to enable significant future cost savings. The business redesign focus of the 1990s demands the hybrid manager who is, not narrowly specialist, but capable of seeing the broad picture and the opportunities present in this total view. Hybrid managers will be critical to the survival of the IS

*function into the next decade. The continuing devolution of many IS areas requires a
hybrid manager to manage the 'new' IS and indeed even the act of assessing the relative
merits of different paths to devolution, and judging what not to devolve requires the
skills as defined to be of a hybrid manager."*

The notion of hybrid management is an essentially British one and a significant
amount of work on the concept of hybrid management has been done. Earl provided the
initial working definition of hybrid managers that has been subsequently adopted by
other works. Hybrid managers are a high risk, high cost, people infrastructure that
enables the organizational integration of IS and business. This integration ensures both
business-consistent IS and IS-exploitative business, and so hybrid managers straddle
two, previously disparate, disciplines. No amount of communication, or translation
bridges, between the two separate disciplines can achieve the same degree of integration.

Whilst there is no theoretical reason why hybrid managers cannot be drawn from
any discipline, experience seems to show that it proves easier to add IS technical
knowledge to a base of business awareness than to inculcate IS technicians with a
broader, organizational vision. The primary benefits of hybrid managers are that they
create islands of true business/IS understanding; these islands then provide the catalyst
that leads to an organizational integration. Even from the earliest stages of integration
programs, organizational gains in flexibility and effectiveness are reported.

Since developing hybrid managers (or any form of management development) is a
costly and uncertain exercise it can be a problem notion in economic depression times.
The long-term benefits, rather than short-term gains, of such a development program can
look easy to trim out of recession-hit budgets. And yet, it is precisely this type of people
infrastructure that supports the cross-boundary, radical re-works typically associated
with business process redesign to enable significant future cost savings. The business
redesign focus of the future demands the hybrid manager who is, not narrowly specialist,
but capable of seeing the broad picture and the opportunities present in this total view.

According to Robson (1997, p. 368), hybrid managers will be critical to the survival
of the IT function in the future. The continuing devolution of many IS areas requires a
hybrid manager to manage the new IS and indeed even the act of assessing the relative
merits of different paths to devolution, and judging what not to devolve requires the skills
as defined to be of a hybrid manager.

This is certainly true if the company is to succeed in Knowledge Management.
Knowledge Management requires not only business literacy and technical competency;
it requires first and foremost an ability to combine the two. Sometimes information
technology is (part of) the solution to Knowledge Management challenges, sometimes
it not. Only business literacy combined with technical competency can enable a CIO to
make an optimal judgment.

Although it was originally expected that the CIO would have high levels of influence
within the firm, as the definition of job responsibilities would suggest, recent surveys
indicate that this may not be the case. CIOs may not actually possess strategic influence
with top management, and they may lack operational and tactical influence with users.
Some specific problems include higher-than-average corporate dismissal rates compared
with other top executives, diminished power with belt tightening and budget cuts, high
expectations of new strategic systems that CIOs may not be able to deliver, lack of secure

Figure 5. CIO Reporting in the US and Norway Over Time

Chief Information Officer (CIO) reporting to:	USA 1992	USA 1997	Norway 1997	Norway 1999	Norway 2000
Chief Executive Officer (CEO)	27%	43%	48%	44%	41%
Chief Financial Officer (CFO)	44%	32%	21%	23%	16%
Other top executive in the company	29%	25%	31%	33%	43%

power bases due to the fact that CIOs are viewed as outsiders by top management, and the fact that few CIOs take part in strategic planning, and many do not report to the CEO.

Over time, the number of CIOs reporting to CEOs seems to increase. In 1992, only 27% of surveyed CIOs in the US reported to CEOs, while this number had increased to 43% five years later, as listed in Figure 5. In Norway, the numbers in Figure 5 seem to indicate a stable level above forty percent or maybe an insignificant decline in the fraction of CIOs reporting to the CEO. An interesting development is indirect reports moving from CFOs to other top executives.

The CIO's pivotal responsibility of aligning business and technology direction presents a number of problems. Moreover, rapid changes in business and information environments have resulted in corresponding changes at the IT function helm. This role has become increasingly complex, causing many firms to look outside the organization for the right qualifications. Characteristics such as professional background, educational background, and current length of tenure have been examined in previous research. CIO problems seem to indicate that, when compared with other senior executives, CIOs do not have the authority or ability to achieve the kind of changes that were promised when the position was initially proposed. A second and possibly related explanation is that CIOs are experiencing managerial role conflicts that prevent them from meeting those expectations as originally envisioned in the CIO position.

One approach to understanding the CIO position is to study managerial roles. Mintzberg (1994) notes a number of different and sometimes conflicting views of the manager's role. He finds that it is a curiosity of the management literature that its best-known writers all seem to emphasize one particular part of the manager's job to the exclusion of the others. Together, perhaps, they cover all the parts, but even that does not describe the whole job of managing.

Based on an observational study of chief executives, Mintzberg (1994) concluded that a manager's work could be described in terms of 10 job roles. As managers take on these roles, they perform management functions. These ten roles consist of three interpersonal roles (figurehead, leader and liaison), three informational roles (monitor, disseminator, and spokesman), and four decisional roles (entrepreneur, disturbance handler, resource allocator, and negotiator):

- *Figurehead* performs some duties of a ceremonial nature. Examples are greeting visitors, responding to journalists' questions, and visiting customers and allies.
- *Personnel leader* is responsible for motivation of subordinates and for staffing and training. Examples are most activities involving subordinates, such as settling disagreements between subordinates.

- *Liaison* establishes a web of external relationships. Examples are attending conferences and giving presentations.
- *Monitor* seeks and receives information to understand and learn from the environment. Examples are reading journals and listening to external experts.
- *Disseminator* transmits information to other organizational members. Examples include forwarding reports and memos, making phone calls to present information, and holding informational meetings.
- *Spokesman* involves the communication of information and ideas. Examples are speaking to the board of directors and top management, and talking to users.
- *Entrepreneur* acts as initiator and designer of much of the controlled change in the organization. Examples are user ideas converted to systems proposals and management objectives transformed to infrastructure actions.
- *Resource allocator* is responsible for allocation of human, financial, material, and other resources. Examples are working on budgets, developing project proposals, and monitoring information technology projects.
- *Negotiator* is responsible for representing the organization in negotiations. Examples are negotiations with unions concerning wages and with vendors concerning procurements.

According to Mintzberg (1994), these ten roles are common in all managerial jobs regardless of the functional or hierarchical level. However, differences do exist in the importance and effort dedicated to each managerial role based on job content, different skill levels, and expertise. Mintzberg (1994) states that managers are in fact specialists, required to perform a particular set of specialized managerial roles that are dependent upon the functional area and hierarchical level in which they work.

Grover et al. (1993) used the Mintzberg framework to study CIO roles. They selected six of ten roles, which they found relevant for CIOs: personnel leader, liaison, monitor, spokesman, entrepreneur and resource allocator. The four other roles (figurehead, disseminator, disturbance handler, and negotiator) were not operationalized because Grover et al. (1993) found that the activities constituting these roles were correlated with the activities of the other six roles and because they found that the activities that comprised those four roles were consistently important only for certain functions and levels of management. The six selected roles were related to information technology management by rephrasing them:

- As the *personnel leader*, the IS manager is responsible for supervising, hiring, training, and motivating a cadre of specialized personnel. Literature has emphasized the impact of this role on IS personnel. This role is mainly internal to the IS organization.
- The *spokesman* role incorporates activities that require the IS manager to extend organizational contacts outside the department to other areas of the organization. Frequently, he or she must cross traditional departmental boundaries and become involved in affairs of production, distribution, marketing, and finance. This role is mainly external in relation to the intra-organizational environment.

- As the *monitor*, the IS manager must scan the external environment to keep up with technical changes and competition. In acting as the firm's technical innovator, the IS manager uses many sources including vendor contacts, professional relationships, and a network of personal contacts. This role is mainly external in relation to the inter-organizational environment.

- As the *liaison*, the IS manager must communicate with the external environment including exchanging information with IS suppliers, customers, buyers, market analysts, and the media. This role is mainly external in relation to the inter-organizational environment.

- As the *entrepreneur*, the IS manager identifies business needs and develops solutions that change business situations. A major responsibility of the IS manager is to ensure that rapidly evolving technical opportunities are understood, planned, implemented, and strategically exploited in the organization.

- As the *resource allocator*, the IS manager must decide how to allocate human, financial, and information resources. The litany of past discussion on charge-back systems (users have to pay for IT services) and the importance of "fairness" in IS resource allocation decisions speak to the importance of this role. This role is mainly internal to the IS organization.

In Figure 6, the selected six CIO roles are illustrated. The roles of personnel leader and resource allocator are both internal to IT functions. The entrepreneur absorbs ideas from the intra-organizational environment, while the spokesman influences the intra-organizational environment. The liaison informs the external environment, while the monitor absorbs ideas from the external environment.

A survey was conducted in Norway to investigate CIO roles. CIOs were asked questions about the importance of the different roles. Survey results indicate some variation in the importance of roles. Responding CIOs found the role of entrepreneur

Figure 6. CIO Roles on Different Arenas

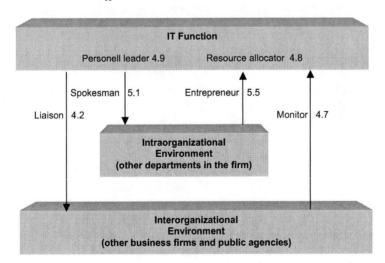

most important and the role of liaison least important. This is indicated with numbers in Figure 6, where the scale went from 1 (not important) to 6 (very important).

In the US, Chatterjee et al. (2001) conducted an investigation to study if newly created CIO positions have any impact. The study's findings provide strong support for the proposition that announcements of newly created CIO positions do indeed provoke positive reactions from the marketplace, but primarily for firms competing in industries with high levels of IT-driven transformation. Within such industries, IT is being applied in innovative ways for competitive purposes. For firms to engage in such strategic behaviors, they must first develop and then effectively exploit an appropriate set of IT capabilities. Strong executive leadership, as reflected in the CIO role, is likely to play a crucial enabling role in the effective deployment of these IT capabilities, and hence be highly valued by a firm's shareholders. Just how valuable is a newly created CIO role? One way to consider the magnitude of the stock market reaction is to compute the impact on each firm's market valuation of common equity.

A conservative approach would calculate this effect through the median statistic (multiplying the median stock market reaction by the median market valuation of common equity); a less conservative approach would use the mean statistic (multiplying the mean stock market reaction by the mean market valuation of common equity). For the entire sample of firms, the net impact per firm of a newly created CIO position was in a range from $7.5 million (median approach) to $76 million (mean approach). If only the IT-driven transformation subgroup is considered, the net impact was in a range from $8 million (median approach) to $297 million (mean approach). Even with the trend in escalated executive salaries, the expected return from such an investment in IT capability appears quite reasonable (Chatterjee et al., 2001, p. 59):

"This study's findings provide strong support for the proposition that announcements of newly created CIO positions do indeed provoke positive reactions from the marketplace, but primarily for firms competing in industries with high levels of IT-driven transformation. Within such industries, IT is being applied in innovative ways for competitive purposes. For firms to engage in such strategic behaviors, they must first develop and then effectively exploit an appropriate set of IT capabilities. Strong executive leadership, as reflected in the CIO role, is likely to play a crucial enabling role in the effective deployment of these IT capabilities, and hence be highly valued by a firm's shareholders.

Just how valuable is a newly created CIO role? One way to consider the magnitude of the stock market reaction is to compute the impact on each firm's market valuation of common equity. A conservative approach would calculate this effect through the median statistic (multiplying the median stock market reaction by the median market valuation of common equity); a less conservative approach would use the mean stratistic (multiplying the mean stock market reaction by the mean market valuation of common equity). For our entire sample of firms, the net impact per firm of a newly created CIO position is in a range from $7.5 million (median approach) to $76 million (mean approach). If only the IT-driven transformation subgroup is considered, the net impact is in a range from $8 million (median approach) to $297 million (mean approach). Even with the trend in escalated executive salaries, the expected return from such an investment in IT capability appears quite reasonable!"

Computer Science Corporation (CSC, 1996) has suggested an alternative set of leadership roles to Mintzberg (1994). These six leadership roles are specifically tailored to information technology executives:

- The *chief architect* designs future possibilities for the business. The primary work of the chief architect is to design and evolve the IT infrastructure so that it will expand the range of future possibilities for the business, not define specific business outcomes. The infrastructure should provide not just today's technical services, such as networking, databases and desktop operating systems, but an increasing range of business level services, such as workflow, portfolio management, scheduling, and specific business components or objects.

- The *change leader* orchestrates resources to achieve optimal implementation of the future. The essential role of the change leader is to orchestrate all those resources that will be needed to execute the change program. This includes providing new IT tools, but it also involves putting in place teams of people who can redesign roles, jobs and workflow, who can change beliefs about the company and the work people do, and who understand human nature and can develop incentive systems to coax people into new and different ways of acting.

- The *product developer* helps define the company's place in the emerging digital economy. For example, a product developer might recognize the potential for performing key business processes (perhaps order fulfillment, purchasing or delivering customer support) over electronic linkages such as the Internet. The product developer must "sell" the idea to a business partner, and together they can set up and evaluate business experiments, which are initially operated out of IS. Whether the new methods are adopted or not, the company will learn from the experiments and so move closer to commercial success in emerging digital markets.

- The *technology provocateur* embeds IT into the business strategy. The technology provocateur works with senior business executives to bring IT and realities of the IT marketplace to bear on the formation of strategy for the business. The technology provocateur is a senior business executive who understands both the business and IT at a deep enough level to integrate the two perspectives in discussions about the future course of the business. Technology provocateurs have a wealth of experience in IS disciplines, so they understand at a fundamental level the capabilities of IT and how IT impacts the business.

- The *coach* teaches people to acquire the skills they will need for the future. Coaches have two basic responsibilities: teaching people how to learn, so that they can become self-sufficient, and providing team leaders with staff able to do the IT-related work of the business. A mechanism that assists both is the center of excellence - a small group of people with a particular competence or skill, with a coach responsible for their growth and development. Coaches are solid practitioners of the competence that they will be coaching, but need not be the best at it in the company.

- The *chief operating* strategist invents the future with senior management. The chief operating strategist is the top IS executive who is focused on the future agenda of the IS organization. The strategist has parallel responsibilities related to helping the business design the future, and then delivering it. The most important, and least understood, parts of the role have to do with the interpretation of new technologies and the IT marketplace, and the bringing of this understanding into the development of the digital business strategy for the organization.

These roles were applied in a survey in Norway. CIOs were asked to rate the importance of each leadership role. The roles were rated on a scale from 1 (not important) to 6 (very important). The role of change leader received the highest score of 4.6, while the role of product developer received the lowest score of 3.3.

The Harvard Business Review invited leading scholars to answer the question: Are CIOs obsolete? They all responded with a "no" answer. Rockart found that all good CIOs today are business executives first, and technologists second (Maruca, 2000, p. 57). Earl paid attention to recruiting new CIOs. His scenario suggests an acid test for selecting the new CIO. Does he or she have the potential to become CEO? If we could develop and appoint such executives, not only will we have CIOs fit for today's challenges, we may be lining up our future CEOs (Maruca, 2000, p. 60).

KEY ISSUES IN IT MANAGEMENT

IT functions face many challenges in today's rapidly changing environment. One approach to understanding these challenges is to survey CIOs to elicit what they consider are key issues. The purpose of such studies is to determine the IT management issues expected to be most important over the next three to five years and thus most deserving of time and resource investments.

A survey was conducted among CIOs in Norway, and their average ranking of key issues in IT management is listed in Figure 7.

Improving links between information systems strategy and business strategy was ranked as most important key issue in IT management in Norway. We have seen how these links can be improved using the Y model and measuring strategic integration.

Implementing and managing knowledge work systems was ranked only sixteenth. That may seem surprising. However, in a separate investigation in knowledge-intensive firms such as law firms, this key issue was ranked sixth.

In Figure 7, scores are listed in a separate column. The scale went from great importance (+4) to little importance (-4). Improving links between information systems strategy and business strategy had an average score of 3.28.

In Figure 7, key issues are classified according to the following dimensions:
- Management (M) versus technology (T).
- Planning (P) versus control (C).
- External (E) versus internal (I).

Improving links between information systems strategy and business strategy is classified as M, P, and E. Improving links is a management task, the task is conducted through planning, and the issue covers more than the IT function. External means external to the IT function, while internal means an issue that is mainly solved within the IT function. Brancheau et al. (1996, p. 233) stressed the importance of alignment:

Figure 7.   Key Issues Ranking in Norway

Rank	Key Issue in IT Management	Score	M/IT	P/C	E/I
1	Improving links between information systems strategy and business strategy	3.28	M	P	E
2	Planning information technology projects for competitive advantage	2.00	M	P	E
3	Improving interorganizational information systems planning	1.05	T	P	E
4	Developing and implementing an information architecture	1.02	T	P	I
5	Controlling a responsive information technology infrastructure	1.02	T	C	I
6	Recruiting and developing IS human resources	0.90	M	P	I
7	Assuring software quality	0.86	T	C	I
8	Ensuring quality with information systems	0.36	T	P	I
9	Reducing IT projects completion time	0.34	M	C	I
10	Making effective use of data and information systems resource	0.31	M	C	E
11	Measuring benefits from information technology applications	0.16	M	C	I
12	Managing Internet applications	-0.02	M	P	E
13	Managing application architecture planning	-0.10	M	P	I
14	Improving control, security and recovery capabilities	-0.21	T	C	E
15	Improving computer operations planning	-0.21	T	P	I
16	Implementing and managing knowledge work systems	-0.34	T	C	E
17	Improving information technology infrastructure planning	-0.47	M	P	I
18	Planning information technology for electronic commerce	-0.78	T	P	E
19	Improving software engineering practices	-1.00	T	C	I
20	Implementing information technology for electronic commerce	-1.10	T	C	E
21	Improving availability of national and international networks	-1.41	M	C	E
22	Managing the technical foundation of information systems	-1.67	M	C	I
23	Managing and controlling end-user computing	-1.78	M	C	E
24	Scanning emerging technologies	-2.21	T	P	E

"The importance of aligning long-range IS plans with strategic business plans has always been high. Rapidly changing business environments, increased involvement of end users, and accelerated technology change make this difficult. Shorter planning cycles require a great deal of flexibility in any plan."

Implementing and managing knowledge work systems is classified as T, C, and E. It is a technology issue, it is to be carried out, and it is to be implemented in the organization.

At the top of the list we find two issues that are M, P, and E. This implies that CIOs in Norway currently struggle with management issues that involve planning and that cannot be solved within the IT function.

Key issues in IT management surveys have been conducted for some years in many nations and regions. Most surveys have used the so-called Delphi method, while the results in Figure 7 are based on the so-called Q Method as presented in the appendix. Key issues studies are of interest to many stakeholders (Niederman et al., 1991, p. 476):

"Vendors can use this information to develop and market products and services. Professional societies can use this information to plan conferences and seminars as well as disseminate knowledge through their publications. Consultants can use this information to help accelerate the transfer of technology and management skills among their clients. Educators can use this information to develop programs and place their graduates. Finally, researchers can use this information to guide their inquiry and improve understanding of critical managerial issues. Thus, the entire IS community needs to be aware of the issues that are judged to be of critical concern by its leading practitioners."

Figure 7 gives a static picture of key issues as it ranks key issues at a specific point in time. Over time, key issues change. Some key issues become more important and some issues become less important. Reduced importance can occur when the issue is becoming solved or when an issue is less relevant than before. The Year 2000 issue is an example, where this issue was extremely important in 1999, but lost its importance after the beginning of the new millennium.

In the US, key issues studies have been conducted several times, enabling comparison of key issues in a time perspective. In 1986, 'improving strategic IS planning' was ranked first. In 1989, this issue had dropped to third rank, and in 1994, this issue had dropped to tenth rank. An interpretation of this result is that strategic IS/IT planning became quite successful in the early 1990s. However, the formulation of the issue is limited to strategic IS planning; it does not cover links between information systems planning and business strategy.

In the same time frame, information architecture rose from eighth rank in 1986 to first rank in 1989 in the US, and then dropped again to fourth rank. In 1994 the issue of infrastructure was at the first rank, climbing from sixth rank in 1989. In 1986, the issue of infrastructure was nonexistent on the key issues list. Brancheau et al. (1996, p. 229) discussed the importance of infrastructure based on the top ranking in 1994:

"Building a technology infrastructure that supports existing applications while remaining responsive to change is a key to long-term enterprise productivity. This task is made difficult by the continuing rapid changes in infrastructure technology and the increasing breadth and depth of applications needing support. More than any other, this issue captures an important contemporary thrust of enterprise IS management: providing the processor power, network connectivity, and application framework required to support core business activities and unknown future ventures."

Key issues studies can be done at the national and regional level. Key issues studies can also be done within a firm. By surveying people in the firm, it is possible to get a picture of ranking that is shared by employees and management. When using the dimensions of M/T, P/C, and E/I, such surveys can provide management with guidance on management or technology focus, planning or control focus, and external or internal focus to solve firm challenges in information technology management.

Ranking of key issues can also be useful as a communication tool for top management and IT management. Often, top management can have a different agenda and different opinions about IT in the organization than IT management. One way of making different opinions visible is to let both top management and IT management produce ranking lists. A comparison of the two lists can spark communication and enable clarification, so that top and IT management have the same priorities in the future.

SUMMARY

CIOs find the managerial role of entrepreneur most important. As the entrepreneur, the CIO identifies business needs and develops solutions that change business situations. A major responsibility of the CIO is to ensure that rapidly evolving technical opportunities are understood, planned, implemented, and strategically exploited in the organization.

Among IS leadership roles, the role of change leader was found most important. The change leader orchestrates resources to achieve optimal implementation of the future. The essential role of the change leader is to orchestrate all those resources that will be needed to execute the change program. This includes providing new IT tools, but it also involves putting in place teams of people who can redesign roles, jobs and workflow, who can change beliefs about the company and the work people do, and who understand human nature and can develop incentive systems to coax people into new and different ways of acting.

In professional service firms, such as law firms, we will find new organizational structures to enable Knowledge Management initiatives. A law firm may have a CEO with two important persons working in KM: the CKO and the CIO. While the CKO is responsible for knowledge architecture and organizational development, the CIO is responsible for information architecture and systems development.

REFERENCES

Applegate, L. M., McFarlan, F. W., & McKenney, J. L. (1996). *Corporate information systems management, Fourth Edition.* Irwin.

Brancheau, J. C., Janz, B. D., & Wetherbe, J. C. (1996). Key issues in information systems management: 1994-95 SIM Delphi results. *MIS Quarterly, 20*(2), 225-242.

Chatterjee, D., Richardson, V. J., & Zmud, R. W. (2001) Examining the shareholder wealth effects of announcements of newly created CIO positions. *MIS Quarterly, 25*(1), 43-70.

CSC. (1996). *New IS leaders, CSC Index Research.* Computer Science Corporation, UK: London.

Earl, M. J., & Feeny, D. F. (1994). Is your CIO adding value? *Sloan Management Review, 35*(3), 11-20.

Grover, V., Jeong, S. R., Kettinger, W. J., & Lee, C. C. (1993). The Chief Information Officer: A study of managerial roles. *Journal of Management Information Systems, 10*(2), 107-130.

Karimi, J., Somers, T. M., & Gupta, Y. P. (2001). Impact of Information Technology management practices on customer service. *Journal of Management Information Systems, 17*(4), 125-148.

Maruca, R. F. (2000). Are CIOs obsolete? *Harvard Business Review,* (March/April),, 55-63.

Mintzberg, H. (1994). Rounding out the manager's job. *Sloan Management Review, 36*(1), 11-26.

Niederman, F., Brancheau, J. C., & Wetherbe, J. C. (1991). Information systems management issues for the 1990s. *MIS Quarterly, 17*(4), 475-500.

Robson, W. (1997). *Strategic management & information systems, Second Edition.* UK: Prentice Hall.

Rockart, J. F., Earl, M. J., & Ross, J. W. (1996). Eight imperatives for the new IT organization. *Sloan Management Review, 38*(1), 43-55.

Sambamurthy, V., Straub, D. W., & Watson, R. T. (2001) Managing IT in the digital era. In G. W. Dickson & G. DeSanctis (Eds.), *Information Technology and the Future Enterprise: New Models for Managers* (pp. 282-305). Prentice Hall.

Chapter XI

Governing Information Technology through COBIT*

Erik Guldentops
IT Governance Institute, USA

ABSTRACT

Board oversight of information technology has not kept pace with the rapid growth of IT as a critical driver of business success. However, this is shortsighted, since effective governance over IT Governance protects shareholder value; makes clear that IT risks are quantified and understood; directs and controls IT investment, opportunity, benefits and risks; aligns IT with the business while accepting IT as a critical input to and component of the strategic plan; sustains current operations and prepares for the future; and is an integral part of a global governance structure.

Like most other governance activities, IT Governance engages both board and executive management. Among the board's responsibilities are reviewing and guiding corporate strategy, setting and monitoring achievement of management's performance objectives, and ensuring the integrity of the organisation's systems. Management's focus is generally on cost-efficiency, revenue enhancement and building capabilities, all of which are enabled by information, knowledge and the IT infrastructure.

The four main focus areas for IT Governance are driven by stakeholder value. Two are outcomes: value delivery and risk mitigation. Two are drivers: strategic alignment and performance measurement.

Action plans for implementing effective IT Governance, from both a board and an executive management point of view, consist of activities, outcome measures, best practices, critical success factors and performance drivers. In addition, organisations must assess how well they are currently performing and be able to identify where and how improvements can be made. The use of maturity models simplifies this task and provides a pragmatic, structured approach for measurement.

Control Objectives for Information and related Technology (COBIT), a third edition of which was issued by the IT Governance Institute in 2000, incorporates material on IT Governance and a Management Guidelines *component. COBIT presents an international and generally accepted IT control framework enabling organisations to implement an IT Governance structure throughout the enterprise.*

The Management Guidelines consist of maturity models, critical success factors, key goal indicators and key performance indicators. This structure delivers a significantly improved framework responding to management's need for control and measurability of IT by providing tools to assess and measure the organisation's IT environment against COBIT's 34 IT processes.

WHAT IS IT GOVERNANCE AND WHY IS IT IMPORTANT?

As information technology has become a critical driver of business success, boards of directors have not kept pace. IT demands thorough and thoughtful board governance, yet such oversight has often been lacking because IT has been seen as an operations matter best left to management, and board members lacked interest or expertise in technology issues.

While boards have always scrutinized business strategy and strategic risks, IT has tended to be overlooked, despite the fact that it involves large investments and huge risks. Reasons include:

- The technical insight required to understand how IT enables the enterprise — and creates risks and opportunities
- The tradition of treating IT as an entity separate to the business
- The complexity of IT, even more apparent in the extended enterprise operating in a networked economy

Closing the IT Governance gap has become imperative as it becomes more difficult to separate an organisation's overall strategic mission from the underlying IT strategy that enables that mission to be fulfilled.

IT Governance is ultimately important because expectations and reality often do not match. Boards expect management to juggle a myriad of responsibilities: deliver quality IT solutions on time and on budget, harness and exploit IT to return business value and leverage IT to increase efficiency and productivity while managing IT risks. However, boards frequently see business losses, damaged reputations or weakened competitive positions, unmet deadlines, higher-than-expected costs, lower-than-expected quality and failures of IT initiatives to deliver promised benefits.

IT Governance extends the board's mission of defining strategic direction and ensuring that objectives are met, risks are managed and resources are used responsibly. Pervasive use of technology has created a critical dependency on IT that calls for a specific focus on IT Governance. Such governance should ensure that an organization's IT sustains and extends its strategies and objectives.

Effective IT Governance:

- Protects shareholder value
- Makes clear that IT risks are quantified and understood
- Directs and controls IT investment, opportunity, benefits and risks
- Aligns IT with the business while accepting IT as a critical input to and component of the strategic plan, influencing strategic opportunities
- Sustains current operations and prepares for the future
- Is an integral part of a global governance structure

WHOM DOES IT CONCERN?

Like most other governance activities, IT Governance intensively engages both board and executive management in a cooperative manner. However, due to complexity and specialisation, this governance layer must rely heavily on the lower layers in the enterprise to provide the information needed in its decision-making and evaluation activities. To have effective IT Governance in the enterprise, the lower layers need to apply the same principles of setting objectives, providing and getting direction, and providing and evaluating performance measures. As a result, good practices in IT Governance need to be applied throughout the enterprise.

WHAT CAN THEY DO ABOUT IT?

Among the board's responsibilities are reviewing and guiding corporate strategy, setting and monitoring achievement of management's performance objectives, and ensuring the integrity of the organisation's systems.

How Should the Board Address the Challenges?

The board should *drive enterprise alignment* by:

- Ascertaining that IT strategy is aligned with enterprise strategy
- Ascertaining that IT delivers against the strategy through clear expectations and measurement
- Directing IT strategy to balance investments between supporting and growing the enterprise
- Making considered decisions about where IT resources should be focused

The board should direct management to *deliver measurable value* through IT by:
- Delivering on time and on budget
- Enhancing reputation, product leadership and cost-efficiency
- Providing customer trust and competitive time-to-market

The board should also *measure performance* by:
- Defining and monitoring measures together with management to verify that objectives are achieved and to measure performance to eliminate surprises
- Leveraging a system of Balanced Business Scorecards maintained by management that form the basis for executive management compensation

The board should *manage enterprise risk* by:
- Ascertaining that there is transparency about the significant risks to the organisation
- Being aware that the final responsibility for risk management rests with the board
- Being conscious that risk mitigation can generate cost-efficiencies
- Considering that a proactive risk management approach can create competitive advantage
- Insisting that risk management be embedded in the operation of the enterprise
- Ascertaining that management has put processes, technology and assurance in place for information security to ensure that:
  - Business transactions can be trusted
  - IT services are usable, can appropriately resist attacks and recover from failures
  - Critical information is withheld from those who should not have access to it

How Should Executive Management Address the Expectations?

The executive's focus is generally on cost-efficiency, revenue enhancement and building capabilities, all of which are enabled by information, knowledge and the IT infrastructure. Because IT is an integral part of the enterprise, and as its solutions become more and more complex (outsourcing, third-party contracts, networking, etc.), adequate governance becomes a critical factor for success. To this end, management should:
- *Embed clear accountabilities* for risk management and control over IT into the organisation
- *Cascade strategy*, *policies* and *goals* down into the enterprise and *align the IT organisation* with the enterprise goals
- *Provide organisational structures* to support the implementation of IT strategies and an *IT infrastructure* to facilitate the creation and sharing of business information
- *Measure performance* by having outcome measures[3] for business value and competitive advantage that IT delivers and performance drivers to show how well IT performs

- *Focus on core business competencies IT must support*, i.e., those that add customer value, differentiate the enterprise's products and services in the marketplace, and add value across multiple products and services over time
- *Focus on important IT processes* that improve business value, such as change, applications and problem management. Management must become aggressive in defining these processes and their associated responsibilities.
- *Focus on core IT competencies* that usually relate to planning and overseeing the management of IT assets, risks, projects, customers and vendors
- *Have clear external sourcing strategies*, focussing on the management of third-party contracts and associated service level and on building trust between organisations, enabling interconnectivity and information sharing

WHAT DOES IT COVER?

Fundamentally, IT Governance is concerned about two things: that IT delivers value to the business and that IT risks are mitigated. The first is driven by strategic alignment of IT with the business. The second is driven by embedding accountability into the enterprise. Both need measurement, for example, by a Balanced Scorecard. This leads to the four main focus areas for IT Governance, all driven by stakeholder value. Two of them are outcomes: value delivery and risk mitigation. Two of them are drivers: strategic alignment and performance measurement.

IT Strategic Alignment: "IT Alignment is a Journey, Not a Destination."

The key question is whether a firm's investment in IT is in harmony with its strategic objectives (intent, current strategy and enterprise goals) and thus building the capabilities necessary to deliver business value. This state of harmony is referred to as "alignment". It is complex, multifaceted and never completely achieved. It is about continuing to move in the right direction and being better aligned than competitors. This may not be attainable for many enterprises because enterprise goals change too quickly, but is nevertheless a worthwhile ambition because there is real concern about the value of IT investment.

Alignment of IT has been synonymous with IT strategy, i.e., does the IT strategy support the enterprise strategy? For IT Governance, alignment encompasses more than strategic integration between the (future) IT organisation and the (future) enterprise organisation. It is also about whether IT operations are aligned with the current enterprise operations. Of course, it is difficult to achieve IT alignment when enterprise units are misaligned.

IT Value Delivery: "IT Value is in the Eye of the Beholder."

The basic principles of IT value are delivery on time, within budget and with the benefits that were promised. In business terms, this is often translated into: competitive advantage, elapsed time for order/service fulfillment, customer satisfaction, customer wait time, employee productivity and profitability. Several of these elements are either subjective or difficult to measure, something all stakeholders need to be aware of.

The value that IT adds to the business is a function of the degree to which the IT organisation is aligned with the business and meets the expectations of the business. The business has expectations relative to the contents of the IT deliverable:

- Fit for purpose, meeting business requirements
- Flexibility to adopt future requirements
- Throughput and response times
- Ease of use, resiliency and security
- Integrity, accuracy and currency of information

The business also has expectations regarding the method of working:

- Time-to-market
- Cost and time management
- Partnering success
- Skill set of IT staff

To manage these expectations, IT and the business should use a common language for value which translates business and IT terminology and is based wholly on fact.

Performance Measurement: "In IT, if You're Playing the Game and Not Keeping Score, You're Just Practising."

Strategy has taken on a new urgency as enterprises mobilise intangible and hidden assets to compete in an information-based global economy. Balanced Scorecards translate strategy into action to achieve goals with a performance measurement system that goes beyond conventional accounting, measuring those relationships and knowledge-based assets necessary to compete in the information age: *customer* focus, *process* efficiency and the ability to *learn* and grow. At the heart of these scorecards is management information supplied by the IT infrastructure. IT also enables and sustains solutions for the actual goals set in the financial (enterprise resource management), customer (customer relationship management), process (intranet and workflow tools) and learning (Knowledge Management) dimensions of the scorecard.

IT needs its own scorecard. Defining clear goals and good measures that unequivocally reflect the business impact of the IT goals is a challenge and needs to be resolved in co-operation among the different governance layers within the enterprise. The linkage between the Business Balanced Scorecard and the IT Balanced Scorecard is a strong method of alignment.

Risk Management: "It's the IT Alligators You Don't See that Will Get You."

Enterprise risk comes in many varieties, not only financial risk. Regulators are specifically concerned about operational and systemic risk, within which technology risk and information security issues are prominent. Infrastructure protection initiatives in the U.S. and the UK point to the utter dependence of all enterprises on IT infrastructures and the vulnerability to new technology risks. The first recommendation these initiatives make is for risk awareness of senior corporate officers.

Therefore, the board should manage enterprise risk by:

- Ascertaining that there is *transparency* about the significant risks to the organisation and clarifying the risk-taking or risk-avoidance policies of the enterprise
- Being aware that the final *responsibility* for risk management rests with the board, so, when delegating to executive management, making sure the constraints of that delegation are communicated and clearly understood
- Being conscious that the system of internal control put in place to manage risks often has the capacity to generate *cost-efficiency*
- Considering that a transparent and proactive risk management approach can create *competitive advantage* that can be exploited
- Insisting that risk management is *embedded in the operation* of the enterprise, responds quickly to changing risks and reports immediately to appropriate levels of management, supported by agreed principles of escalation (what to report, when, where and how)

WHAT QUESTIONS SHOULD BE ASKED?

While it is not the most efficient IT Governance process, asking tough questions is an effective way to get started. Of course, those responsible for governance want good answers to these questions. Then they want action. Then they need follow-up. It is essential to determine, along with the action, *who* is responsible to deliver *what* by *when*.

An extensive checklist of questions is provided in *Board Briefing on IT Governance*. The questions focus on three objectives: questions asked to discover IT issues, to find out what management is doing about them, and to self-assess the board's governance over them. For example:

To Uncover IT Issues

- How often do IT projects fail to deliver what they promised?
- Are end users satisfied with the quality of the IT service?
- Are sufficient IT resources, infrastructure and competencies available to meet strategic objectives?

To Find Out How Management Addresses the IT Issues

- How well are enterprise and IT objectives aligned?
- How is the value delivered by IT being measured?
- What strategic initiatives has executive management taken to manage IT's criticality relative to maintenance and growth of the enterprise, and are they appropriate?

To Self-Assess IT Governance Practices

- Is the board regularly briefed on IT risks to which the enterprise is exposed?
- Is IT a regular item on the agenda of the board and is it addressed in a structured manner?
- Does the board articulate and communicate the business objectives for IT alignment?

HOW IS IT ACCOMPLISHED?

Action plans for implementing effective IT Governance, from both a board and an executive management point of view, are provided in detail in *Board Briefing on IT Governance*. These plans consist of the following elements:

- *Activities* list what is done to exercise the IT Governance responsibilities and the *subjects* identify those items that typically get onto an IT Governance agenda.
- *Outcome measures* relate directly to the subjects of IT Governance, such as the alignment of business and IT objectives, cost-efficiencies realised by IT, capabilities and competencies generated and risks and opportunities addressed.
- *Best practices* list examples of how the activities are being performed by those who have established leadership in governance of technology.
- *Critical success factors* are conditions, competencies and attitudes that are critical to being successful in the practices.
- *Performance drivers* provide indicators on *how* IT Governance is achieving, as opposed to the outcome measures that measure *what* is being achieved. They often relate to the critical success factors.

The plans list IT Governance activities and link a set of subjects and practices to them. Practices are classified to reflect the IT Governance area(s) to which they provide the greatest contribution: value delivery, strategic alignment, risk management and/or performance (V, A, R, P). A list of critical success factors is provided in support of the practices. Finally, two sets of measures are listed: outcome measures that relate to the IT Governance subjects and performance drivers that relate to how activities are performed and the associated practices and critical success factors.

HOW DOES YOUR ORGANISATION COMPARE?

For effective governance of IT to be implemented, organisations need to assess how well they are currently performing and be able to identify where and how improvements can be made. This applies to both the IT Governance process itself and to all the processes that need to be managed within IT.

The use of maturity models greatly simplifies this task and provides a pragmatic and structured approach for measuring how well developed your processes are against a consistent and easy-to-understand scale:

0 = Non-existent. Management processes are not applied at all.
1 = Initial. Processes are ad hoc and disorganised.
2 = Repeatable. Processes follow a regular pattern.
3 = Defined. Processes are documented and communicated.
4 = Managed. Processes are monitored and measured.
5 = Optimised. Best practices are followed and automated.

(For a complete description of the various maturity levels, see *Board Briefing on IT Governance*.)

Using this technique the organisation can:

- Build a view of current practices by discussing them in workshops and comparing to example models
- Set targets for future development by considering model descriptions higher up the scale and comparing to best practices
- Plan projects to reach the targets by defining the specific changes required to improve management
- Prioritise project work by identifying where the greatest impact will be made and where it is easiest to implement

INTRODUCING COBIT

Control Objectives for Information and related Technology (COBIT) was initially published by the Information Systems Audit and Control Foundation™ (ISACF™ in 1996, and was followed by a second edition in 1998. The third edition, which incorporates all-new material on IT Governance and Management Guidelines, was issued by the IT Governance Institute in 2000. COBIT presents an international and generally accepted IT control framework enabling organisations to implement an IT Governance structure throughout the enterprise.

Since its first issuance, COBIT has been adopted in corporations and by governmental entities throughout the world.

All portions of COBIT, except the Audit Guidelines, are considered an open standard and may be downloaded on a complimentary basis from the Information Systems Audit and Control Association's web site, www.isaca.org/cobit.htm. The Audit Guidelines are available on a downloadable basis to ISACA members only.

THE COBIT FRAMEWORK

Business orientation is the main theme of COBIT. It begins from the premise that IT needs to deliver the information that the enterprise needs to achieve its objectives. It is designed to be employed as comprehensive guidance for management and business process owners. Increasingly, business practice involves the full empowerment of business process owners so they have total responsibility for all aspects of the business process. In particular, this includes providing adequate controls. COBIT promotes a process focus and process ownership.

The COBIT Framework provides a tool for the business process owner that facilitates the discharge of this responsibility. The Framework starts from a simple and pragmatic premise:

In order to provide the information that the organisation needs to achieve its objectives, IT resources need to be managed by a set of naturally grouped processes.

The Framework continues with a set of 34 high-level Control Objectives, one for each of the IT processes, grouped into four domains:

- **Planning and Organisation:** This domain covers strategy and tactics, and concerns the identification of the way IT can best contribute to the achievement of the business objectives. Furthermore, the realisation of the strategic vision needs to be planned, communicated and managed for different perspectives. Finally, a proper organisation as well as technological infrastructure must be put in place.

- **Acquisition and Implementation:** To realise the IT strategy, IT solutions need to be identified, developed or acquired, as well as implemented and integrated into the business process. In addition, changes in and maintenance of existing systems are covered by this domain to make sure that the lifecycle is continued for these systems.

- **Delivery and Support:** This domain is concerned with the actual delivery of required services, which range from traditional operations over security and continuity aspects to training. In order to deliver services, the necessary support processes must be set up. *This domain includes the actual processing of data by application systems, often classified under application controls.*

- **Monitoring:** All IT processes need to be regularly assessed over time for their quality and compliance with control requirements. This domain thus addresses management's oversight of the organisation's control process and independent assurance provided by internal and external audit or obtained from alternative sources.

Corresponding to each of the 34 high-level control objectives is an Audit Guideline to enable the review of IT processes against COBIT's 318 recommended detailed control objectives to provide management assurance and/or advice for improvement.

The Management Guidelines further enhance and enable enterprise management to deal more effectively with the needs and requirements of IT Governance. The guidelines are action-oriented and generic and provide management direction for getting the enterprise's information and related processes under control, for monitoring achievement of organisational goals, for monitoring performance within each IT process and for benchmarking organisational achievement.

COBIT also contains an Implementation Tool Set that provides lessons learned from those organisations that quickly and successfully applied COBIT in their work environments. It has two particularly useful tools — Management Awareness Diagnostic and IT Control Diagnostic — to assist in analyzing an organisation's IT control environment.

Over the next few years, the management of organisations will need to demonstrably attain increased levels of security and control. COBIT is a tool that allows managers to bridge the gap with respect to control requirements, technical issues and business risks and communicate that level of control to stakeholders. COBIT enables the development of clear policy and good practice for IT control throughout organisations worldwide. Thus, COBIT is designed to be *the* break-through IT Governance tool that helps in understanding and managing the risks and benefits associated with information and related IT.

THE COBIT CONTROL OBJECTIVES

For the purposes of COBIT, the following definitions are provided. "Control" is adapted from the COSO Report *(Internal Control — Integrated Framework*, Committee of Sponsoring Organisations of the Treadway Commission, 1992) and "IT Control Objective" is adapted from the SAC Report *(Systems Auditability and Control Report*, The Institute of Internal Auditors Research Foundation, 1991 and 1994).

Control is defined as the policies, procedures, practices and organisational structures designed to provide reasonable assurance that business objectives will be achieved and that undesired events will be prevented or detected and corrected.

IT Control Objective is a statement of the desired result or purpose to be achieved by implementing control procedures in a particular IT activity.

To satisfy business objectives, information needs to conform to certain criteria, which COBIT refers to as business requirements for information. In establishing the list of requirements, COBIT combines the principles embedded in existing and known reference models:

- **Quality requirements** — Quality, Cost, Delivery
- **Fiduciary requirements** (COSO Report) — Effectiveness and Efficiency of operations; Reliability of Information; Compliance with laws and regulations
- **Security requirements** — Confidentiality; Integrity; Availability

Starting the analysis from the broader Quality, Fiduciary and Security requirements, seven distinct, certainly overlapping, categories were extracted. COBIT's working definitions are as follows:

- **Effectiveness** deals with information being relevant and pertinent to the business process as well as being delivered in a timely, correct, consistent and usable manner.
- **Efficiency** concerns the provision of information through the optimal (most productive and economical) use of resources.
- **Confidentiality** concerns the protection of sensitive information from unauthorised disclosure.
- **Integrity** relates to the accuracy and completeness of information as well as to its validity in accordance with business values and expectations.
- **Availability** relates to information being available when required by the business process now and in the future. It also concerns the safeguarding of necessary resources and associated capabilities.
- **Compliance** deals with complying with those laws, regulations and contractual arrangements to which the business process is subject, i.e., externally imposed business criteria.

- **Reliability of Information** relates to the provision of appropriate information for management to operate the entity and for management to exercise its financial and compliance reporting responsibilities.

The IT resources identified in COBIT can be explained/defined as follows:

- **Data** are objects in their widest sense (i.e., external and internal), structured and non-structured, graphics, sound, etc.
- **Application Systems** are understood to be the sum of manual and programmed procedures.
- **Technology** covers hardware, operating systems, database management systems, networking, multimedia, etc.
- **Facilities** are all the resources to house and support information systems.
- **People** include staff skills, awareness and productivity to plan, organise, acquire, deliver, support and monitor information systems and services.

COBIT consists of high-level control objectives for each process which identify which information criteria are most important in that IT process, state which resources will usually be leveraged and provide considerations on what is important for controlling that IT process. The underlying theory for the classification of the control objectives is that there are, in essence, three levels of IT efforts when considering the management of IT resources. Starting at the bottom, there are the activities and tasks needed to achieve a measurable result. Activities have a lifecycle concept while tasks are more discrete. The lifecycle concept has typical control requirements different from discrete activities. Processes are then defined one layer up as a series of joined activities or tasks with natural (control) breaks. At the highest level, processes are naturally grouped together into domains. Their natural grouping is often confirmed as responsibility domains in an organisational structure and is in line with the management cycle or lifecycle applicable to IT processes.

Thus, the conceptual framework can be approached from three vantage points: (1) information criteria, (2) IT resources and (3) IT processes.

It is clear that all control measures will not necessarily satisfy the different business requirements for information to the same degree.

- **Primary** is the degree to which the defined control objective directly impacts the information criterion concerned.
- **Secondary** is the degree to which the defined control objective satisfies only to a lesser extent or indirectly the information criterion concerned.
- **Blank** could be applicable; however, requirements are more appropriately satisfied by another criterion in this process and/or by another process.

Similarly, all control measures will not necessarily impact the different IT resources to the same degree. Therefore, the COBIT Framework specifically indicates the applicability of the IT resources that are specifically managed by the process under consideration (not those that merely take part in the process). This classification is made within the COBIT Framework, based on a rigorous process of input from researchers, experts and reviewers, using the strict definitions previously indicated.

Each high-level control objective is accompanied by detailed control objectives, 318 in all, providing additional detail on how control should be exercised over that particular process. In addition, extensive audit guidelines are included for building on the objectives.

Sample high-level control objectives, with their related detailed control objectives, are provided at the end of the chapter for PO9, the Assess Risks process in the Planning and Organisation domain, and DS5, the Ensure System Security process in the Delivery and Support domain.

COBIT'S MANAGEMENT GUIDELINES

COBIT's Management Guidelines consist of maturity models, critical success factors (CSFs), key goal indicators (KGIs) and key performance indicators (KPIs). This structure delivers a significantly improved framework responding to management's need for control and measurability of IT by providing management with tools to assess and measure their organisation's IT environment against COBIT's 34 IT processes.

COBIT's Management Guidelines are generic and action-oriented for the purpose of addressing the following types of management concerns:

• Performance measurement — What are the indicators of good performance?

• IT control profiling — What's important? What are the critical success factors for control?

• Awareness — What are the risks of not achieving our objectives?

• Benchmarking — What do others do? How do we measure and compare?

An answer to these requirements of determining and monitoring the appropriate IT security and control level is the definition of specific:

• **Benchmarking** of IT control practices (expressed as maturity models)

• **Performance indicators** of the IT processes — for their outcome and their performance

• **Critical success factors** for getting these processes under control

The Management Guidelines are consistent with and build upon the principles of the Balanced Business Scorecard.[4] In "simple terms", these measures will assist management in monitoring their IT organisation by answering the following questions:

1. What is the management concern?
   Make sure that the enterprise needs are fulfilled.
2. Where is it measured?
   On the Balanced Business Scorecard as a key goal indicator, representing an outcome of the business process.
3. What is the IT concern?
   That the IT processes deliver on a timely basis the right information to the enterprise, enabling the business needs to be fulfilled. This is a critical success factor for the enterprise.
4. Where is that measured?
   On the IT Balanced Scorecard, as a key goal indicator representing the outcome for

IT, which is that information is delivered with the right criteria (effectiveness, efficiency, confidentiality, integrity, availability, compliance and reliability).
5. What else needs to be measured?
  Whether the outcome is positively influenced by a number of critical success factors that need to be measured as key performance indicators of how well IT is doing.

Each element of the Management Guidelines will be examined in further detail.

Maturity Models

IT management is constantly on the lookout for benchmarking and self-assessment tools in response to the need to know what to do in an efficient manner. Starting from COBIT's processes and high-level control objectives, the process owner should be able to incrementally benchmark against that control objective. This creates three needs:

- A relative measure of where the organisation is
- A manner to decide efficiently where to go
- A tool for measuring progress against the goal

The approach to maturity models for control over IT processes consists of developing a method of scoring so that an organisation can grade itself from non-existent to optimised (from 0 to 5). This approach is based on the maturity model that the Software Engineering Institute defined for the maturity of the software development capability.[5] Whatever the model, the scales should not be too granular, as that would render the system difficult to use and suggest a precision that is not justifiable.

In contrast, one should concentrate on maturity levels based on a set of conditions that can be unambiguously met. Against levels developed for each of COBIT's 34 IT processes, management can map:

- The current status of the organisation — where the organisation is today
- The current status of (best-in-class in) the industry — the comparison
- The current status of international standard guidelines — additional comparison
- The organisation's strategy for improvement — where the organisation wants to be

For each of the 34 IT processes, there is an incremental measurement scale, based on a rating of 0 through 5. The scale is associated with generic qualitative maturity model descriptions ranging from Non-existent to Optimised as follows:

- **0 Non-existent.** Complete lack of any recognisable processes. The organisation has not even recognised that there is an issue to be addressed.
- **1 Initial.** There is evidence that the organisation has recognised that the issues exist and need to be addressed. There are no standardised processes but instead there are ad hoc approaches that tend to be applied on an individual or case-by-case basis. The overall approach to management is disorganised.
- **2 Repeatable.** Processes have developed to the stage where similar procedures are followed by different people undertaking the same task. There is no formal training

or communication of standard procedures and responsibility is left to the individual. There is a high degree of reliance on the knowledge of individuals and therefore errors are likely.

- **3 Defined.** Procedures have been standardised and documented, and communicated through training. It is, however, left to the individual to follow these processes, and it is unlikely that deviations will be detected. The procedures themselves are not sophisticated but are the formalisation of existing practices.
- **4 Managed.** It is possible to monitor and measure compliance with procedures and to take action where processes appear not to be working effectively. Processes are under constant improvement and provide good practice. Automation and tools are used in a limited or fragmented way.
- **5 Optimised.** Processes have been refined to a level of best practice, based on the results of continuous improvement and maturity modelling with other organisations. IT is used in an integrated way to automate the workflow, providing tools to improve quality and effectiveness, making the enterprise quick to adapt.

The maturity model scales help professionals explain to managers where IT management shortcomings exist and set targets for where they need to be by comparing their organisation's control practices to the best practice examples. The right maturity level will be influenced by the enterprise's business objectives and operating environment. Specifically, the level of control maturity depends on the enterprise's dependence on IT, its technology sophistication and, most importantly, the value of its information.

A strategic reference point for an organisation to improve security and control could also consist of looking at emerging international standards and best-in-class practices. The emerging practices of today may become the expected level of performance of tomorrow and are therefore useful for planning where an organisation wants to be over time.

In summary, maturity models:

- Refer to business requirements and the enabling aspects at the different maturity levels
- Are a scale that lends itself to pragmatic comparison, where differences can be made measurable in an easy manner
- Help setting "as-is" and "to-be" positions relative to IT Governance, security and control maturity
- Lend themselves to gap analysis to determine what needs to be done to achieve a chosen level
- Avoid, where possible, discrete levels that create thresholds that are difficult to cross
- Increasingly apply critical success factors
- Are not industry-specific nor always applicable. The type of business defines what is appropriate.

Critical Success Factors

Critical success factors provide management with guidance for implementing control over IT and its processes. They are the most important things to do that contribute to the IT process achieving its goals. They are activities that can be of a strategic, technical, organisational, process or procedural nature. They are usually dealing with capabilities and skills and have to be short, focused and action-oriented, leveraging the resources that are of primary importance in the process under consideration.

A number of critical success factors can be deduced that apply to most IT processes:

Applying to IT in General

- IT processes are defined and aligned with the IT strategy and the business goals.
- The customers of the process and their expectations are known.
- Processes are scalable and their resources are appropriately managed and leveraged.
- The required quality of staff (training, transfer of information, morale, etc.) and availability of skills (recruit, retain, retrain) exist.
- IT performance is measured in financial terms, in relation to customer satisfaction, for process effectiveness and for future capability. IT management is rewarded based on these measures.
- A continuous quality improvement effort is applied.

Applying to Most IT Processes

- All process stakeholders (users, management, etc.) are aware of the risks, of the importance of IT and the opportunities it can offer, and provide strong commitment and support.
- Goals and objectives are communicated across all disciplines and understood; it is known how processes implement and monitor objectives, and who is accountable for process performance.
- People are goal-focused and have the right information on customers, on internal processes and on the consequences of their decisions.
- A business culture is established, encouraging cross-divisional co-operation, teamwork and continuous process improvement.
- There is integration and alignment of major processes, e.g., change, problem and configuration management.
- Control practices are applied to increase efficient and optimal use of resources and improve the effectiveness of processes.

Applying to IT Governance

- Control practices are applied to increase transparency, reduce complexity, promote learning, provide flexibility and scalability, and avoid breakdowns in internal control and oversight.
- Practices that enable sound oversight are applied: a control environment and culture; a code of conduct; risk assessment as a standard practice; self-assessments; formal compliance on adherence to established standards; monitoring and follow-up of control deficiencies and risk.

- IT Governance is recognised and defined, and its activities are integrated into the enterprise governance process, giving clear direction for IT strategy, a risk management framework, a system of controls and a security policy.
- IT Governance focuses on major IT projects, change initiatives and quality efforts, with awareness of major IT processes, the responsibilities and the required resources and capabilities.
- An audit committee is established to appoint and oversee an independent auditor, drive the IT audit plan and review the results of audits and third party opinions.

In summary, critical success factors are:
- Essential enablers focused on the process or supporting environment
- A thing or a condition that is required to increase the probability of success of the process
- Observable — usually measurable — characteristics of the organisation and process
- Either strategic, technological, organisational or procedural in nature
- Focused on obtaining, maintaining and leveraging capability and skills
- Expressed in terms of the process, not necessarily the business

Key Goal Indicators

A key goal indicator, representing the process goal, is a measure of *what* has to be accomplished. It is a measurable indicator of the process achieving its goals, often defined as a target to achieve. By comparison, a key performance indicator is a measure of *how well* the process is performing.

How are business and IT goals and measures linked? The COBIT Framework expresses the objectives for IT in terms of the information criteria that the business needs in order to achieve the business objectives, which will usually be expressed in terms of:
- Availability of systems and services
- Absence of integrity and confidentiality risks
- Cost-efficiency of processes and operations
- Confirmation of reliability, effectiveness and compliance

The goal for IT can then be expressed as delivering the information that the business needs in line with these criteria. These information criteria are provided in the Management Guidelines with an indication whether they have primary or secondary importance for the process under review. In practice, the information criteria profile of an enterprise would be more specific. The degree of importance of each of the information criteria is a function of the business and the environment in which the enterprise operates.

Key goal indicators are *lag* indicators, as they can be measured only after the fact, as opposed to key performance indicators, which are *lead* indicators, giving an indication of success before the fact. They also can be expressed negatively, i.e., in terms of the impact of not reaching the goal.

Key goal indicators should be measurable as a number or percentage. These measures should show that information and technology are contributing to the mission and strategy of the organisation. Because goals and targets are specific to the enterprise and its environment, many key goal indicators have been expressed with a direction, e.g., increased availability, decreased cost. In practice, management has to set specific targets which need to be met, taking into account past performance and future goals.

In summary, key goal indicators are:

- A representation of the process goal, i.e., a measure of *what*, or a target to achieve
- The description of the outcome of the process and therefore lag indicators, i.e., measurable after the fact
- Immediate indicators of the successful completion of the process or indirect indicators of the value the process delivered to the business
- Possibly descriptions of a measure of the impact of not reaching the process goal
- Focused on the customer and financial dimensions of the Balanced Business Scorecard
- IT-oriented but business-driven
- Expressed in precise, measurable terms wherever possible
- Focused on those information criteria that have been identified as most important for this process

Key Performance Indicators

Key performance indicators are measures that tell management that an IT process is achieving its business requirements by monitoring the performance of the enablers of that IT process. Building on Balanced Business Scorecard principles, the relationship between key performance indicators and key goal indicators is as follows: key performance indicators are short, focused and measurable indicators of performance of the enabling factors of the IT processes, indicating how well the process enables the goal to be reached. While key goal indicators focus on *what*, the key performance indicators are concerned with *how*. They often are a measure of a critical success factor and, when monitored and acted upon, identify opportunities for the improvement of the process. These improvements should positively influence the outcome and, as such, key performance indicators have a cause-effect relationship with the key goal indicators of the process.

While key goal indicators are business-driven, key performance indicators are process-oriented and often express how well the processes and the organisation leverage and manage the needed resources. Similar to key goal indicators, they often are expressed as a number or percentage. A good test of a key performance indicator is to see whether it really does predict success or failure of the process goal and whether or not it assists management in improving the process.

Some generic key performance indicators follow that usually are applicable to all IT processes:

Applying to IT in General

- Reduced cycle times (i.e., responsiveness of IT production and development)
- Increased quality and innovation
- Utilisation of communications bandwidth and computing power
- Service availability and response times

- Satisfaction of stakeholders (survey and number of complaints)
- Number of staff trained in new technology and customer service skills

Applying to most IT Processes
- Improved cost-efficiency of the process (cost vs. deliverables)
- Staff productivity (number of deliverables) and morale (survey)
- Amount of errors and rework

Applying to IT Governance
- Benchmark comparisons
- Number of non-compliance reportings

In summary, key performance indicators:
- Are measures of how well the process is performing
- Predict the probability of success or failure in the future, i.e., are lead indicators
- Are process-oriented, but IT-driven
- Focus on the process and learning dimensions of the Balanced Business Scorecard
- Are expressed in precisely measurable terms
- Help in improving the IT process when measured and acted upon
- Focus on those resources identified as the most important for this process

MANAGEMENT GUIDELINES FOR SELECTED COBIT PROCESSES

Although COBIT consists of 34 high-level IT control practices, through extensive testing and surveying, the 15 most important have been identified. At the end of the chapter, COBIT's Management Guideline for seven of these 15 processes is included, outlining critical success factors, key goal indicators, key performance indicators and a maturity model for each.

REFERENCES

Control Objectives for Information and related Technology (COBIT) 3rd Edition, IT Governance Institute, 1998, www.isaca.org/cobit.htm. (All sections of COBIT, except the Audit Guidelines, can be downloaded on a complimentary basis.)

Board Briefing on IT Governance, IT Governance Institute, 2001, www.ITgovernance.org/resources.htm.

Information Security Governance: Guidance for Boards of Directors and Executive Management, IT Governance Institute, 2001, www.ITgovernance.org/resources.htm.

ENDNOTES

[1]   In this document, "stakeholder" is used to indicate anyone who has either a responsibility for or an expectation from the enterprise's IT, e.g., shareholders, directors, executives, business and technology management, users, employees, governments, suppliers, customers and the public.

[2]   In this document, "board of directors" and "board" are used to indicate the body that is ultimately accountable to the stakeholders of the enterprise.

[3]   The COBIT control framework refers to key goal indicators (KGIs) and key performance indicators (KPIs) for the Balanced Business Scorecard concepts of outcome measures and performance drivers.

[4]   "The Balanced Business Scorecard — Measurements that Drive Performance," Robert S. Kaplan and David P. Norton, *Harvard Business Review*, January-February 1992.

[5]   "Capability Maturity Model SM for Software," Version 1.1. Technical Report CMU/SEI-93-TR-024, Software Engineering Institute, Carnegie Mellon University, Pittsburgh, PA, February 1993.

[*]   The information in this chapter is based primarily on *Control Objectives for Information and Related Technology (COBIT)*, published by the IT Governance Institute. © Control Objectives for Information and Related Technology (COBIT) 3rd Edition, IT Governance Institute, 2000. Reprinted by permission.

PO9 Planning & Organisation
Assess Risks

COBIT

HIGH-LEVEL CONTROL OBJECTIVE

P | S | P | P | P | S | S

Planning &
Organisation

Acquisition &
Implementation

Delivery &
Support

Monitoring

Control over the IT process of

assessing risks

that satisfies the business requirement

of supporting management decisions through achieving IT objectives
and responding to threats by reducing complexity, increasing
objectivity and identifying important decision factors

is enabled by

the organisation engaging itself in IT risk-identification and
impact analysis, involving multi-disciplinary functions and taking
cost-effective measures to mitigate risks

and takes into consideration

- risk management ownership and accountability
- different kinds of IT risks (technology, security, continuity,
  regulatory, etc.)
- defined and communicated risk tolerance profile
- root cause analyses and risk brainstorming sessions
- quantitative and/or qualitative risk measurement
- risk assessment methodology
- risk action plan
- timely reassessment

IT GOVERNANCE INSTITUTE

DETAILED CONTROL OBJECTIVES

9   ASSESS RISKS

9.1   Business Risk Assessment
CONTROL OBJECTIVE
Management should establish a systematic risk assessment framework. Such a framework should incorporate a regular assessment of the relevant information risks to the achievement of the business objectives, forming a basis for determining how the risks should be managed to an acceptable level. The process should provide for risk assessments at both the global level and system specific level, for new projects as well as on a recurring basis, and with cross-disciplinary participation. Management should ensure that reassessments occur and that risk assessment information is updated with results of audits, inspections and identified incidents.

9.2   Risk Assessment Approach
CONTROL OBJECTIVE
Management should establish a general risk assessment approach which defines the scope and boundaries, the methodology to be adopted for risk assessments, the responsibilities and the required skills. Management should lead the identification of the risk mitigation solution and be involved in identifying vulnerabilities. Security specialists should lead threat identification and IT specialists should drive the control selection. The quality of the risk assessments should be ensured by a structured method and skilled risk assessors.

9.3   Risk Identification
CONTROL OBJECTIVE
The risk assessment approach should focus on the examination of the essential elements of risk and the cause/effect relationship between them. The essential elements of risk include tangible and intangible assets, asset value, threats, vulnerabilities, safeguards, consequences and likelihood of threat. The risk identification process should include qualitative and, where appropri-

ate, quantitative risk ranking and should obtain input from management brainstorming, strategic planning, past audits and other assessments. The risk assessment should consider business, regulatory, legal, technology, trading partner and human resources risks.

9.4   Risk Measurement
CONTROL OBJECTIVE
The risk assessment approach should ensure that the analysis of risk identification information results in a quantitative and/or qualitative measurement of risk to which the examined area is exposed. The risk acceptance capacity of the organisation should also be assessed.

9.5   Risk Action Plan
CONTROL OBJECTIVE
The risk assessment approach should provide for the definition of a risk action plan to ensure that cost-effective controls and security measures mitigate exposure to risks on a continuing basis. The risk action plan should identify the risk strategy in terms of risk avoidance, mitigation or acceptance.

9.6   Risk Acceptance
CONTROL OBJECTIVE
The risk assessment approach should ensure the formal acceptance of the residual risk, depending on risk identification and measurement, organisational policy, uncertainty incorporated in the risk assessment approach itself and the cost effectiveness of implementing safeguards and controls. The residual risk should be offset with adequate insurance coverage, contractually negotiated liabilities and self-insurance.

continued on next page

IT GOVERNANCE INSTITUTE

PO9 Planning & Organisation
Assess Risks

COBIT

DETAILED CONTROL OBJECTIVES *continued*

9.7 **Safeguard Selection**
CONTROL OBJECTIVE
While aiming for a reasonable, appropriate and proportional system of controls and safeguards, controls with the highest return on investment (ROI) and those that provide quick wins should receive first priority. The control system also needs to balance prevention, detection, correction and recovery measures. Furthermore, management needs to communicate the purpose of the control measures, manage conflicting measures and monitor the continuing effectiveness of all control measures.

9.8 **Risk Assessment Commitment**
CONTROL OBJECTIVE
Management should encourage risk assessment as an important tool in providing information in the design and implementation of internal controls, in the definition of the IT strategic plan and in the monitoring and evaluation mechanisms.

DS5 Delivery & Support
Ensure Systems Security

HIGH-LEVEL CONTROL OBJECTIVE

Control over the IT process of

ensuring systems security

that satisfies the business requirement

to safeguard information against unauthorised use, disclosure or
modification, damage or loss

is enabled by

logical access controls which ensure that access to systems, data and
programmes is restricted to authorised users

and takes into consideration

- confidentiality and privacy requirements
- authorisation, authentication and access control
- user identification and authorisation profiles
- need-to-have and need-to-know
- cryptographic key management
- incident handling, reporting and follow-up
- virus prevention and detection
- firewalls
- centralised security administration
- user training
- tools for monitoring compliance,
  intrusion testing and reporting

IT GOVERNANCE INSTITUTE

CONTROL OBJECTIVES   DS5

DETAILED CONTROL OBJECTIVES

5   ENSURE SYSTEMS SECURITY

5.1   Manage Security Measures
CONTROL OBJECTIVE
IT security should be managed such that security measures are in line with business requirements. This includes:
- Translating risk assessment information to the IT security plans
- Implementing the IT security plan
- Updating the IT security plan to reflect changes in the IT configuration
- Assessing the impact of change requests on IT security
- Monitoring the implementation of the IT security plan
- Aligning IT security procedures to other policies and procedures

5.2   Identification, Authentication and Access
CONTROL OBJECTIVE
The logical access to and use of IT computing resources should be restricted by the implementation of adequate identification, authentication and authorisation mechanisms, linking users and resources with access rules. Such mechanisms should prevent unauthorised personnel, dial-up connections and other system (network) entry ports from accessing computer resources and minimise the need for authorised users to use multiple sign-ons. Procedures should also be in place to keep authentication and access mechanisms effective (e.g., regular password changes).

5.3   Security of Online Access to Data
CONTROL OBJECTIVE
In an online IT environment, IT management should implement procedures in line with the security policy that provides access security control based on the individual's demonstrated need to view, add, change or delete data.

5.4   User Account Management
CONTROL OBJECTIVE
Management should establish procedures to ensure timely action relating to requesting, establishing, issuing, suspending and closing of user accounts. A formal approval procedure outlining the data or system owner granting the access privileges should be included. The security of third-party access should be defined contractually and address administration and non-disclosure requirements. Outsourcing arrangements should address the risks, security controls and procedures for information systems and networks in the contract between the parties.

5.5   Management Review of User Accounts
CONTROL OBJECTIVE
Management should have a control process in place to review and confirm access rights periodically. Periodic comparison of resources with recorded accountability should be made to help reduce the risk of errors, fraud, misuse or unauthorised alteration.

5.6   User Control of User Accounts
CONTROL OBJECTIVE
Users should systematically control the activity of their proper account(s). Also information mechanisms should be in place to allow them to oversee normal activity as well as to be alerted to unusual activity in a timely manner.

5.7   Security Surveillance
CONTROL OBJECTIVE
IT security administration should ensure that security activity is logged and any indication of imminent security violation is reported immediately to all who may be concerned, internally and externally, and is acted upon in a timely manner.

continued on next page

IT GOVERNANCE INSTITUTE

DS5 Delivery & Support
Ensure Systems Security

COBIT

DETAILED CONTROL OBJECTIVES *continued*

5.8 Data Classification
CONTROL OBJECTIVE
Management should implement procedures to ensure that all data are classified in terms of sensitivity by a formal and explicit decision by the data owner according to the data classification scheme. Even data needing "no protection" should require a formal decision to be so designated. Owners should determine disposition and sharing of data, as well as whether and when programs and files are to be maintained, archived or deleted. Evidence of owner approval and data disposition should be maintained. Policies should be defined to support reclassification of information, based on changing sensitivities. The classification scheme should include criteria for managing exchanges of information between organisations, addressing both security and compliance with relevant legislation.

5.9 Central Identification and Access Rights Management
CONTROL OBJECTIVE
Controls are in place to ensure that the identification and access rights of users as well as the identity of system and data ownership are established and managed in a unique and central manner to obtain consistency and efficiency of global access control.

5.10 Violation and Security Activity Reports
CONTROL OBJECTIVE
IT security administration should ensure that violation and security activity is logged, reported, reviewed and appropriately escalated on a regular basis to identify and resolve incidents involving unauthorised activity. The logical access to the computer resources accountability information (security and other logs) should be granted based upon the principle of least privilege, or need-to-know.

5.11 Incident Handling
CONTROL OBJECTIVE
Management should establish a computer security incident handling capability to address security incidents by providing a centralised platform with sufficient expertise and equipped with rapid and secure communication facilities. Incident management responsibilities and procedures should be established to ensure an appropriate, effective and timely response to security incidents.

5.12 Reaccreditation
CONTROL OBJECTIVE
Management should ensure that reaccreditation of security (e.g., through "tiger teams") is periodically performed to keep up-to-date the formally approved security level and the acceptance of residual risk.

5.13 Counterparty Trust
CONTROL OBJECTIVE
Organisational policy should ensure that control practices are implemented to verify the authenticity of the counterparty providing electronic instructions or transactions. This can be implemented through trusted exchange of passwords, tokens or cryptographic keys.

5.14 Transaction Authorisation
CONTROL OBJECTIVE
Organisational policy should ensure that, where appropriate, controls are implemented to provide authenticity of transactions and establish the validity of a user's claimed identity to the system. This requires use of cryptographic techniques for signing and verifying transactions.

IT GOVERNANCE INSTITUTE

CONTROL OBJECTIVES   **DS5**

5.15 Non-Repudiation
Control Objective
Organisational policy should ensure that, where appropriate, transactions cannot be denied by either party, and controls are implemented to provide non-repudiation of origin or receipt, proof of submission, and receipt of transactions. This can be implemented through digital signatures, time stamping and trusted third-parties, with appropriate policies that take into account relevant regulatory requirements.

5.16 Trusted Path
Control Objective
Organisational policy should ensure that sensitive transaction data is only exchanged over a trusted path. Sensitive information includes security management information, sensitive transaction data, passwords and cryptographic keys. To achieve this, trusted channels may need to be established using encryption between users, between users and systems, and between systems.

5.17 Protection of Security Functions
Control Objective
All security related hardware and software should at all times be protected against tampering to maintain their integrity and against disclosure of secret keys. In addition, organisations should keep a low profile about their security design, but should not base their security on the design being secret.

5.18 Cryptographic Key Management
Control Objective
Management should define and implement procedures and protocols to be used for generation, change, revocation, destruction, distribution, certification, storage, entry, use and archiving of cryptographic keys to ensure the protection of keys against modification and unauthorised dis-

closure. If a key is compromised, management should ensure this information is propagated to any interested party through the use of Certificate Revocation Lists or similar mechanisms.

5.19 Malicious Software Prevention, Detection and Correction
Control Objective
Regarding malicious software, such as computer viruses or trojan horses, management should establish a framework of adequate preventative, detective and corrective control measures, and occurrence response and reporting. Business and IT management should ensure that procedures are established across the organisation to protect information systems and technology from computer viruses. Procedures should incorporate virus protection, detection, occurrence response and reporting.

5.20 Firewall Architectures and Connections with Public Networks
Control Objective
If connection to the Internet or other public networks exists, adequate firewalls should be operative to protect against denial of services and any unauthorised access to the internal resources; should control any application and infrastructure management flows in both directions; and should protect against denial of service attacks.

5.21 Protection of Electronic Value
Control Objective
Management should protect the continued integrity of all cards or similar physical mechanisms used for authentication or storage of financial or other sensitive information, taking into consideration the related facilities, devices, employees and validation methods used.

PO1 Planning & Organisation
Define a Strategic Information Technology Plan

 COBIT

Control over the IT process **Define a Strategic IT Plan** with the business goal of *striking an optimum balance of information technology opportunities and IT business requirements as well as ensuring its further accomplishment*

ensures delivery of information to the business that addresses the required Information Criteria and is measured by Key Goal Indicators

is enabled by *a strategic planning process undertaken at regular intervals giving rise to long-term plans; the long-term plans should periodically be translated into operational plans setting clear and concrete short-term goals*

considers Critical Success Factors that leverage specific IT Resources and is measured by Key Performance Indicators

Information Criteria
- P effectiveness
- S efficiency
- confidentiality
- integrity
- availability
- compliance
- reliability

(P) primary (S) secondary

IT Resources
- ✓ people
- ✓ applications
- ✓ technology
- ✓ facilities
- ✓ data

(✓) applicable to

Key Goal Indicators
- Percent of IT and business strategic plans that are aligned and cascaded into long- and short-range plans leading to individual responsibilities
- Percent of business units that have clear, understood and current IT capabilities
- Management survey determines clear link between responsibilities and the business and IT strategic goals
- Percent of business units using strategic technology covered in the IT strategic plan
- Percent of IT budget championed by business owners
- Acceptable and reasonable number of outstanding IT projects

Critical Success Factors
- The planning process provides for a prioritisation scheme for the business objectives and quantifies, where possible, the business requirements
- Management buy-in and support is enabled by a documented methodology for the IT strategy development, the support of validated data and a structured, transparent decision-making process
- The IT strategic plan clearly states a risk position, such as leading edge or road-tested, innovator or follower, and the required balance between time-to-market, cost of ownership and service quality
- All assumptions of the strategic plan have been challenged and tested
- The processes, services and functions needed for the outcome are defined, but are flexible and changeable, with a transparent change control process
- A reality check of the strategy by a third party has been conducted to increase objectivity and is repeated at appropriate times
- IT strategic planning is translated into roadmaps and migration strategies

Key Performance Indicators
- Currency of IT capabilities assessment (number of months since last update)
- Age of IT strategic plan (number of months since last update)
- Percent of participant satisfaction with the IT strategic planning process
- Time lag between change in the IT strategic plans and changes to operating plans
- Index of participants involved in strategic IT plan development, based on size of effort, ratio of involvement of business owners to IT staff and number of key participants
- Index of quality of the plan, including timelines of development effort, adherence to structured approach and completeness of plan

IT GOVERNANCE INSTITUTE

MANAGEMENT GUIDELINES   PO1

PO1 Maturity Model

Control over the IT process **Define a Strategic IT Plan** with the business goal of *striking an optimum balance of information technology opportunities and IT business requirements as well as ensuring its further accomplishment*

0   **Non-existent** IT strategic planning is not performed. There is no management awareness that IT strategic planning is needed to support business goals.

1   **Initial/Ad Hoc** The need for IT strategic planning is known by IT management, but there is no structured decision process in place. IT strategic planning is performed on an *as needed* basis in response to a specific business requirement and results are therefore sporadic and inconsistent. IT strategic planning is occasionally discussed at IT management meetings, but not at business management meetings. The alignment of business requirements, applications and technology takes place reactively, driven by vendor offerings, rather than by an organisation-wide strategy. The strategic risk position is identified informally on a project-by-project basis.

2   **Repeatable but Intuitive** IT strategic planning is understood by IT management, but is not documented. IT strategic planning is performed by IT management, but only shared with business management on an as needed basis. Updating of the IT strategic plan occurs only in response to requests by management and there is no proactive process for identifying those IT and business developments that require updates to the plan. Strategic decisions are driven on a project-by-project basis, without consistency with an overall organisation strategy. The risks and user benefits of major strategic decisions are being recognised, but their definition is intuitive.

3   **Defined Process** A policy defines when and how to perform IT strategic planning. IT strategic planning follows a structured approach, which is documented and known to all staff. The IT planning process is reasonably sound and ensures that appropriate planning is likely to be performed. However, discretion is given to individual managers with respect to implementation of the process and there are no procedures to examine the process on a regular basis. The overall IT strategy includes a consistent definition of risks that the organisation is willing to take as an innovator or follower. The IT financial, technical and human resources strategies increasingly drive the acquisition of new products and technologies.

4   **Managed and Measurable** IT strategic planning is standard practice and exceptions would be noticed by management. IT strategic planning is a defined management function with senior level responsibilities. With respect to the IT strategic planning process, management is able to monitor it, make informed decisions based on it and measure its effectiveness. Both short-range and long-range IT planning occurs and is cascaded down into the organisation, with updates done as needed. The IT strategy and organisation-wide strategy are increasingly becoming more coordinated by addressing business processes and value-added capabilities and by leveraging the use of applications and technologies through business process re-engineering. There is a well-defined process for balancing the internal and external resources required in system development and operations. Benchmarking against industry norms and competitors is becoming increasingly formalised.

5   **Optimised** IT strategic planning is a documented, living process, is continuously considered in business goal setting and results in discernable business value through investments in IT. Risk and value added considerations are continuously updated in the IT strategic planning process. There is an IT strategic planning function that is integral to the business planning function. Realistic long-range IT plans are developed and constantly being updated to reflect changing technology and business-related developments. Short-range IT plans contain project task milestones and deliverables, which are continuously monitored and updated, as changes occur. Benchmarking against well-understood and reliable industry norms is a well-defined process and is integrated with the strategy formulation process. The IT organisation identifies and leverages new technology developments to drive the creation of new business capabilities and improve the competitive advantage of the organisation.

PO9 Planning & Organisation
Assess Risks

COBIT

Control over the IT process **Assess Risks** with the business goal of *supporting management decisions in achieving IT objectives and responding to threats by reducing complexity, increasing objectivity and identifying important decision factors*

ensures delivery of information to the business that addresses the required Information Criteria and is measured by Key Goal Indicators

is enabled by *the organisation engaging itself in IT risk-identification and impact analysis, involving multi-disciplinary functions and taking cost-effective measures to mitigate risks*

considers Critical Success Factors that leverage specific IT Resources and is measured by Key Performance Indicators

Information Criteria	
P	effectiveness
S	efficiency
P	confidentiality
P	integrity
P	availability
S	compliance
S	reliability

(P) primary (S) secondary

IT Resources	
✓	people
✓	applications
✓	technology
✓	facilities
✓	data

(✓) applicable to

Key Goal Indicators
- Increased degree of awareness of the need for risk assessments
- Decreased number of incidents caused by risks identified after the fact
- Increased number of identified risks that have been sufficiently mitigated
- Increased number of IT processes that have formal documented risk assessments completed
- Appropriate percent or number of cost effective risk assessment measures

Critical Success Factors
- There are clearly defined roles and responsibilities for risk management ownership and management accountability
- A policy is established to define risk limits and risk tolerance
- The risk assessment is performed by matching vulnerabilities, threats and the value of data
- Structured risk information is maintained, fed by incident reporting
- Responsibilities and procedures for defining, agreeing on and funding risk management improvements exist
- Focus of the assessment is primarily on real threats and less on theoretical ones
- Brainstorming sessions and root cause analyses leading to risk identification and mitigation are routinely performed
- A reality check of the strategy is conducted by a third party to increase objectivity and is repeated at appropriate times

Key Performance Indicators
- Number of risk management meetings and workshops
- Number of risk management improvement projects
- Number of improvements to the risk assessment process
- Level of funding allocated to risk management projects
- Number and frequency of updates to published risk limits and policies
- Number and frequency of risk monitoring reports
- Number of personnel trained in risk management methodology

IT GOVERNANCE INSTITUTE

MANAGEMENT GUIDELINES   PO9

PO9 Maturity Model

Control over the IT process **Assess Risks** with the business goal of *supporting management decisions in achieving IT objectives and responding to threats by reducing complexity, increasing objectivity and identifying important decision factors*

0   **Non-existent** Risk assessment for processes and business decisions does not occur. The organisation does not consider the business impacts associated with security vulnerabilities and with development project uncertainties. Risk management has not been identified as relevant to acquiring IT solutions and delivering IT services.

1   **Initial/Ad Hoc** The organisation is aware of its legal and contractual responsibilities and liabilities, but considers IT risks in an ad hoc manner, without following defined processes or policies. Informal assessments of project risk take place as determined by each project. Risk assessments are not likely to be identified specifically within a project plan or to be assigned to specific managers involved in the project. IT management does not specify responsibility for risk management in job descriptions or other informal means. Specific IT-related risks such as security, availability and integrity are occasionally considered on a project-by-project basis. IT-related risks affecting day-to-day operations are infrequently discussed at management meetings. Where risks have been considered, mitigation is inconsistent.

2   **Repeatable but Intuitive** There is an emerging understanding that IT risks are important and need to be considered. Some approach to risk assessment exists, but the process is still immature and developing. The assessment is usually at a high-level and is typically applied only to major projects. The assessment of on-going operations depends mainly on IT managers raising it as an agenda item, which often only happens when problems occur. IT management has not generally defined procedures or job descriptions dealing with risk management.

3   **Defined Process** An organisation-wide risk management policy defines when and how to conduct risk assessments. Risk assessment follows a defined process that is documented and available to all staff through training. Decisions to follow the process and to receive training are left to the individual's discretion. The methodology is convincing and sound, and ensures that key risks to the business are likely to be identified. Decisions to follow the process are left to individual IT managers and there is no procedure to ensure that all projects are covered or that the ongoing operation is examined for risk on a regular basis.

4   **Managed and Measurable** The assessment of risk is a standard procedure and exceptions to following the procedure would be noticed by IT management. It is likely that IT risk management is a defined management function with senior level responsibility. The process is advanced and risk is assessed at the individual project level and also regularly with regard to the overall IT operation. Management is advised on changes in the IT environment which could significantly affect the risk scenarios, such as an increased threat from the network or technical trends that affect the soundness of the IT strategy. Management is able to monitor the risk position and make informed decisions regarding the exposure it is willing to accept. Senior management and IT management have determined the levels of risk that the organisation will tolerate and have standard measures for risk/return ratios. Management budgets for operational risk management projects to reassess risks on a regular basis. A risk management database is established.

5   **Optimised** Risk assessment has developed to the stage where a structured, organisation-wide process is enforced, followed regularly and well managed. Risk brainstorming and root cause analysis, involving expert individuals, are applied across the entire organisation. The capturing, analysis and reporting of risk management data are highly automated. Guidance is drawn from leaders in the field and the IT organisation takes part in peer groups to exchange experiences. Risk management is truly integrated into all business and IT operations, is well accepted and extensively involves the users of IT services.

PO10 Planning & Organisation
Manage Projects

CobiT

Control over the IT process **Manage Projects** with the business goal of *setting priorities and delivering on time and within budget*

ensures delivery of information to the business that addresses the required Information Criteria and is measured by Key Goal Indicators

is enabled by *the organisation identifying and prioritising projects in line with the operational plan and the adoption and application of sound project management techniques for each project undertaken*

considers Critical Success Factors that leverage specific IT Resources and is measured by Key Performance Indicators

Information Criteria	
P effectiveness	
P efficiency	
confidentiality	
integrity	
availability	
compliance	
reliability	

(P) primary (S) secondary

IT Resources	
✓ people	
✓ applications	
✓ technology	
✓ facilities	
data	

(✓) applicable to

Key Goal Indicators

- Increased number of projects completed on time and on budget
- Availability of accurate project schedule and budget information
- Decrease in systemic and common project problems
- Improved timeliness of project risk identification
- Increased organisation satisfaction with project delivered services
- Improved timeliness of project management decisions

Critical Success Factors

- Experienced and skilled project managers are available
- An accepted and standard programme management process is in place
- There is senior management sponsorship of projects, and stakeholders and IT staff share in the definition, implementation and management of projects
- There is an understanding of the abilities and limitations of the organisation and the IT function in managing large, complex projects
- An organisation-wide project risk assessment methodology is defined and enforced
- All projects have a plan with clear traceable work breakdown structures, reasonably accurate estimates, skill requirements, issues to track, a quality plan and a transparent change process
- The transition from the implementation team to the operational team is a well-managed process
- A system development life cycle methodology has been defined and is used by the organisation

Key Performance Indicators

- Increased number of projects delivered in accordance with a defined methodology
- Percent of stakeholder participation in projects (involvement index)
- Number of project management training days per project team member
- Number of project milestone and budget reviews
- Percent of projects with post-project reviews
- Average number of years of experience of project managers

IT GOVERNANCE INSTITUTE

PO10 Planning & Organisation
Manage Projects

COBIT

Control over the IT process **Manage Projects** with the business goal of *setting priorities and delivering on time and within budget*

Information Criteria		IT Resources
P effectiveness		✓ people
P efficiency		✓ applications
confidentiality		✓ technology
integrity		✓ facilities
availability		data
compliance		
reliability		

(P) primary   (S) secondary   (✓) applicable to

ensures delivery of information to the business that addresses the required Information Criteria and is measured by Key Goal Indicators

is enabled by *the organisation identifying and prioritising projects in line with the operational plan and the adoption and application of sound project management techniques for each project undertaken*

considers Critical Success Factors that leverage specific IT Resources and is measured by Key Performance Indicators

Key Goal Indicators

- Increased number of projects completed on time and on budget
- Availability of accurate project schedule and budget information
- Decrease in systemic and common project problems
- Improved timeliness of project risk identification
- Increased organisation satisfaction with project delivered services
- Improved timeliness of project management decisions

Critical Success Factors

- Experienced and skilled project managers are available
- An accepted and standard programme management process is in place
- There is senior management sponsorship of projects, and stakeholders and IT staff share in the definition, implementation and management of projects
- There is an understanding of the abilities and limitations of the organisation and the IT function in managing large, complex projects
- An organisation-wide project risk assessment methodology is defined and enforced
- All projects have a plan with clear traceable work breakdown structures, reasonably accurate estimates, skill requirements, issues to track, a quality plan and a transparent change process
- The transition from the implementation team to the operational team is a well-managed process
- A system development life cycle methodology has been defined and is used by the organisation

Key Performance Indicators

- Increased number of projects delivered in accordance with a defined methodology
- Percent of stakeholder participation in projects (involvement index)
- Number of project management training days per project team member
- Number of project milestone and budget reviews
- Percent of projects with post-project reviews
- Average number of years of experience of project managers

IT GOVERNANCE INSTITUTE

AI6 Acquisition & Implementation
Manage Changes

COBIT

Control over the IT process **Manage Changes** with the business goal of *minimising the likelihood of disruption, unauthorised alterations and errors*

ensures delivery of information to the business that addresses the required Information Criteria and is measured by Key Goal Indicators

is enabled by *a management system which provides for the analysis, implementation and follow-up of all changes requested and made to the existing IT infrastructure*

considers Critical Success Factors that leverage specific IT Resources and is measured by Key Performance Indicators

Information Criteria	
P	effectiveness
P	efficiency
	confidentiality
P	integrity
P	availability
	compliance
S	reliability

(P) primary (S) secondary

IT Resources	
✓	people
✓	applications
✓	technology
✓	facilities
✓	data

(✓) applicable to

Key Goal Indicators

- Reduced number of errors introduced into systems due to changes
- Reduced number of disruptions (loss of availability) caused by poorly managed change
- Reduced impact of disruptions caused by change
- Reduced level of resources and time required as a ratio to number of changes
- Number of emergency fixes

Critical Success Factors

- Change policies are clear and known and they are rigorously and systematically implemented
- Change management is strongly integrated with release management and is an integral part of configuration management
- There is a rapid and efficient planning, approval and initiation process covering identification, categorisation, impact assessment and prioritisation of changes
- Automated process tools are available to support workflow definition, pro-forma workplans, approval templates, testing, configuration and distribution
- Expedient and comprehensive acceptance test procedures are applied prior to making the change
- A system for tracking and following individual changes, as well as change process parameters, is in place
- A formal process for hand-over from development to operations is defined
- Changes take the impact on capacity and performance requirements into account
- Complete and up-to-date application and configuration documentation is available
- A process is in place to manage co-ordination between changes, recognising interdependencies
- An independent process for verification of the success or failure of change is implemented
- There is segregation of duties between development and production

Key Performance Indicators

- Number of different versions installed at the same time
- Number of software release and distribution methods per platform
- Number of deviations from the standard configuration
- Number of emergency fixes for which the normal change management process was not applied retroactively
- Time lag between the availability of the fix and its implementation
- Ratio of accepted to refused change implementation requests

IT GOVERNANCE INSTITUTE

MANAGEMENT GUIDELINES   AI6

AI6 Maturity Model
Control over the IT process **Manage Changes** with the business goal of *minimising the likelihood of disruption, unauthorised alterations and errors*

0   Non-existent There is no defined change management process and changes can be made with virtually no control. There is no awareness that change can be disruptive for both IT and business operations, and no awareness of the benefits of good change management.

1   Initial/Ad Hoc It is recognised that changes should be managed and controlled, but there is no consistent process to follow. Practices vary and it is likely that unauthorised changes will take place. There is poor or non-existent documentation of change and configuration documentation is incomplete and unreliable. Errors are likely to occur together with interruptions to the production environment caused by poor change management.

2   Repeatable but Intuitive There is an informal change management process in place and most changes follow this approach; however, it is unstructured, rudimentary and prone to error. Configuration documentation accuracy is inconsistent and only limited planning and impact assessment takes place prior to a change. There is considerable inefficiency and rework.

3   Defined Process There is a defined formal change management process in place, including categorisation, prioritisation, emergency procedures, change authorisation and release management, but compliance is not enforced. The defined process is not always seen as suitable or practical and, as a result, workarounds take place and processes are bypassed. Errors are likely to occur and unauthorised changes will occasionally occur. The analysis of the impact of IT changes on business operations is becoming formalised, to support planned rollouts of new applications and technologies.

4   Managed and Measurable The change management process is well developed and consistently followed for all changes and management is confident that there are no exceptions. The process is efficient and effective, but relies on considerable manual procedures and controls to ensure that quality is achieved. All changes are subject to thorough planning and impact assessment to minimise the likelihood of post-production problems. An approval process for changes is in place. Change management documentation is current and correct, with changes formally tracked. Configuration documentation is generally accurate. IT change management planning and implementation is becoming more integrated with changes in the business processes, to ensure that training, organisational changes and business continuity issues are addressed. There is increased co-ordination between IT change management and business process re-design.

5   Optimised The change management process is regularly reviewed and updated to keep in line with best practices. Configuration information is computer based and provides version control. Software distribution is automated and remote monitoring capabilities are available. Configuration and release management and tracking of changes is sophisticated and includes tools to detect unauthorised and unlicensed software. IT change management is integrated with business change management to ensure that IT is an enabler in increasing productivity and creating new business opportunities for the organisation.

DS5 Delivery & Support
Ensure Systems Security

COBIT

Control over the IT process **Ensure Systems Security** with the business goal of *safeguarding information against unauthorised use, disclosure or modification, damage or loss*

ensures delivery of information to the business that addresses the required Information Criteria and is measured by Key Goal Indicators

is enabled by *logical access controls which ensure that access to the systems, data and programmes is restricted to authorised users*

considers Critical Success Factors that leverage specific IT Resources and is measured by Key Performance Indicators

Information Criteria	
	effectiveness
	efficiency
P	confidentiality
P	integrity
S	availability
S	compliance
S	reliability

(P) primary (S) secondary

IT Resources	
✓	people
✓	applications
✓	technology
✓	facilities
✓	data

(✓) applicable to

Key Goal Indicators

- No incidents causing public embarrassment
- Immediate reporting on critical incidents
- Alignment of access rights with organisational responsibilities
- Reduced number of new implementations delayed by security concerns
- Full compliance, or agreed and recorded deviations from minimum security requirements
- Reduced number of incidents involving unauthorised access, loss or corruption of information

Critical Success Factors

- An overall security plan is developed that covers the building of awareness, establishes clear policies and standards, identifies a cost-effective and sustainable implementation, and defines monitoring and enforcement processes
- There is awareness that a good security plan takes time to evolve
- The corporate security function reports to senior management and is responsible for executing the security plan
- Management and staff have a common understanding of security requirements, vulnerabilities and threats, and they understand and accept their own security responsibilities
- Third-party evaluation of security policy and architecture is conducted periodically
- A "building permit" programme is defined, identifying security baselines that have to be adhered to
- A "drivers licence" programme is in place for those developing, implementing and using systems, enforcing security certification of staff
- The security function has the means and ability to detect, record, analyse significance, report and act upon security incidents when they do occur, while minimising the probability of occurrence by applying intrusion testing and active monitoring
- A centralised user management process and system provides the means to identify and assign authorisations to users in a standard and efficient manner
- A process is in place to authenticate users at reasonable cost, light to implement and easy to use

Key Performance Indicators

- Reduced number of security-related service calls, change requests and fixes
- Amount of downtime caused by security incidents
- Reduced turnaround time for security administration requests
- Number of systems subject to an intrusion detection process
- Number of systems with active monitoring capabilities
- Reduced time to investigate security incidents
- Time lag between detection, reporting and acting upon security incidents
- Number of IT security awareness training days

IT GOVERNANCE INSTITUTE

MANAGEMENT GUIDELINES   DS5

DS5 Maturity Model

Control over the IT process **Ensure Systems Security** with the business goal of *safeguarding information against unauthorised use, disclosure or modification, damage or loss*

0   Non-existent The organisation does not recognise the need for IT security. Responsibilities and accountabilities are not assigned for ensuring security. Measures supporting the management of IT security are not implemented. There is no IT security reporting and no response process to IT security breaches. There is a complete lack of a recognisable system security administration process.

1   Initial/Ad Hoc The organisation recognises the need for IT security, but security awareness depends on the individual. IT security is addressed on a reactive basis and not measured. IT security breaches invoke "finger pointing" responses if detected, because responsibilities are unclear. Responses to IT security breaches are unpredictable.

2   Repeatable but Intuitive Responsibilities and accountabilities for IT security are assigned to an IT security co-ordinator with no management authority. Security awareness is fragmented and limited. IT security information is generated, but is not analysed. Security solutions tend to respond reactively to IT security incidents and by adopting third-party offerings, without addressing the specific needs of the organisation. Security policies are being developed, but inadequate skills and tools are still being used. IT security reporting is incomplete, misleading or not pertinent.

3   Defined Process Security awareness exists and is promoted by management. Security awareness briefings have been standardised and formalised. IT security procedures are defined and fit into a structure for security policies and procedures. Responsibilities for IT security are assigned, but not consistently enforced. An IT security plan exists, driving risk analysis and security solutions. IT security reporting is IT focused, rather than business focused. Ad hoc intrusion testing is performed.

4   Managed and Measurable Responsibilities for IT security are clearly assigned, managed and enforced. IT security risk and impact analysis is consistently performed. Security policies and practices are completed with specific security baselines. Security awareness briefings have become mandatory. User identification, authentication and authorisation are being standardised. Security certification of staff is being established. Intrusion testing is a standard and formalised process leading to improvements. Cost/benefit analysis, supporting the implementation of security measures, is increasingly being utilised. IT security processes are co-ordinated with the overall organisation security function. IT security reporting is linked to business objectives.

5   Optimised IT security is a joint responsibility of business and IT management and is integrated with corporate security business objectives. IT security requirements are clearly defined, optimised and included in a verified security plan. Security functions are integrated with applications at the design stage and end users are increasingly accountable for managing security. IT security reporting provides early warning of changing and emerging risk, using automated active monitoring approaches for critical systems. Incidents are promptly addressed with formalised incident response procedures supported by automated tools. Periodic security assessments evaluate the effectiveness of implementation of the security plan. Information on new threats and vulnerabilities is systematically collected and analysed, and adequate mitigating controls are promptly communicated and implemented. Intrusion testing, root cause analysis of security incidents and pro-active identification of risk is the basis for continuous improvements. Security processes and technologies are integrated organisation wide.

DS10 Delivery & Support
Manage Problems and Incidents

COBIT

Control over the IT process **Manage Problems and Incidents** with the business goal of *ensuring that problems and incidents are resolved, and the cause investigated to prevent any recurrence*

ensures delivery of information to the business that addresses the required Information Criteria and is measured by Key Goal Indicators

is enabled by *a problem management system which records and progresses all incidents*

considers Critical Success Factors that leverage specific IT Resources and is measured by Key Performance Indicators

Information Criteria	IT Resources
P effectiveness	✓ people
P efficiency	✓ applications
confidentiality	✓ technology
integrity	✓ facilities
S availability	✓ data
compliance	
reliability	

(P) primary (S) secondary   (✓) applicable to

Key Goal Indicators

- A measured reduction of the impact of problems and incidents on IT resources
- A measured reduction in the elapsed time from initial symptom report to problem resolution
- A measured reduction in unresolved problems and incidents
- A measured increase in the number of problems avoided through pre-emptive fixes
- Reduced time lag between identification and escalation of high-risk problems and incidents

Critical Success Factors

- There is clear integration of problem management with availability and change management
- Accessibility to configuration data, as well as the ability to keep track of problems for each configuration component, is provided
- An accurate means of communicating problem incidents, symptoms, diagnosis and solutions to the proper support personnel is in place
- Accurate means exist to communicate to users and IT the exceptional events and symptoms that need to be reported to problem management
- Training is provided to support personnel in problem resolution techniques
- Up-to-date roles and responsibilities charts are available to support incident management
- There is vendor involvement during problem investigation and resolution
- Post-facto analysis of problem handling procedures is applied

Key Performance Indicators

- Elapsed time from initial symptom recognition to entry in the problem management system
- Elapsed time between problem recording and resolution or escalation
- Elapsed time between evaluation and application of vendor patches
- Percent of reported problems with already known resolution approaches
- Frequency of coordination meetings with change management and availability management personnel
- Frequency of component problem analysis reporting
- Reduced number of problems not controlled through formal problem management

IT GOVERNANCE INSTITUTE

MANAGEMENT GUIDELINES   DS10

DS10 Maturity Model
Control over the IT process **Manage Problems and Incidents** with the business goal of *ensuring that problems and incidents are resolved, and the cause investigated to prevent any recurrence*

0   **Non-existent** There is no awareness of the need for managing problems and incidents. The problem-solving process is informal and users and IT staff deal individually with problems on a case-by-case basis.

1   **Initial/Ad Hoc** The organisation has recognised that there is a need to solve problems and evaluate incidents. Key knowledgeable individuals provide some assistance with problems relating to their area of expertise and responsibility. The information is not shared with others and solutions vary from one support person to another, resulting in additional problem creation and loss of productive time, while searching for answers. Management frequently changes the focus and direction of the operations and technical support staff.

2   **Repeatable but Intuitive** There is a wide awareness of the need to manage IT related problems and incidents within both the business units and information services function. The resolution process has evolved to a point where a few key individuals are responsible for managing the problems and incidents occurring. Information is shared among staff; however, the process remains unstructured, informal and mostly reactive. The service level to the user community varies and is hampered by insufficient structured knowledge available to the problem solvers. Management reporting of incidents and analysis of problem creation is limited and informal.

3   **Defined Process** The need for an effective problem management system is accepted and evidenced by budgets for the staffing, training and support of response teams. Problem solving, escalation and resolution processes have been standardised, but are not sophisticated. Nonetheless, users have received clear communications on where and how to report on problems and incidents. The recording and tracking of problems and their resolutions is fragmented within the

response team, using the available tools without centralisation or analysis. Deviations from established norms or standards are likely to go undetected.

4   **Managed and Measurable** The problem management process is understood at all levels within the organisation. Responsibilities and ownership are clear and established. Methods and procedures are documented, communicated and measured for effectiveness. The majority of problems and incidents are identified, recorded, reported and analysed for continuous improvement and are reported to stakeholders. Knowledge and expertise are cultivated, maintained and developed to higher levels as the function is viewed as an asset and major contributor to the achievement of IT objectives. The incident response capability is tested periodically. Problem and incident management is well integrated with interrelated processes, such as change, availability and configuration management, and assists customers in managing data, facilities and operations.

5   **Optimised** The problem management process has evolved into a forward-looking and proactive one, contributing to the IT objectives. Problems are anticipated and may even be prevented. Knowledge is maintained, through regular contacts with vendors and experts, regarding patterns of past and future problems and incidents. The recording, reporting and analysis of problems and resolutions is automated and fully integrated with configuration data management. Most systems have been equipped with automatic detection and warning mechanism, which are continuously tracked and evaluated.

IT GOVERNANCE INSTITUTE

DS11 Delivery & Support
Manage Data

COBIT

Control over the IT process **Manage Data** with the business goal of *ensuring that data remains complete, accurate and valid during its input, update and storage*

ensures delivery of information to the business that addresses the required Information Criteria and is measured by Key Goal Indicators

is enabled by *an effective combination of application and general controls over the IT operations*

considers Critical Success Factors that leverage specific IT Resources and is measured by Key Performance Indicators

Information Criteria	IT Resources
effectiveness	people
efficiency	applications
confidentiality	technology
P integrity	facilities
availability	✓ data
compliance	
P reliability	

(P) primary (S) secondary   (✓) applicable to

Key Goal Indicators

- A measured reduction in the data preparation process and tasks
- A measured improvement in the quality, timeline and availability of data
- A measured increase in customer satisfaction and reliance upon the data
- A measured decrease in corrective activities and exposure to data corruption
- Reduced number of data defects, such as redundancy, duplication and inconsistency
- No legal or regulatory data compliance conflicts

Critical Success Factors

- Data entry requirements are clearly stated, enforced and supported by automated techniques at all levels, including database and file interfaces
- The responsibilities for data ownership and integrity requirements are clearly stated and accepted throughout the organisation
- Data accuracy and standards are clearly communicated and incorporated into the training and personnel development processes
- Data entry standards and correction are enforced at the point of entry
- Data input, processing and output integrity standards are formalised and enforced
- Data is held in suspense until corrected
- Effective detection methods are used to enforce data accuracy and integrity standards
- Effective translation of data across platforms is implemented without loss of integrity or reliability to meet changing business demands
- There is a decreased reliance on manual data input and re-keying processes
- Efficient and flexible solutions promote effective use of data
- Data is archived and protected and is readily available when needed for recovery

Key Performance Indicators

- Percent of data input errors
- Percent of updates reprocessed
- Percent of automated data integrity checks incorporated into the applications
- Percent of errors prevented at the point of entry
- Number of automated data integrity checks run independently of the applications
- Time interval between error occurrence, detection and correction
- Reduced data output problems
- Reduced time for recovery of archived data

IT GOVERNANCE INSTITUTE

MANAGEMENT GUIDELINES   DS11

DS11 Maturity Model

Control over the IT process **Manage Data** with the business goal of *ensuring that data remains complete, accurate and valid during its input, update and storage*

0   **Non-existent** Data is not recognised as a corporate resource and asset. There is no assigned data ownership or individual accountability for data integrity and reliability. Data quality and security is poor or non-existent.

1   **Initial/Ad Hoc** The organisation recognises a need for accurate data. Some methods are developed at the individual level to prevent and detect data input, processing and output errors. The process of error identification and correction is dependent upon manual activities of individuals, and rules and requirements are not passed on as staff movement and turnover occur. Management assumes that data is accurate because a computer is involved in the process. Data integrity and security are not management requirements and, if security exists, it is administered by the information services function.

2   **Repeatable but Intuitive** The awareness of the need for data accuracy and maintaining integrity is prevalent throughout the organisation. Data ownership begins to occur, but at a department or group level. The rules and requirements are documented by key individuals and are not consistent across the organisation and platforms. Data is in the custody of the information services function and the rules and definitions are driven by the IT requirements. Data security and integrity are primarily the information services function's responsibilities, with minor departmental involvement.

3   **Defined Process** The need for data integrity within and across the organisation is understood and accepted. Data input, processing and output standards have been formalised and are enforced. The process of error identification and correction is automated. Data ownership is assigned, and integrity and security are controlled by the responsible party. Automated techniques are utilised to prevent and detect errors and inconsistencies. Data definitions, rules and requirements

are clearly documented and maintained by a database administration function. Data becomes consistent across platforms and throughout the organisation. The information services function takes on a custodian role, while data integrity control shifts to the data owner. Management relies on reports and analyses for decisions and future planning.

4   **Managed and Measurable** Data is defined as a corporate resource and asset, as management demands more decision support and profitability reporting. The responsibility for data quality is clearly defined, assigned and communicated within the organisation. Standardised methods are documented, maintained and used to control data quality, rules are enforced and data is consistent across platforms and business units. Data quality is measured and customer satisfaction with information is monitored. Management reporting takes on a strategic value in assessing customers, trends and product evaluations. Integrity of data becomes a significant factor, with data security recognised as a control requirement. A formal, organisation-wide data administration function has been established, with the resources and authority to enforce data standardisation.

5   **Optimised** Data management is a mature, integrated and cross-functional process that has a clearly defined and well-understood goal of delivering quality information to the user, with clearly defined integrity, availability and reliability criteria. The organisation actively manages data, information and knowledge as corporate resources and assets, with the objective of maximising business value. The corporate culture stresses the importance of high quality data that needs to be protected and treated as a key component of intellectual capital. The ownership of data is a strategic responsibility with all requirements, rules, regulations and considerations clearly documented, maintained and communicated.

IT GOVERNANCE INSTITUTE

Chapter XII

Governance in IT Outsourcing Partnerships[1]

Erik Beulen
Tilburg University, The Netherlands

ABSTRACT

An IT outsourcing partnership consists of an outsourcing relationship and one or more external IT suppliers and the relationship between them. Alignment of mutually set goals of the IT outsourcing relationship is a prerequisite to achieve governance. In order to achieve governance the management of IT outsourcing partnerships is also essential. Managing an IT outsourcing relationship requires substantial effort from both the outsourcing organisation and the IT supplier.

This chapter is based on 11 international IT outsourcing partnerships, five expert interviews and on literature. Three dimensions are described in a descriptive IT outsourcing partnership governance framework: outsourcing organization, the maintenance of the relationship, and the IT supplier. Eleven governance factors are detailed in the framework. These governance factors include guidelines for the implementation of the IT strategy and the information management. Furthermore, this chapter focuses on the IT outsourcing contract. The role of the contract management and account management of the IT suppliers and the implementation of global service delivery processes is also detailed in this chapter.

INTRODUCTION

Just as it is necessary to align the business strategy with the IT strategy (Rockart & Morton, 1984; Henderson & Venkatraman, 1992; Brown & Magioll, 1994), it is also necessary to align the IT strategy with the sourcing strategy (Quinn & Hilmer, 1995). Only a sound sourcing strategy will result in proper IT Governance (Dreyfuss, 2002) and solid IT outsourcing partnerships (Lacity & Hirschheim, 1993; Willcocks & Fitzgerald, 1994; Earl, 1996).

Organizations basically have two options for sourcing IT services: they can either provide the IT service themselves through their own internal IT division — insourcing — or use external IT suppliers — outsourcing. These choices are based on a "make or buy" decision whose essence is described by Coase (1937) and Williamson (1975).

The focus will be on the governance of IT outsourcing partnerships, not insourcing, and covers the governance of IT outsourcing partnerships from the perspective of both outsourcing organization and IT supplier. It aims to provide a better understanding of governing IT outsourcing partnerships by proposing a descriptive framework with governance factors.

The definitions of the most important concepts related to the governance of IT outsourcing partnerships and the positioning of IT outsourcing partnerships are detailed in first section. The second section sets out the research framework used. The third section explains the management of IT outsourcing partnership issues. A descriptive framework for the management of the IT outsourcing partnership is described in the fourth section and is further developed in the fifth through seventh sections. Future trends are developed in the eighth section. The final section contains the conclusions.

DEFINITIONS

The literature first devoted attention to IT outsourcing partnerships in 1990 (Gantz, 1990; Rochester & Douglas, 1990). An IT outsourcing partnership consists of an outsourcing organization and one or more external IT suppliers and the relationship between them. This definition is based on the work of Lacity and Hirschheim (1993), Willcocks and Chio (1995a) and Currie (1998). The IT services outsourcing market is still growing every year. Trend analysts such as Morgan Chambers, IDC and Gartner predict annual growth figures of approximately 10% (Morgan & Chambers, 2001; Lamy, 2001; Cox, 2002a). This growth is also confirmed by publications such as the OutsourcingProject (2002). In addition, market analyses carried out by Gartner indicate that more and more organizations are outsourcing IT services related to their primary business processes. Furthermore, outsourcing organizations are outsourcing complete business processes (Brown & Scholl, 2002). This increases the impact of the IT outsourcing partnership for outsourcing organizations and is further discussed. All of these factors make it essential that sufficient attention be devoted to the governance of IT outsourcing partnerships. Governance will be defined.

To position the governance of IT outsourcing partnerships, it is essential to properly understand the advantages and disadvantages of outsourcing and to further develop the elements of an IT outsourcing partnership.

Governance

Governance of IT outsourcing partnerships is, amongst others, defined by Lacity and Hirschheim (1993), Willcocks and Fitzgerald (1994), Corbett (1994) and Klepper (1995, 1998). Based on their definitions the governance of IT outsourcing partnership results in realizing the mutually set goals of the IT outsourcing relationship. These goals may be contrary, such as the cost saving goal of the outsourcing organisation versus the return on investment goals of the IT supplier. The alignment of the goals is a precondition for governance (Beulen, 2000). To achieve alignment governance the contract between the outsourcing organisation and the IT supplier is essential. In order to achieve governance the management of IT outsourcing partnerships is also essential. Management of IT outsourcing is defined as the activities that the outsourcing organisation and the IT supplier take to achieve governance. This definition is based on the work of McFarlan and Nolan (1995), Cox (1999), Heckmann (1999) and Kern and Willcocks (2002). Managing an IT outsourcing requires substantial effort from both the outsourcing organisation and the IT supplier. The management attention for the outsourcing organisation will be detailed. The management attention for the IT supplier will also be detailed.

Outsourcing Benefits and Disadvantages

Outsourcing decisions are strongly situation-dependent (Lacity & Hirschheim, 1993; Beulen, Ribbers, & Roos, 1994; de Looff, 1996) and this section consequently summarizes the general advantages and disadvantages of outsourcing. Much was published on this subject during the nineties. This resulted in the perception that outsourcing is only cheaper due to the fact that IT suppliers are able to generate economies of scale and that outsourcing creates a large dependency on the IT supplier. The following summary illustrates this perception.

Outsourcing Benefits:

- Access to expertise and the deployment of new technologies (KPMG Impact, 1995; International Data Corporation, 1998): rapid technological developments require a significant portion of the human resources capacity of internal IT divisions and require high investments in the training of IT professionals. An IT supplier whose core business consists of the delivery of IT services is able to keep the level of knowledge of its IT professionals up to date more effectively and efficiently.

- Increase in the level of flexibility (Lacity & Hirschheim, 1993; Klepper, 1995): due to the fact that an IT supplier has several customers, the IT supplier is better able to absorb the peaks and valleys in the demand for IT services than the internal IT division, which generally only provides services to its parent organization.

- Decrease in costs (Apte, 1990; Willcocks & Fitzgerald, 1994): due to their scale and ability to share production resources, IT suppliers are able to provide more efficient and effective IT services.

- Increase the predictability of costs (Lacity, Willcocks, & Feeny, 1995b; Currie & Willcocks, 1998): outsourcing contracts are generally multi-year contracts. This increases the predictability of costs for the outsourcing organization. This is an important advantage, particularly for investors.

- The generation of cash flows (Earl, 1996; Willcocks & Lester, 1997): through the sale of assets — hardware and immovable property — the outsourcing organization is able to generate a one-time cash flow by outsourcing its IT services.

Outsourcing Disadvantages:

- Management of IT supplier(s) (KPMG Impact, 1995; Feeny, 1997a): the management of IT suppliers requires the attention of the management of the outsourcing organization and this carries its own costs. Furthermore, many organizations have difficulty finding qualified managers to assume this role.
- Confidentiality (Willcocks & Fitzgerald, 1994; Klepper & Jones, 1998): outsourcing arrangements cause the outsourcing organization's confidential data to be accessible to the IT supplier's employees. This constitutes a risk that must be considered when the decision to outsource is taken.
- Dependency on the IT supplier(s) (Terdiman, 1993; Lacity & Hirschheim, 1993): by entering into a multi-year contract, outsourcing organizations become dependent on their IT suppliers, particularly when there are changes in IT services required by the outsourcing organization.

Outsourcing Organization

Outsourcing organizations use one of three organization structures[2] (Duncan, 1979; Daft, 1998): the functional organization structure — denoted by Burns and Stalker as mechanistic organizations (1961); the divisional organization structure — the organization structure here follows the organization's business strategy, which leads to the creation of autonomous business units within the enterprise (Chandler, 1962; Lawrence & Lorsch, 1967; Mintzberg, 1979); and the matrix organization structure (David & Laurance, 1977).

The position of the Chief Information Officer (CIO) and the information managers within the organization structure of the outsourcing organization are related to the organization structure of the outsourcing organization (Dickson, Leitheiser, & Wetherbe, 1984; Niederman, Brancheau, & Wetherbe, 1991), see Table 1, and influences the manner in which the governance of the IT outsourcing partnership is set up (Earl & Feeny, 1997a; Beulen, 2000). It is essential for outsourcing organizations to staff the CIO and information manager positions in order to be able to implement proper governance of the IT outsourcing partnership (Earl, 1996; Lacity, Willcocks, & Feeny, 1996). The roles of the CIO and the information managers are further described.

The IT Supplier

The IT supplier is responsible for the delivery of the IT services on the basis of a contractual agreement. A distinction is made here between the internal IT supplier — the internal IT division — and external IT suppliers. Cases where the internal IT division provides the IT services are not considered outsourcing arrangements: these are referred to as insourcing.

External IT suppliers may be subdivided into specialized companies and full-service suppliers (Pinnington & Woolcock, 1997; Dataquest, 1998). Full service suppliers are

Table 1.   The Position of the Chief Information Officer (CIO) and the Information Managers Within the Organization Structure of the Outsourcing Organization (Earl, Edwards, & Feeny, 1997b)

Functional Organization Structure: The CIO and the information managers occupy a centralized position within the organization. The information managers functionally report to the business unit managers of the outsourcing organization and, hierarchically, to the CIO. Earl refers to this as the Corporate Service IS function (Earl, Edwards, & Feeny, 1997b).

Divisional Organization Structure: The CIO occupies a centralized position within the organization, and the information managers have a decentralized position and are part of the business units. The information managers therefore report hierarchically to the business unit managers and functionally to the CIO. Earl refers to this as the Decentralized IS function (Earl, Edwards, & Feeny, 1997b).

Matrix Organization Structure: The CIO occupies a centralized position within the organization and a number of the information managers occupy a centralized position within the organization. These information managers functionally report to the business unit managers of the outsourcing organization and hierarchically to the CIO. The remaining information managers occupy a decentralized position within the organization. They report hierarchically to the business unit managers and functionally to the CIO. Earl refers to this as the Federal IS function (Earl, Edwards, & Feeny, 1997b).

---

further subdivided into first and second tier suppliers. Second tier suppliers provide more limited geographic coverage than first tier suppliers, which are able to provide a consistent worldwide portfolio of IT services (Dataquest, 1998). Specialized companies may be further subdivided into companies that provide specific IT services and companies that operate in vertical markets.

In selecting an IT supplier, the outsourcing organization must choose an IT supplier with a profile that fits the requested IT services (Willcocks & Fitzgerald, 1994; KPMG Impact, 1995; Lacity & Willcocks, 2001). It is important here to select an IT supplier that is comparable to the outsourcing organization in relation to its relative size. If the IT supplier is relatively far larger than the outsourcing organization the attention may not be as appropriate as necessary over the contract period. In the case that the IT supplier is relatively smaller than the outsourcing organization there might be a chance that the outsourcing company cannot benefit from economies of scale. Furthermore, IT suppliers may find difficulties in implementing innovations and flexibility.

There are four variants that may be used by an outsourcing organization to have its external IT suppliers provide the required IT services. A choice must be made in this regard between outsourcing the entire IT service and partial outsourcing. Currie and Willcocks (1998) refer to this as "total outsourcing" and "selective outsourcing" respectively. If an organization opts for selective sourcing, a portion of the IT services will be provided by the internal IT division. The IT literature raises many questions in relation to total outsourcing due to its large dependency on IT suppliers. This includes Lacity, Willcocks and Feeny (1995b) and Huber (1993). In addition, the outsourcing organization must make a choice between outsourcing to a single vendor or to multiple vendors. Currie and Willcocks (1998) refer to this as "single sourcing" and "multiple outsourcing" respectively. "Multiple outsourcing" requires the outsourcing organization to provide greater coordination between IT suppliers and consequently generally leads to higher quality IT services at more competitive prices (Cox, 2002b).

The choice of "single sourcing" or "multiple sourcing" in combination with the choice of "total outsourcing" and "selective outsourcing" yields the following four variations:

1.   Selective Single Outsourcing: the outsourcing organization only outsources a single component to a single external IT supplier. The other components are provided by the internal IT division.
2.   Selective Multiple Outsourcing: the outsourcing organization outsources a portion of its IT services to multiple external IT suppliers in combination with portions that are insourced to the internal IT division.
3.   Total Single Outsourcing: the outsourcing organization has outsourced all of its IT services to a single external IT supplier.
4.   Total Multiple Outsourcing: the outsourcing organization has outsourced all of its IT services to multiple external IT suppliers.

The Relationship

The relationship consists of the contractual link between the outsourcing organization and the IT suppliers for the delivery of the IT services.

The contracts in all cases include a framework agreement (Willcocks & Fitzgerald, 1994; KMPG Impact, 1995). The framework agreement includes the Service Level Agreements. The contractual relationship can essentially be set up in one of two ways. By linking the method of grouping the SLAs to the geographical location and the business units of the outsourcing organization it is possible to attain a much higher degree of contract effectiveness. This is due to the fact that this approach makes it possible to align the IT services included in the contract with the local information needs of the business units of the outsourcing organization (Beulen, 2000). The grouping of SLAs by the type of IT service makes it possible to build greater efficiency into the contracts. This is so because this approach makes it possible for the IT supplier to standardize its IT services, which in turn enables the IT supplier to realize economies of scale, including the inherent cost savings (Lacity & Hirschheim, 1993).

There are three types of IT outsourcing: Information Systems Outsourcing, Processing Outsourcing and Business Process Outsourcing (International Data Corporation, 1998)[3] — see Table 2. Information Systems Outsourcing is the most commonly occurring form of outsourcing (Cox, 2002b). This chapter will briefly delve into the subject of Business Process Outsourcing.

The most important difference here in relation to Information Systems Outsourcing and Processing Outsourcing is that in the case of Business Process Outsourcing the primary emphasis is on the performance of the outsourced process, rather than on the performance of the information systems.

RESEARCH FRAMEWORK

This analysis is based on the literature, supplemented with 25 interviews with business and IT executives in outsourcing organizations, executives of IT suppliers[4]

Table 2. Types of IT Services (International Data Corporation, 1998)

Information Systems Outsourcing: A long-term contract, including facilities management, in which the IT supplier assumes responsibility, or part of the responsibility, for providing the IT services and where there is a possibility that the IT supplier takes over the property, or parts of the property, of the internal IT division, and also takes over its personnel.

Processing Outsourcing: IT services for specific processing functions that include a high degree of standardization. The IT supplier carries responsibility here for executing a process that includes IT and non-IT related elements. There is no transfer of personnel or transfer of property to the IT supplier.

Business Process Outsourcing: A specific set of activities and knowledge that is required by an IT supplier for carrying out the activities of a division, process or function of the outsourcing organization. The execution of these activities in support of an IT system or application forms part of the services provided whereby the IT supplier also carries responsibility for non-IT related activities.

The most important difference here in relation to Information Systems Outsourcing and Processing Outsourcing is that in the case of Business Process Outsourcing the primary emphasis is on the performance of the outsourced process rather than on the performance of the information systems.

---

related to 11 international outsourcing situations, see Table 3, and interviews with five experts with broad outsourcing experience as consultants and/or academics, see Table 4. The scope, geographic coverage and the nature of the IT services covered vary significantly. In view of the limited number of cases that were analyzed, it is not possible to draw specific conclusions about the impact of the characteristics of the case studies on the governance factors. This research only provides a general image of the governance factors related to IT outsourcing partnerships. Case studies and quotes from the interviews conducted have been included in order to provide insight into the issue of the governance of IT outsourcing partnerships. In addition to the literature and professional judgments, these case studies and quotes form a supporting structure for the descriptive framework for the governance of IT outsourcing partnerships.

The nature of this type of research is explorative. Case studies can be used for so-called analytical generalization, not statistical generalization (Yin, 1989). The case study method has been used because it enables "reality" to be captured in considerably greater detail than other methods, and it also allows for the analysis of a considerable greater number of variables.

The same research protocol was followed in all interviews conducted for the case studies. Furthermore, all interviews were fully transcribed and submitted to the interviewees for approval. All interviews started with open questions related to the governance of the IT outsourcing partnership. In the second part of the interview, all governance factors of the framework were verified and clarified by the interviewees. For each case study at least two interviews, one with a representative of the outsourcing organization and one with a representative of the IT supplier, were conducted, except for case studies seven and 10. This enabled us to crosscheck the opinions of the interviewees. The interviews with the experts completed the investigation. This enabled us to check

Table 3. IT Outsourcing Partnerships Analyzed

Firm	Sector	# of Employees[5]	Region(s)	Total Contract Value in US $	Contract Start Date	Contract Duration (Years)
1[6]	Discrete Manufacturing	6,200	Europe	30 million	1992 (already renewed)	5
2[7]	Utilities	1,800	Europe	23 million	1996	5
3[8]	Discrete Manufacturing	200,000	Asia	21 million	1998	5
4[9]	Services	200,000	Europe	40 million	1997	5
5[10]	Process Industry	68,000	Europe/Asia/ North America	Yearly revenues 90 million (750 employees)	1999 onwards	Purchase of internal IT division
6	Process Industry	68,000	Asia	0.4 million	1999	3
7[11]	Tele-communications	100,000	Europe/ Middle East/Asia	Confidential (> 20 million)	1997	5
8	Media	8,000	Europe	4 million	1995 (already renewed)	5
9	Discrete Manufacturing	2,000	Asia	1 million	2000	2
10[12]	Utilities	2,000	Europe	100 million	2000	5
11	Discrete Manufacturing	200,000	Europe/Asia/ Americas	Yearly revenues 550 million (initially 1,500 employees)	1990 onwards	Purchase of internal IT division (various contracts)

for possible influences associated with the fact that all outsourcing organizations in the case studies investigated used the same IT supplier. The experts had access to a large number of complex IT outsourcing partnerships involving other IT suppliers. This helped us to avoid drawing conclusions based only on the best practices of one IT supplier. In addition to the interviews, supporting documentation for the case studies analyzed was collected, including annual reports, organization charts and research reports on IT outsourcing.

ISSUES

As indicated in the first section, the outsourcing market is still growing. It is therefore essential for both the outsourcing organization as well as the IT supplier that the governance issue is dealt with. Yet not all organizations that have outsourced their IT services are satisfied (Lacity & Hirschheim, 1995a).

Expert 5: "Over the past 10 years I have advised many organizations on outsourcing and I guided eight organizations in the selection of an IT supplier and the implementation of the outsourcing contract. Let me say this carefully: not all outsourcing partnerships resulted in the realization of the high expectations. The expectations for most contracts were only partially met by IT suppliers."

The reasons why outsourcing does not always result in success are legion. "Successful" here is defined as the realization of prior set objectives by both the outsourcing organization as well as the IT supplier. This definition fits with the definition of Governance as mentioned earlier. This definition of "Successful" is also consistent with the research into outsourcing conducted respectively by the Nolan Norton Institute, Morgan Chambers and Gartner (van der Zee, 1997; Morgan & Chambers, 2001; Ackerman, 2002). On the basis of both the literature as well as actual practice, three

Table 4. Profiles of Experts Interviewed

No.	Analysts/Opinion Leaders	Business Profile
1.	Director	International research institute of leading consultancy firm
2.	European Expertise Center Manager	International market research institute
3.	Analyst	International market research institute
4.	Research Consultant	International market research institute
5.	Consultant	Leading international consultancy firm

important factors that influence the success of an IT outsourcing partnership have been identified — see Table 5: maturity of the outsourcing organization (Klepper & Hartog, 1991; Willcocks & Fitzgerald, 1994; McFarland & Nolan, 1995), the degree of flexibility inherent in the contracts (Lacity & Hirschheim, 1993; Klepper, 1995; Burnett, 1998) and the ability of the IT suppliers to integrate the IT services taken over into their own organization (Terdiman, 1991; Feeny & Willcocks, 1997c; Lacity & Willcocks, 2001). This obviously does not constitute a comprehensive summary of the influencing factors, but it does provide an overview of the governance issue.

The descriptive framework includes some guidelines for setting up the IT outsourcing partnership such that it will become a successful partnership for both the outsourcing organization as well as the IT supplier(s).

FRAMEWORK

In line with the definition of IT outsourcing partnerships, the descriptive framework for the governance of IT outsourcing partnerships contains three dimensions. The framework includes governance factors for the outsourcing organization, the maintenance of the relationship, and the IT suppliers.

The outsourcing organizations and the IT suppliers may use the governance factors to provide substance to the IT outsourcing partnership. The governance factors do not constitute a comprehensive inventory of the essential measures that should be adopted, but simply constitute specific points of importance that can be implemented by outsourcing organizations and IT suppliers.

Table 6 contains a summary of the governance factors. These are further explained. The governance factors are based on the case studies analyzed, the interviews with the experts, professional judgment and a study of the literature.

GOVERNANCE FACTORS RELATED TO THE OUTSOURCING ORGANIZATION

There are four management factors related to the outsourcing organization: attention to IT within business units, a clear IT strategy, Information Management as the link between business functions and IT suppliers and a properly functioning Chief Informa-

Table 5. Issues that Affect the Success of IT Outsourcing Partnerships

Maturity (Klepper & Hartog, 1991; Willcocks & Fitzgerald, 1994; McFarland & Nolan, 1995): Outsourcing organizations with prior outsourcing experience are better able to implement the governance structure required for IT outsourcing partnerships (Klepper & Hartog, 1991; McFarland & Nolan, 1995). However, most organizations have limited experience with outsourcing (Willcocks & Fitzgerald, 1994), particularly outsourcing relationships that involve the management of multiple IT outsourcing suppliers (Currie & Willcocks, 1998). An important recommendation put forth by Gartner is consequently as follows: "develop the required specific management capabilities" for managing the contract (Dreyfuss, 2002).

Chief Information Officer, Firm 5: "..... <My company> had already outsourced their network services before signing the purchasing agreement <the investigated outsourcing contract>. This experience contributed to defining the governance structure of the outsourcing relationship and resulted in reaching fairly quick agreement and contract signature ....."

Flexibility (Lacity & Hirschheim, 1993; Klepper, 1995; Burnett, 1998): The sixth lesson learned from the study of Lacity and Hirschheim is as follows: "If an organization decides to outsource, the contract is the only mechanism to ensure that expectations are realized" (Lacity & Hirschheim, 1993). Most outsourcing contracts are of long duration, however, and are not structured to allow changes in information needs to be easily operationalized (Klepper, 1995). This requires procedures for dealing with changes that lead to situations that are not covered by the contract (Gietzmann, 1996). Burnett (1998) uses the Liaison Model to describe this aspect. This procedure is used as a basis for formulating agreements between both parties that complement the signed IT outsourcing contract. Hart (1995) and Segal (1999) use the concept of ex-post negotiations to describe this aspect.

Business Manager, Firm 1: "...particularly flexibility in terms of volume. We are, of course, a one-product company. It is easy to see that we are tremendously affected by the cyclical character of the economy ... ...This means that you have to be extremely flexible in relation to the marketplace in terms of variations in volume."

Integration (Terdiman, 1991; Feeny & Willcocks, 1997c; Lacity & Willcocks, 2001): The IT suppliers must integrate the IT services, including the IT professionals taken over, into their own organization. Objectives can only be realized through integration (Feeny & Willcocks, 1997c; Lacity & Willcocks, 2001). Many IT suppliers have difficulty with this integration, however (Terdiman, 1991). Even today this still requires much attention from IT supplier management (Beulen & Ribbers, 2002a). In addition to management attention (Young & Cournoyer, 2000), initial investments constitute an important issue related to integration (Outsourcing Transition Management, 1996) and result in an increase in costs.

Expert 2: "...a few years ago <1997> I was involved in some research conducted by my American colleagues. One of the important issues that came to light was the difficulty experienced by IT suppliers in integrating services into their own organization. Fortunately, today we observe that many IT suppliers are dealing much more effectively with this issue."

Table 6. Governance Factors by Dimension

Dimensions	Governance Factors
The outsourcing organization	1.1 Attention to IT within business units
	1.2 A clear IT strategy
	1.3 Information management as the link between business units and IT suppliers
	1.4 A properly functioning Chief Information Officer
The maintenance of the relationship	2.1 Mutual trust between the outsourcing organization and the IT supplier
	2.2 Experience in establishing and maintaining IT outsourcing relationships
	2.3 Efficient and effective IT outsourcing contracts
	2.4 An audit & benchmark process in place
The IT suppliers	3.1 Adequate contract and account management
	3.2 Adequate service delivery processes
	3.3 The availability of human resources to IT suppliers

tion Officer (CIO). These factors are further explained in this section and in Appendix A — Measuring the governance factors for IT outsourcing partnership.

Attention to IT within Business Units

The outsourcing objectives are not always equally clear to everyone involved within the outsourcing organization (KPMG Impact, 1995). This doesn't enable achieving governance in IT outsourcing partnerships. It is therefore the task of the CIO, as well as senior management, to communicate these internally. The outsourcing objectives of outsourcing organizations are often focused on cost savings (Morgan & Chambers, 2001; Outsourcingproject, 2002). This is not necessarily in line with the information needs of the business units of the outsourcing organization.

Expert 5: "...there needs to be awareness within the organization <the outsourcing organization>, of what the objectives are, and probably not only by the CIO, but also by corporate management."

Building on this, business units should no longer judge IT on the basis of costs, but on the basis of its added value. A shift must take place from the minimization of costs to the maximization of business impact (Kotwica & Field, 1999). Lacity and Willcocks (2001) support this transformation, albeit with a somewhat different nuance. This transformation is related to the changing role of IT. The role of IT is shifting from the support of non-primary processes to IT that forms part of the products or services that are provided by the outsourcing organization, such as in-product software. Another example is provided by Willcocks and Plant (2001): UPS has a track and trace Internet application which allows clients to track their shipments during transport. When business management pays proper attention to IT, the organization is able to anticipate needs in a timely fashion. In view of the ever-decreasing time-to-market, it is not an excessive luxury to involve IT proactively in the development of new products and services. This line of thought fits into the observations made by Gerrity and Rockart, who already in the mid-80s concluded that line managers feel themselves to be increasingly responsible for IT (Gerrity & Rockart, 1986). It also fits in with the suggestion made by Earl that IT can be used "to develop new business" (Earl, 1987).

Business Manager, Firm 5: "We specifically involve our information managers in our product creation process. This helps us to prevent IT from becoming a bottleneck when we are ready to introduce new products."

The management of the outsourcing relationship is the responsibility of the CIO and IM. This concept is further developed. However, it is important for business management to devote proper attention to the outsourcing relationship (Willcocks & Fitzgerald, 1994; Klepper & Jones, 1998). Attention to the outsourcing relationship on the part of business management has a positive influence on the management of outsourcing relationships (Quinn & Hilmer, 1995). Gartner supports this position and adds: [business management] "must actively manage their side of the partnership" (Terdiman, 1991). To retain business management's attention, the type of reporting about the IT services delivered is very important and must be consistent with the type of reporting units that business management is familiar with (Katz & Katz, 1966; Feeny, Willcocks, & Core, 1998).

A Clear IT Strategy

IT strategy is defined as the strategy of the outsourcing organization in relation to its information technology and IT services, and the role these play, or will play, within the outsourcing organization. This is consistent with the definitions put forth by Zani (1970), McLean and Soden (1977) and King (1978). The development and implementation of the IT strategy is the responsibility of the Chief Information Officer (CIO). An IT strategy is essential because organizations are able to implement new technologies within their organization on the basis of this strategy, and will subsequently be able to derive strategic benefits from this (Porter & Millar, 1985; Earl, 1987). From a governance perspective it is important to create alignment between the business and IT. The IT strategy can be a facilitating factor in accomplishing this (Henderson & Venkatraman, 1993). Furthermore, organizations are well advised to incorporate their sourcing strategy into the IT strategy: how does the organization intend to deliver the required IT services in support of business operations (Currie & Willcocks, 1998)? This issue requires continuous attention from the CIO (Grigg & Block, 2002).

The development and implementation of an IT strategy is not easy. Through the development of different scenarios it is possible to reduce uncertainty (Emery & Trist, 1965; van der Heijden, 1996). Scenarios can also be used for the development of an IT strategy (Rosser, 1998; van der Zee & van Wijngaarden, 1999). Among others, Gartner and IDC also work with scenarios.

It is important to emphasize that responsibility for the development of the IT strategy lies with the outsourcing organization itself (KPMG Impact, 1995). The capacity for executing this responsibility is denoted as IT/IS leadership by Feeny and Willcocks (1997c). Gartner also views the transfer of responsibility for developing an IT strategy to someone else as a high risk: "...retain control of, or be able to influence, strategic technology issues and directions" (Terdiman, 1991). It is possible, however, to involve external consultants in the formulation of an IT strategy. The use of external consultants may be considered as a means of supplementing a shortage in required capacity and for hiring specific knowledge that the organization is not able to develop and/or maintain itself.

Business Director, Firm 1: "You do the following <when developing an IT strategy>, you throw out some ad hoc questions and you ask for some research to be conducted. In some instances a project manager from the consultancy firm may be involved. You use this person to test your ideas. However, the situation must not be such that the external consultant jointly determines your IT strategy."

Information Management as the Link between Business Units and IT Suppliers

The information manager is responsible for the alignment of the demand for IT services by the outsourcing organization's business units with the services provided by the IT suppliers (Quinn, Doorley, & Paquette, 1990). The information manager is furthermore responsible for the IT outsourcing partnership and supports the Chief Information Office (CIO) in the implementation of the IT strategy (Corbett, 1994; McFarland & Nolan, 1995). The effort required on the part of outsourcing organizations to manage the IT outsourcing partnership is substantial. The cost of the effort expended by the information management function is between 2% and 10% of the contract value (Aylott, 2002).

In order to be able to properly carry out these tasks, information managers require knowledge of the business operations, as well as IT (Willcocks & Fitzgerald, 1994; Kitzis, 1998). Many outsourcing organizations have difficulty finding qualified candidates to carry out these functions (Heckmann, 1999).

Expert 3: Three months ago <2000> I was doing...a presentation at a conference...and resourcing the information management office was one of the major discussions that took place. There were a lot of vendors <IT suppliers> and major user organizations <outsourcing organizations>. What most of the user organizations came up with, they were saying it's so hard.... To find people who have experience or expertise with this supplier management model, that are able to manage multiple suppliers in particular."

It is important to ensure there exists a strong information management function that can provide a counterbalance to business unit management. Lacity and Hirschheim (1995a) refer to this as "senior management must empower IS <meaning information management> to implement changes."

A Properly Functioning Chief Information Officer

The Chief Information Officer (CIO) is responsible for the development and implementation of the IT strategy and carries final responsibility for the IT outsourcing partnership relationships (Earl & Feeny, 1997a; Kotwica & Fields, 1999; Lacity & Willcocks, 2001).

In order to be able to provide proper direction to their responsibilities, which includes safeguarding the governance of the IT outsourcing partnership, it is important that in addition to their knowledge of information technologies, both the CIO and the information managers also understand the business operations and developments in the markets in which the outsourcing organization is active or intends to become active (Willcocks & Fitzgerald, 1994; de Looff, 1996; Kitzis, 1998). A CIO could also play an important role in the alignment of business operations with IT. Many organizations set up an IT Board for this purpose. Gartner refers to these boards as coordinating committees: the business-IT strategy committee (Dreyfuss, 2002). This is an organization-wide steering committee in which all business units and information management is represented and which is chaired by the CIO. The IT Board is able to explore the political field of influence and decisions concerning the IT services to be provided can be prepared and discussed here. All case studies analyzed included the use of some kind of IT Board.

Account Manager IT Supplier, Firm 2: "My customer has an IT Board that functions as kind of an awareness club for different business functions, including purchasing, sales and transportation. Due to the fact that all divisions are represented, there exists strong support for preparing decisions about the IT services to be provided."

To ensure that the CIO is able to function properly, it is also essential that the CIO function be positioned at the proper level within the organization. Outsourcing organizations are ill advised to place the CIO on the Board of Directors. Responsibility for providing IT services is an integral responsibility of the entire Board of Directors. The CIO should therefore report to one of the members of the Board of Directors (Earl & Feeny, 2000). A trend is being observed in this regard which shows that an increasing number of CIOs are directly reporting to the Chief Executive Officer instead of to the Chief Financial Officer or to the Chief Operating Officer (Kotwica & Fields, 1998).

Expert 1: "Anyone who'd like to play the role of an entrepreneur, and this should include all members of the Board of Directors, must simply possess understanding and knowledge <about the IT function> and maybe then you can simply turn over the service aspects to the CIO. But the understanding of technology, what that <technology> can do for the business, I think that this will become an integral part of entrepreneurship."

GOVERNANCE FACTORS FOR THE MAINTENANCE OF THE RELATIONSHIP

There are four factors that govern the maintenance of the relationship: mutual trust between the outsourcing organization and the IT supplier; experience in establishing and maintaining IT outsourcing relationships; efficient and effective outsourcing contracts; and an audit and benchmark process in place. These governance factors are described in further detail in this section and in Appendix A — Measuring the governance factors for IT outsourcing partnership.

Mutual Trust between the Outsourcing Organization and the IT Supplier

Trust is a particularly important criterion for the selection process used by outsourcing organizations (Apte, 1990; Lacity & Hirschheim, 1993; Willcocks & Fitzgerald, 1994; Earl, 1996).

Account Representative IT Supplier, Firm 8: "We obtained the contract at the time <1995> because they had confidence in the management of our organization. This helps us in maintaining that relationship to this very day. Trust is indispensable to an outsourcing relationship."

Country Manager IT Supplier, Firm 6: "Our customer has chosen us because of our track record and cultural fit, especially here in X <Asia>, this is really important."

But trust is also necessary for maintaining a relationship. Lacity and Hirschheim (1995a) state that mutual trust can be increased by exchanging strategies with the objective of aligning mutual strategies: complementary or shared goals. This is not easy in actual practice. There does not always exist equality, a true partnership, between the outsourcing organization and the IT supplier. This issue is not specific to the IT services industry. Kraljic (1983) has been pleading for the creation of partnerships since the 80s: "Purchasing must become Supplier Management". In the case of IT services there often exists a client-supplier relationship (Kanter, 1994). A client-supplier relationship is referred to by Gartner as buying technology. When this joint relationship evolves into a partnership, the relationship is referred to as buying services (Cox, 2002b).

Furthermore the culture factor is also important for creating trust (Hofstede, 1980). Grönroos states that the culture of a service provider is focused on creating service-oriented attitudes and proactive behaviors (Grönroos, 1990). This is consistent with the opinions of Aylott, an outsourcing specialist with Orbys (Aylott, 2002): "...a culture that is proactive, collaborative and supportive." It is not essential for the cultures of the outsourcing organization and the IT supplier to be identical. It is more important that they not be in conflict (Eggleton & Otter, 1991) — cultural convergence (Kern & Willcocks, 2000), and this becomes particularly important when a global IT outsourcing partnership is involved (Beulen & Ribbers, 2002b). In addition, it is important that the outsourcing organization and the IT supplier are both clearly aware of the differences in their respective cultures.

Expert 5: "I recently attended a meeting of the Board of Directors of a large multinational. They were in the process of negotiating the outsourcing of their worldwide infrastructure with four IT suppliers. To provide you with an indication of their size, their yearly IT budget is approximately US$200 million. The CFO had a strong preference for one of the IT suppliers because its culture was a good fit with the culture of his own organization. By way of background to this outsourcing initiative, there was a desire on the part of the outsourcing organization for things to change dramatically. During the discussions I suggested that the organization avoid selecting an IT supplier with a similar culture. Otherwise you'll never be able to make a break with established patterns."

Experience Establishing and Maintaining IT Outsourcing Relationships

Experience is an important selection criterion for choosing an IT supplier (Klepper & Hartog, 1991; McFarland & Nolan, 1995). By providing references, IT suppliers are able to provide the outsourcing organization with an overview of their track record. In addition, mature IT suppliers use a standard methodology for setting up and managing IT outsourcing partnerships in order to achieve governance. Examples include Hewlett-Packard (1999) and Atos Origin (Outsourcing Transition Management, 1996). Research conducted by Gartner indicates that there are no significant differences between these methodologies (Young & Cournoyer, 2000). However, outsourcing organizations should use a standardized process for the implementation of outsourcing (Chaderton & van de Wittenboer, 2002). Aside from the experiences of the IT supplier, it is also important for the outsourcing organization to build up experience in outsourcing. Lacity and Hirschheim (1995a) state that the lack of experience with outsourcing is an important argument for deciding not to outsource: "customers' inexperience with IS outsourcing". It is not essential that the experience with outsourcing involves the outsourcing of IT services. Experience with the outsourcing of other business processes, such as document reproduction, the treasury or logistics services also contributes to the experience with outsourcing (Beulen, 2000).

The method used for reporting about the IT services provided is particularly important in maintaining sound outsourcing relationships (Willcocks & Fitzgerald, 1994; KPMG Impact, 1995; Lacity & Willcocks, 2001). The reports provided by the IT supplier must be consistent with the reporting units used by the outsourcing organization. This makes it possible for the outsourcing organization to provide direction to the delivery of IT services (Terdiman, 1993; Klepper & Jones, 1998). In addition, the frequency of reporting and the review milestones and reporting levels linked to it are also important for maintaining a sound outsourcing relationship. The choices made in this regard depend on the impact of the outsourced IT services on business operations. The greater the impact, the more frequent the reporting and the higher the level within the organization where reporting should take place (Terdiman, 1991; KPMG Impact, 1995).

Purchasing Manager, Firm 5: "We outsourced portions of our IT services to multiple IT suppliers. Reporting is of vital importance to us for keeping things under control. Every month we receive an elaborate overview of all delivered services and the service levels attained for the major portion of the IT services provided. This reporting is based on exception reporting. We receive a weekly update for some business-critical IT services. This particularly concerns itself with the progress of projects.... Supplementary agreements are formulated and other measures are discussed with the IT suppliers on the basis of this reporting."

Efficient and Effective Outsourcing Contracts

Many authors, such as Lacity and Hirschheim (1993), and Kern and Willcocks (2002) emphasize the importance of a contract — "...the centrality of the contract." Aside from a precise description of the IT services to be provided and the agreed upon service levels, it is important for contracts to be flexible (Klepper & Jones, 1998; Kitzis, 1998). Inflexible contracts carry additional costs for the outsourcing organization (Cox, 1999). One way of achieving flexibility is to manage IT as a portfolio and to enter into contract with multiple suppliers: "...avoid being locked into a single supplier" (KPMG Impact, 1995). This also requires additional attention from information management to manage the relationship (Cross, 1995). It furthermore carries additional risks. The responsibilities of the different IT suppliers must be clear (Lacity, Willcocks, & Feeny, 1995b).

Description of contract clauses, Firm 7: "The contracts do not include a guarantee of exclusivity '...neither grants to suppliers an exclusive right of privilege to sell to X any or all service of the type described in this framework agreement which X may require, nor requires the purchase of other materials or services from the supplier by X.' and no guarantee of purchase '...shall neither restrict the right of X to cease purchasing nor require X to continue any level of such purchase.'"

Also individual IT suppliers in outsourcing partnerships can make use of subcontractors. This impacts the governance of IT outsourcing partnerships (Willcocks & Choi, 1995a). Outsourcing organizations have to ensure that their IT suppliers are able to be held responsible for the contracted services, including the services that are provided by their subcontractors (Burnett, 1998). Also involving subcontractors carries additional risks (de Looff, 1996). In order to limit these risks the involved subcontracts have to be agreed with the outsourcing organization.

Account Manager, Firm 11: "In order to provide the services we make use of a limited number of subcontracts. Sometimes is it necessary to make use of subcontracts that are not on the list. But all these subcontractors are prior to providing the services approved by the outsourcing organization. But my company is anyway responsible of providing the services to our customer."

In addition to completeness, which can be addressed by creating a procedure for situations that are not covered by the contract, the termination of a contract constitutes an important area requiring attention, because it means that the responsibility for the delivery of the IT services must once again be transferred. Responsibilities are transferred not only when an IT outsourcing contract is initially signed, but also when the IT outsourcing contract is assigned to an IT supplier other than the IT supplier currently performing the IT services. This also involves costs that are considered to be coordination costs. A transfer demands much attention from the outsourcing organization, the current IT supplier and the new IT supplier (Outsourcing Transition Management, 1996; Anderson & Christensen, 2002). In particular, the knowledge that the current IT supplier has built up over the years, also referred to as tacit knowledge, will need to be transferred to the new IT supplier. The only interest that the current IT supplier has in carefully transferring this knowledge is the preservation of its reputation (Beulen, 2000). The IT outsourcing contract must therefore include agreements concerning the transfer process at the end of the contract.

Contract Manager, Firm 1: "This contract is almost coming to an end. I am not sure that it will be extended once again. I am already thinking about the resources that are available for the preparation of a transition plan and for the execution of that plan. The current resources that carry responsibility for the delivery of the IT service play an important role in this regard, but a strong project manager and a few consultants are essential for ensuring that this effort proceeds along the proper lines. Obviously I am hoping that it will not come to this and that we will be able to sign an extension of a few more years."

An Audit & Benchmark Process in Place

It is important for the outsourcing organization to formulate agreements about the conduct of audits (Allen, 1975; Willcocks & Fitzgerald, 1994; Parikh, 2002): a verification of the IT supplier's processes. In addition, agreements must be made about establishing regular benchmarks (Lacity & Hirschheim, 1995a; Lacity & Willcocks, 1998; Cox, 1999): is the price/quality ratio of the IT services provided in conformance with the marketplace?

IT suppliers that deliver their services in a process-oriented manner, for example through means of ISO certification, the implementation of ITIL or CMM, are generally in a better position to provide the services contracted for. Experts 1, 2 and 5 consider this a critical success factor for the IT supplier. This is further explained in the next section. The outsourcing organization must audit the IT service delivery processes on a regular basis.

Expert 2: "The process-oriented delivery of their services is no longer a differentiating factor for IT suppliers, but a necessity. Even so, it is important for outsourcing organization to keep a finger on the pulse. Furthermore, there are many outsourcing organizations that must regularly audit their suppliers on the basis of their own procedures or certification."

Gartner states: "Enterprises and their External Services Providers must acquire the monitoring and measurement tools required to keep their outsourcing agreements technically and financially on track, and also to demonstrate that they are on track" (Cox, 1999). This ongoing process (Linsenmeyer, 1991) must be the subject of prior agreements between the outsourcing organization and the IT supplier. Morgan Chambers recognizes four phases as part of the benchmarking process: organization and planning, data analysis, action and review (Eliades, 2002). It is important in this regard that the benchmark be carried out by an independent third party (Ackerman, 2002) and that the IT supplier be involved in this process in order to avoid differences in interpretation and to provide supplementary information to the independent third party.

Expert 3: "It is essential for the IT supplier to be involved in the benchmarking process.... The IT supplier may be able to provide additional information...and even more important is likely able to accept the conclusions of the benchmark analysis more quickly due to its involvement. This makes any discussions on this subject far easier for the outsourcing organization."

GOVERNANCE FACTORS
FOR IT SUPPLIERS

There are three governance factors related to the IT supplier: adequate contract and account management, adequate service delivery processes and the availability of human resources. These factors are explained in further detail here and in Appendix A — Measuring the governance factors for IT outsourcing partnership.

Adequate Contract and Account Management

Contract and account management make up the front office of the IT supplier and are denoted as the "customer outsourcing interface" by MacFarland and Nolan (1995). It is important in this regard to make a clear distinction between contract management and account management, due to the fact that there is a big difference between the primary tasks performed by an account manager and a contract manager (Beulen, 2000).

Contract management is responsible for the operational management of the relationship and therefore the direction of the service delivery processes. Contract management is therefore focused on the effectiveness and efficiency of the agreed upon contractual commitments.

Account management is responsible for maintaining the relationship with the outsourcing organization (Holden, 1990) and is focused on obtaining an extension of existing contracts and on expanding the services provided through means of new contracts. This requires the account manager to understand both new business developments and technological developments.

Expert 4: "An anecdotal example...an outsourcing organization was getting desktop hardware from supplier X. At the time they were also talking with this supplier about the use of Lotus Notes, which was going to be used to perform business processes, which is a very different way of working. Now one contract is a hardware acquisition commodity contract, while the other is very innovative and very collaborative and involves how a product is going to be used as part of the business. They <the outsourcing organization> had enormous difficulties in persuading the account manager to cooperate, because this account manager used to work in a commodity environment, and he simply didn't understand what they were talking about. They got him moved before they could get any sense out of supplier X, because he was blocking the line of communication to this supplier."

Expert 1 indicates that particularly smaller outsourcing contracts, with a value of less than U.S.$5 million, combine different functions, because the scope of these contracts does not warrant the involvement of two different contracting officers. For the case studies analyzed here, this is indeed so for case studies 6 and 9, but not for case study 8. The explanation for the latter is likely that the size of this contract was decreased after it was signed.

Adequate Service Delivery Processes

The existence of adequate service delivery processes are considered to be a "core capability" by Feeny (1997b): "delivery of IS services" and using Mintzberg's (1979) terminology, the "operating core". When the service delivery processes are set up it is essential to make a distinction between the IT supplier's business units, which are involved in the delivery of the IT services that have been contractually agreed upon with the IT supplier's customers: the so-called service delivery units, and the IT supplier's business units that are involved in researching the potential of new technological developments and which are responsible for building up knowledge about these new technologies: the so-called competence centers (Cash, McFarlan, & McKenney, 1988; Markus, 1996). The service delivery units must be assessed in terms of their degree of effectiveness and efficiency and are directly controlled by the contract managers responsible for the contracts signed with the outsourcing organization. It is important in this regard that the service delivery units use a standardized methodology for delivering their services. For this purpose, IT suppliers could make use of methodologies such as ITIL (CCTA, 1993) for setting up the management organization, and CMM for software development (Paulk, Curtis, Chrisses, & Weber, 1993). These industry standards help ensure uniformity, which is certainly essential when multiple IT suppliers are involved.

Contract manager, Firm 3: "A portion of the software is developed in India. We use a software development process, which is CMM Level 5 certified. This certification was the deciding factor for company X to select us as their supplier. This certification only carries advantages with it for performing our activities. Because of the certification our processes have been set up such that the quality of our services is high and that we are able to work efficiently."

The competence centers are directed by the IT supplier's general management and fall within Mintzberg's (1979) so-called "technostructure". New technologies are researched on the basis of proposals and projects. Feeny refers to this as "making the technology work" (Feeny & Willcocks, 1997c). It is evident that it is more difficult to measure the output of a competence center than the results produced by a service delivery unit. An important factor here is the development of an innovative capability. The competence centers develop the IT services of the future.

Account Manager, Firm 11: "I maintain intensive contacts with the managers of the competence centers. I very much want to discuss the innovative things that are being developed there, with my customers. However, it is the service delivery units that deliver the actual services when a contract is signed."

The Availability of Human Resources to IT Suppliers

From the mid-90s to the year 2000, the shortage of IT professionals was seen as a factor that would limit the growth of the IT services industry (Kitzis, 1998). Due to economic developments there has been a decrease in this shortage in the labor market (Hirschheim & Lacity, 2000). In spite of this, IT suppliers must continue to pay attention to the availability of IT professionals for delivering their IT services. A staff disposition plan involves the grouping of all IT professionals into expertise groupings, on the basis of their current expertise. Based on the projected IT services to be delivered in the future, a plan is prepared for updating the current expertise groupings. There are three factors that play a role in this plan: staff turnover, training and recruitment. By preparing a staff disposition plan, it becomes possible to ensure that the required IT services can continue to be delivered in the future (Outsourcing Transition Management, 1996; Young & Cournoyer, 2000).

Contract Manager, Firm 2: "Due to the migration from a mainframe environment to a client server environment, I required IT professionals with a completely different skill set. The mainframe experts were phased out on the basis of a staff disposition plan. They are now working for customers who are still using mainframe technology. I obtained access to a team that had worked for another one of our customers to provide the new services. I complemented this team with three recruited trainees and an experienced service manager from the responsible service delivery unit. Unfortunately the contract was not extended. However, these IT professionals were able to find employment with my customer."

TRENDS

Trends are not explicitly addressed in the interviews. On the basis of an analysis of the literature, three trends related to IT outsourcing partnerships became evident: Business Process Outsourcing (BPO) (Kern & Willcocks, 2002; Outsourcingproject, 2002; Brown & Scholl, 2002), Multiple Sourcing (Currie & Willcocks, 1998; Parikh, 2002) and Insourcing (Lacity & Hirschheim, 1995a; Cox, 2002a). These trends have already been defined in the first section.

In analyzing these trends, a link was made between these trends and the governance factors for IT outsourcing partnerships. This is shown in Table 7 and is further developed here, primarily on the basis of professional judgment. In addition, it is also possible that these trends may result in the identification of new governance factors. This aspect has not been dealt with in the executed research. Consequently, it is not possible to draw any conclusions about this here.

Business Process Outsourcing

In addition to outsourcing the IT services for their supporting processes, an increasing number of organizations are also deciding to outsource complete business processes. IDC: Another remarkable trend emerging from the analysis of the top 100 deals is the onwards march of BPO contracts (Kern & Willcocks, 2002). Gartner: BPO is the fastest growing segment (Brown & Scholl, 2002). BPO requires a greater level of attention from business management (governance factor 1.1) because it involves a higher level of impact. Business managers must ensure that the business processes that are still carried out by their own organization remain consistent with the business processes outsourced. Furthermore, the sourcing strategy, which is part of the IT strategy (governance factor 1.2), must be defined. Once again, due to the high level of impact of BPO, outsourcing organizations must go through careful deliberations before making a decision to outsource their business processes. This sourcing issue must be anchored within the IT strategy.

Trust is an important factor for the outsourcing of the IT services for supporting processes (governance factor 2.1). Due to the higher level of impact associated with BPO, trust becomes an even more important factor. Suppliers are able to gain trust through proven successes. This is particularly important, because the market is not yet really mature and there are still very few mature providers (Brown & Scholl, 2002). This is also linked to governance factor 2.2: experience in establishing and maintaining IT outsourcing relationships. As indicated earlier, experience on the part of IT suppliers is important for the success of an IT outsourcing partnership (Lacity, Willcocks, & Feeny, 1995a). Finally, BPO requires different skills from IT professionals. IT professionals must have greater knowledge of the business processes and they must possess business knowledge: "Recruitment using traditional technical criteria was seen as particularly inappropriate..." (Feeny, 1997a).

Multiple Sourcing

Many organizations outsource their IT services to more than one IT supplier (Parikh, 2002). This is referred to as multiple sourcing (Currie & Willcocks, 1998). Lacity and Hirschheim (1998) indicate that it is better for outsourcing organizations to outsource their IT services to multiple suppliers. Multiple sourcing affects the role of information management (governance factor 1.3) and the Chief Information Officer (CIO) (governance factor 1.4). The management of multiple IT suppliers requires greater effort from information management and the CIO. The option of using multiple sourcing also carries consequences with it for the individual IT suppliers (governance factor 3.1). IT suppliers will have to fight for their position in the case of competition. This requires higher level account and contract managers. They have to be able to work together, not only with the outsourcing organization, but also with the other IT suppliers. This requires a proper demarcation of responsibilities among IT suppliers (governance factor 2.3). The most important advantage of multiple sourcing for the outsourcing organization is that it promotes competition among IT suppliers. This should lead to a reduction in costs and an increase in the quality of the services provided (Currie & Willcocks, 1998; Lacity & Willcocks, 1998; Parikh, 2002). However, it must be possible for the outsourcing organization to measure this. This makes benchmarking more than essential (governance factor 2.4). Outsourcing organizations must realize, however, that they must not allow themselves to fall into the "divide and conquer" trap: "Enterprises that have embraced the multisourced environment must seriously evaluate the service value lost to ineffective sourcing decisions and management" (Stone, 2002).

Insourcing

Insourcing means that not all of the IT services are performed by one or more IT suppliers and that a portion of the IT services are provided by the internal IT division. Lacity and Hirschheim suggest: "Selective sourcing is right sourcing" (Lacity & Hirschheim, 1995a). This implies that some aspects will continue to be looked after by the internal IT division itself. The internal IT division also requires attention: "Re-shaping the internal team will become a priority in 2002-2003" (Cox, 2002a). Furthermore, the choice to insource will need to be anchored into the organization's IT strategy (governance factor 1.2). In addition, as in the case of multiple sourcing, insourcing has an impact on contracts, since the internal IT division also constitutes an IT supplier with which the external IT suppliers are required to cooperate (governance factor 2.3). Monopolies are no longer the norm. This requires significant effort on the part of the account and contract management team (governance factor 3.1) within the internal IT division. Due to recent labor market developments insourcing has a limited impact on the resourcing of the internal IT division (governance factor 3.3). Until recently, the scarce labor market for IT professionals was a serious threat to the resourcing of internal IT divisions (Hirschheim & Lacity, 2000). Due to recent developments, however, internal IT organizations are now also able to attract IT professionals with the proper qualifications.

Table 7.   Link Between Governance Factors of the IT Outsourcing Partnership Framework and Trends

Governance Factors	Trends		
	BPO	Multiple Sourcing	Insourcing
The Outsourcing Organization			
1.1 Attention to IT within business units	X		
1.2 A clear IT strategy	X		X
1.3 Information management as the link between business units and IT suppliers		X	
1.4 A properly functioning Chief Information Officer		X	
The Maintenance of the Relationship			
2.1 Mutual trust between the outsourcing organization and the IT supplier	X		
2.2 Experience in establishing and maintaining IT outsourcing relationships	X		
2.3 Efficient and effective IT outsourcing contracts		X	X
2.4 An audit & benchmark process in place		X	
1.1 *IT Suppliers*			
3.1 Adequate Contract and Account Management		X	X
3.2 Adequate service delivery processes			
3.3 The availability of human resources to the IT suppliers	X		X

CONCLUSION

Organizations must deliberate carefully before taking a decision to outsource. Outsourcing should certainly not be considered to be superior to insourcing during these deliberations. These careful deliberations should be based on an IT strategy that incorporates the sourcing strategy. If the organization decides to outsource, it must also make the required decisions about the scope and nature of the outsourcing relationship: single/multiple and total/selective. In addition, the type of IT services to be outsourced must be defined: Information Systems Outsourcing, Processing Outsourcing or Business Process Outsourcing. These different types of outsourcing possess an increasing degree of complexity and demand increasing management attention from both the outsourcing organization as well as the IT suppliers.

It is clear that not all IT outsourcing partnerships are a success. There are three key factors that cause IT outsourcing partnerships to be unsuccessful: lack of maturity of the outsourcing organization, contracts that are inflexible and insufficient degree of integration of the IT division taken over into the organization of the IT supplier.

These factors were explored as part of the descriptive framework to achieve governance in IT outsourcing partnerships. Aside from the role of the outsourcing organization and the IT supplier, the maintenance of the relationship also takes on an important role in this framework. In addition to the already mentioned need for a clear IT and sourcing strategy, it is important for outsourcing organizations that their business units are committed to IT. Furthermore, the management of the IT function must be properly anchored within the organization: the information management function and the Chief Information Officer (CIO). The IT supplier must ensure that an interface with the outsourcing organization is created for managing the relationship: the account management and contract management functions. Contract management is responsible for directing the IT supplier's service delivery processes in this regard. The IT supplier must also ensure that sufficient and sufficiently qualified personnel are available. Trust is a key concept for the relationship between the outsourcing organization and IT suppliers, and experience with outsourcing relationships helps in maintaining the outsourcing relationship. Effective and efficient contracts are essential for managing the relationship and audit and benchmarking processes must be set up. Only then does it become possible to guarantee the continuous and consistent governance of the IT outsourcing partnership.

However, time does not stand still. Business Process Outsourcing is expected to become increasingly dominant and increasing numbers of multiple outsourcing relationships will come into being. In addition, outsourcing organizations will also selectively repatriate certain components of their IT services. This will have an impact and will lead to an increase of complexity in governing IT outsourcing partnerships.

ACKNOWLEDGMENTS

I would first like to express my thanks to all persons interviewed and who generously provided their cooperation to this research project. Their valuable insights form the basis of this research project. I am also grateful to Atos Origin for providing me with the opportunity to write this chapter.

The critical comments of Lielle van Laren have been of immeasurable help in improving the content of this chapter. Christine Holdert's critical eye led to significant improvement of the structure and readability of this text.

REFERENCES

Ackerman, D. (2002). Measuring success in outsourcing relationships. *Gartner IT Services and Sourcing Summit 2002, 15–17 May, Nevada USA,* C7, STD5, 2002.

Allen, B. (1975). Guide to computers. *Harvard Business Review,* (July/August).

Anderson, T., & Stampe Christensen, M. (2002). Contract renewal under uncertainty. *Journal of Economic Dynamics & Control, 26*(4), April, 637-652.

Apte, U. (1990). Global outsourcing of information systems and processing services. *The Information Society, 7,* 287-303.

Aylott, B. (2002). Questions and answers. *Montgomery Research Europe.*

Barnard, C. (1938). *The functions of the executive.* Cambridge, MA: Harvard University Press.

Beulen, E. (2000). *Beheersing van IT-outsourcingsrelaties.* Ph.D. thesis, Tilburg University, The Netherlands, (in Dutch).

Beulen, E., & Ribbers, P. (2002). Lessons learned: Managing an IT-partnership in Asia: Case Study: The relationship between a global outsourcing company and their suppliers. *Proceedings of Hawaii International Conference on Systems Sciences 2002.*

Beulen, E., & Ribbers, P. (2002). Managing complex IT outsourcing – partnerships. *Proceedings of Hawaii International Conference on Systems Sciences 2002.*

Beulen, E., Ribbers, P., & Roos, J. (1994). *Outsourcing van IT-dienstverlening, een 'make or buy' beslissing, Kluwer Bedrijfswetenschappen, Deventer, The Netherlands,* (in Dutch).

Brown, C., & Magioll, S. (1994). Alignment of the IS functions with the enterprise: Towards model of antecedents. *MIS Quarterly,* (December).

Brown, R., & Scholl, R. (2002). European business processing outsourcing trends 2001. *Gartner focus report,* (February 19).

Burnett, R. (1998). *Outsourcing IT- the legal aspects.* Aldershot, Gower.

Burns, T., & Stalker, G. (1961). *The management of innovation.* Tavistock, London.

Busher, J. (2002). What is your sourcing strategy? *Montgomery Research Europe, ISSN 1476-2064.*

Cash, J., McFarlan, F., & McKenney, J. *Corporate information systems management, the issues facing senior executives.* Irwin.

CCTA. 1993). *The infrastructure library: An introduction.* CCTA.

Chaderton, R., & van de Wittenboer, J. P. (2002). Eleven steps to partnership heaven. *Montgomery Research Europe.*

Chandler, A. (1962). *Strategy and structure.* Cambridge, MA: MIT Press.

Coase, R. (1937). The nature of the firm. *Economica, 4,* 386-405.

Corbett, M. (1994). Outsourcing and the new IT executive: A trend report. *Information Systems Management,* (Fall).

Cox, R. (1999). Managing outsourcer relationships. *Conference presentation, ESC11Outsourc99RCox.*

Cox, R. (2002). Gartner services and sourcing scenario. *Gartner Symposium ITXPO 2002,* (April 8-10) *Florence, Italy.*

Cox, R. (2002). Strategic sourcing: Shifting the focus from buying service to buying relationships. *Montgomery Research Europe.*

Cross, J. (1995). IT-outsourcing: British Petroleum's competitive approach. *Harvard Business Review,* (May/June).

Currie, W., & L. Willocks, L. (1998). New strategies in IT-outsourcing: Major trends and global best practices. *Business Intelligence.*

Daft, R. (1998). *Organizations theory and design, (6th ed.).* South-Western College Publishing.

Dataquest. (1998). *Worldwide service: Market definitions*, by M. Sadlowski.

David, S., & Laurance, P. (1977). *Matrix.* Addison Wesley.

Dickson, G., Leitheiser, R., & Wetherbe, J. (1984). Key information system issues for the 1980s. *MIS Quarterly,* (September).

Dreyfuss, C. (2002). Sourcing governance: What, where and how. *Gartner IT Services and Sourcing Summit 2002,* (May 15-17) *Nevada USA,* C1, STD5.

Duncan, R. (1979). What is the right organisation structure? Decision tree analysis provide the answer. *Organization Dynamics,* (Winter).

Earl, M. (1987). Information systems strategy formulation. In R. Boland & R. Hirschheim (Eds.), *Critical Issues in Information System Research.* John Wiley & Sons.

Earl, M. (1996). The risks of outsourcing IT. *Sloan Management Review,* (Spring).

Earl, M., & Feeny, D. (1997). Is your CIO Adding value?, In L. Willcocks, D. Feeny, & G. Islei (Eds.), *Managing IT as a Strategic Resource.* McGraw Hill.

Earl, M., & Feeny, D. (2000). Opinion: How to be a CEO for the Information Age. *Sloan Management Review,* (Winter).

Earl, M., Edwards, B., & Feeny, D. (1997). Configuring the IS function in complex organizations. In L. Willcocks, D. Feeny, & G. Islei (Eds.), *Managing IT as a Strategic Resource.* McGraw Hill.

Eggleton, D., & Otter, G. (1991). A directors' briefing: Outsourcing information systems services. *Butler Cox Foundation,* (April).

Eliades, B. (2002). Benchmarking outsourced services. *Montgomery Research Europe.*

Emery, F., & Trist, E. (1965). The causal texture of organisational environment. *Human Relations, 18,* 21-32.

Feeny, D. (1997). The five-year learning of ten IT directors. In L. Willcocks, D. Feeny, & G. Islei, (Eds.), *Managing IT as a Strategic Resource.* McGraw Hill.

Feeny, D., & Willcocks, L. (1997). The IT-function: Changing capabilities and skills. In L. Willcocks, D. Feeny, & G. Islei (Eds.), *Managing IT as a Strategic Resource.* McGraw Hill.

Feeny, D., Earl, M., & Edwards, B. (1997). Information systems organization: The role of users and specialists. In L. Willcocks, D. Feeny, & G. Islei (Eds.), *Managing IT as a Strategic Resource.* McGraw Hill.

Feeny, D., Willcocks, L., & Core, I. (1998). Capabilities for exploiting IT. *Sloan Management Review, 39*(3), 1-26.

Gantz, J. (1990). Outsourcing: Treat or salvation? *Networking Management, 10.*

Gerrity, T., & Rockart, J. (1986). End user computing: Are you a leader of a laggard? *Sloan Management Review, Summer,* 25-34.

Gietzmann, M. (1996). Incomplete contracts and the make or buy decision: Governance design and attainable flexibility. *Accounting, Organization and Society, 21*(6), 611-626.

Grigg, J., & Block, D. (2002). Building a sourcing strategy. *Gartner IT Services and Sourcing Summit 2002,* (May 15-17), *Nevada USA,* B2, STD5.

Grönroos, C. (1990). *Service management and marketing, managing the moment of truth in service competition.* Lexington Books.

Hart, O. (1995). *Contracts and financial structure.* Oxford University Press.

Heckmann, R. (1999). Organizing and managing supplier relationships in information technology procurement. *International Journal of Information Management, 19,* 141-155.

Henderson, J., & Venkatraman, N. (1992). Strategic alignment: A model for organizational transformation through Information Technology. In T. Kochan & M. Usseem (Eds.), *Transforming organizations,* (pp. 97-117). New York: Oxford University Press.

Henderson, J., & Venkatraman, N. (1993). Strategic alignment: Leveraging information technology for transforming organizations. *IBM Systems Journal, 32*(1).

Hewlett-Packard Limited. (1999). *Customer roadmap, 5499-1301-9/00.*

Hirschheim, R., & Lacity, M. (2000). The myths and realities of information technology insourcing. *Communication of the ACM, 43*(2), February.

Hofstede, G. (1980). *Culture's consequences.* Sage Publications.

Holden, J. (1990). *Value based selling.* Holden Corporation.

Huber, R. (1993). How Continental Banks outsourced its crown jewels. *Harvard Business Review,* (January/February).

International Data Corporation, M. Lukacs. (1998). *European consulting and management services: European outsourcing markets and trends, 1996 – 2002 research report.*

Kanter, R. (1994). Collaborative Advantage: The art of alliances. *Harvard Business Review,* (July/ August), 96-108.

Katz, D., & Katz, L. (1966). *The social psychology of organizations.* New York: John Wiley & Sons.

Kern, T., & Willcocks, L. (2000). Exploring Information Technology outsourcing relationships: Theory and practice. *Journal of Strategic Information Systems, 9*, 321-350.

Kern, T., & Willcocks, L. (2002). Exploring relationships in Information Technology outsourcing: The interaction approach. *European Journal of Information Systems, 11*, 3-19.

King, W. (1978). Strategic planning for management, information systems. MIS Quarterly, *2*(2), June, 27-37.

Kitzis, E. (1998). *Report Gartner group/conference presentation, Market definition and forecast*, VEN1SvcsMkt498Ekitzis.

Klepper, R. (1995). The management of partnering development in I/S outsourcing. *Journal of Technology, 10.*

Klepper. R., & Hartog, C. (1991). Some determinants of MIS outsourcing behaviour. *Handbook BIK, D2700,* (November).

Klepper, R., & Jones, W. (1998). *Outsourcing Information Technology systems and services.* Prentice Hall.

Kotwica, K., & Fields, T. (1998). *The CIO executive research centre, What is a CIO?* Available online: www.cio.com, last update 17 December 1998.

Kotwica, K., & Fields, T. (1999). *The changing role of the chief information officer.* Available online: www.CIO.com, last update 11 January 1999.

KPMG Impact. (1995). *IMPACT outsourcing workshop: Best practice guidelines for outsourcing,* handout, (May).

Kraljic, P. (1983). Purchasing must become supply management. *Harvard Business Review,* (September/October), 109-117.

Lacity, M., & Hirschheim, R. (1993). *Information systems outsourcing.* John Wiley & Sons.

Lacity, M., & Hirschheim, R. (1995). *Beyond the information systems outsourcing bandwagon: The insourcing response.* John Wiley & Sons.

Lacity, M., & Willocks, L. (1998). An empirical investigation of Information Technology sourcing practices: Lessons from experience. MIS Quarterly, *22*(3), 363-408.

Lacity, M., & Willcocks, L. (2001). *Global Information Technology outsourcing: In search of business advantage.* John Wiley & Sons.

Lacity, M., Willcocks, L., & Feeny, D. (1995). IT outsourcing: Maximize flexibility and control. *Harvard Business Review,* (May/June).

Lacity, M., Willocks, L., & Feeny, D. (1996). The value of selective sourcing. *Sloan Management Review,* (Spring).

Lamy, L. (2001). *IDCs top 100 outsourcing deals 2000.* IDC research report.

Lawrence, P., & Lorsch, J. (1967). *Organisation and management.* Boston, MA: Harvard Business School Press.

Linsenmeyer, A. (1991). Fad or fundamental: A chat with Bob Camp of Xerox, the man who wrote the book on benchmarking. *FW, 160*(19), September 17, 34-35.

Looff, L de. (1996). A model for information systems outsourcing decision making. *Doctoral Dissertation.* Delft University of Technology.

Markus, M. (1996). The futures of IT management. *Database, 27*(4), Fall.

McFarland, W., & Nolan, R. (1995). How to manage an IT-outsourcing alliance. *Sloan Management Review, 36,* 9-23.

McLean, E., & Soden, J. (1977). *Strategic planning for MIS.* New York: John Wiley & Sons.

Mintzberg, H. (1979). *The structuring of organizations.* Englewood Cliffs, NJ: Prentice Hall.

Morgan & Chambers. (2001). Outsourcing in the FTSE top 100. *Computer Weekly.*

Niederman, F., Brancheau, J., & Wetherbe, J. (1991). Information systems management issues for the 1990s. *MIS Quarterly, 15*(4), December.

Outsourcing Transition Management (1996). *Atos Origin method for implementing outsourcing contracts –internal procedure, version 1.0.*

Outsourcingproject. (2002). *Montgomery Research Europe,.*

Parikh, K. (2002). Strategic sourcing –The art of negotiating performance-based outsourcing contracts. *Gartner IT services and sourcing summit 2002, Las Vegas, Nevada.*

Paulk, M., Curtis, B., Chrisses, M., & C. Weber, C. (1993). *Capability maturity Model for Software, version 1.1.* Software Engineering Institute, , (February).

Pinnington, A., & and Woolcock, P. (1997). The role of vendor companies in IS/IT outourcing. *International Journal of Information Management, 17*(3).

Poisson, J. (2002). The Outsourcing Performance Assessment model: A framework to a relationship management audit. *Montgomery Research Europe.*

Porter, M., & Millar, V. (1985). How information gives you competitive advantage. *Harvard Business Review,* (July/August).

Quinn, J., & Hilmer, F. (1995). Strategic outsourcing. *The McKinsey Quarterly, 1.*

Quinn, J., Doorley, L., & Paquette, P. (1990). Technology in services: Rethinking strategic focus. *Sloan Management Review,* 79-87.

Rochester, J., & Douglas, D. (Eds.) (1990). Taking an objective look at outsourcing. *I/S Analyzer, 28*(9).

Rockart, J., & Scott Morton, S. (1984). Implications of changes in Information Technology for corporate strategy. *Interfaces, 14* (1), January-February, 84-95.

Rosser, B. (1998). *Report Gartnergroup, How is the IT strategic planning changing?* VEN1ITPlan498Brosser.

Segal, I. (1999). Complexity and renegotiation: A foundation for incomplete contracts. *Review of Economic Studies, January.* Oxford.

Stone, L. (2002). Chief Sourcing Officer: New role, new reality. *Gartner research note,* COM-16-1899, (May 9).

Terdiman, R. (1991). Gartner report: Outsourcing: Threat or salvation? *Gartner, IS: R-980-108,* (July).

Terdiman, R. (1993). Outsourcing in the '90s. CT 06904, at *Enterprise '93: Profit through information access, World Trade Center, Boston,* (June 16-18).

van der Heijden, K. (1996). *Scenarios, the art of strategic conversation.* John Wiley & Sons.

van der Zee, H. (1997). *Succesvol outsourcen in Nederland, Ten Hagen Stam (in Dutch).*

van der Zee, H., & van Wijngaarden, P. (1999). Strategic sourcing and partnerships, Challenging scenarios for IT alliances in the network era. Addison Wesley.

Willcocks, L., & Choi, C. (1995). CO-operative partnerships and "total" IT-outsourcing: From contractual obligation to strategic alliance? *European Management Journal, 13*(1), March.

Willcocks, L., & Fitzgerald, G. (1994). A business guide to outsourcing IT. *Business Intelligence.*

Willcocks, L., & Lester, S. (1997). Assessing IT-productivity: Any way out of the Labyrinth? In L. Willcocks, D. Feeny, & G. Islei (Eds.), *Managing IT as a Strategic Resource.* McGraw Hill.

Willcocks, L., & Plant, R. (2001). Pathways to e-business leadership: Getting from bricks to clicks. *MIT Sloan Management Review,* (Spring).

Willcocks, L., Fitzgerald, G., & Feeny, D. (1995). Outsourcing the strategic implications. *Long Range Planning, 28*(5).

Williamson, O. (1975). *Markets and hierarchies.* New York: The Free Press.

Yin, R. (1994). *Case study research, design and methods.* SAGE.

Young, A., & Cournoyer, S. (2000). *Strategic profiles of market leaders: Employee transition in outsourcing engagements, competitive analysis of Gartner Group,* (February 14).

Zani, W. (1970). Blueprint for MIS. *Harvard Business Review, 48*(6), November-December, 95-100.

ENDNOTES

[1]   This chapter is based on a long-term research program into the management of IT outsourcing partnerships conducted by Prof. Dr. Pieter Ribbers, Prof. Jan Roos and the author of this chapter, as part of the Economics branch of the Faculty of Economics of the University of Tilburg in the Netherlands. A large number of outsourcing case studies were analyzed over the last few years as part of this research program. The descriptive framework for the governance of IT outsourcing partnerships developed as part of the author's doctoral thesis (Beu'00) is further refined in this chapter on the basis of an analysis of additional references and case studies.

[2]   In addition, organizations could also adopt a hybrid organization structure in which portions of the organization are structured on the basis of the functional organization structure and portions on the basis of the divisional organization structure. This hybrid organizational structure is not addressed in this chapter.

[3]   Gartner uses the terms IT Management Services, Transaction Processing Services and Business Process Management, respectively (Kitzis, 1998).

[4]   *The IT supplier in all cases was Atos Origin.*

[5]   The data was collected in 2002.

[6]   This case study was investigated in 2000 and 2002.

[7]   This case study was investigated in 2000 and 2002.

[8]   This Firm is part of the Firm described in case study 6, which was investigated in 2000 and 2002.

[9]   This Firm is part of the Firm described in case study 6. Because of the different type of activities of this business unit this case study is classified under the services sector instead of the discrete manufacturing sector.

[10]   This case study was investigated in 1994 and 2000.

[11]   Only the outsourcing contracts were analyzed for this case study and no interviews were conducted.

[12]   Only the outsourcing contracts were analyzed for this case study and no interviews were conducted.

APPENDIX A
Measuring the Governance Factors for IT Outsourcing Partnership (Beulen, 2000)

Table 8. Governance Indicators by Governance Factor for the Outsourcing Organization

Governance Factors for the Outsourcing Organization	Governance Indicators	
1.1 Attention to IT within business units	1.1.1.	Responsibility for IT has been assigned to the management team of the business function
	1.1.2.	IT service delivery is evaluated in terms of its added value and its cost
	1.1.3.	The business functions proactively involve the IT supplier's IT specialists in the development of new products or services
	1.1.4.	Business function play particular attention to the strategic management of the IT outsourcing relationship
1.2 A clear IT strategy	1.2.1	The outsourcing organization's IT strategy is linked to and interacts with outsourcing organization's overall strategy
	1.2.2	The outsourcing organization's supplier strategy is an explicit part of the IT strategy and focused on continuity
	1.2.3	The IT strategy anticipates new developments in the market in which the outsourcing organization operates or is going to operate and which offers opportunities for developing new technologies
	1.2.4	The role of external IT suppliers and consultancy firms in the development and implementation of the IT strategy is limited to a facilitating or supporting role. The outsourcing company is accountable for developing and implementing the IT strategy
	1.2.5	Alignment of the IT strategy with the parent company's IT strategy
	1.2.6	An adequate IT board
1.3 Information management as the link between business units and IT suppliers	1.3.1	The information management comprises the following tasks: development & implementation of IT strategy, contract management, points of contact of the IT supplier and points of contact for the business functions
	1.3.2	Information management possesses both business knowledge and IT knowledge
	1.3.3	Information management has a facilitating role in implementing the IT strategy on behalf of the business functions
1.4 A properly functioning Chief Information Officer	1.4.1	Within the board of management of the outsourcing organization there is attention for IT
	1.4.2	The role of the CIO comprises the following tasks: alignment of business and IT, development & implementation of the IT strategy, managing of IT outsourcing relationships and governance of IT
	1.4.3	The CIO is accountable for the IT outsourcing relationship
	1.4.4	The CIO possesses both business knowledge and IT knowledge

Table 9. Governance Indicators by Governance Factor for the Maintenance of the Relationship

Governance Factors for the Maintenance of the Relationship	Governance Indicators	
2.1 Mutual trust between the outsourcing organization and the IT supplier	2.1.1	Overall strategies are enhanced between the outsourcing organization and the IT suppliers and the objectives of the outsourcing company and the IT suppliers are aligned
	2.1.2	The cultures of the outsourcing organization and the IT suppliers do not clash
	2.1.3	Mutual trust between the staff members of the outsourcing organization and the external IT suppliers
2.2 Experience in establishing and maintaining IT outsourcing relationships	2.2.1	The IT suppliers have an experienced-based methodology for defining IT outsourcing relationships and the outsourcing organization has an experienced-based process description for defining IT outsourcing relationships
	2.2.2	The outsourcing organization's strategy focuses on establishing alliances with suppliers and the strategy of the IT suppliers is focusing on establishing long-term relationships with outsourcing organizations
	2.2.3	The outsourcing organization and the external IT suppliers do not only judge the performance of the outsourcing relationship on the basis of short-term results
	2.2.4	After the IT outsourcing contract has been signed, a planned and successful transition was carried out
2.3 Efficient and effective IT outsourcing contracts	2.3.1	The IT outsourcing contracts between the outsourcing organization and the IT suppliers are flexible
	2.3.2	The descriptions in the outsourcing contracts as well as any other mutual obligations are in line with the spirit of the relationship between the outsourcing organization and the IT suppliers: completeness is not an issue
	2.3.3	The IT outsourcing contracts clearly specify the reporting frequency and the content of the reporting
	2.3.4	The outsourcing contracts explicitly specify the consequences of not or not fully meeting the terms of the contract and the maximum amount of damages to be paid for not or not fully meeting the terms of the contract
	2.3.5	IT outsourcing contract have a satisfactory profit margin
2.4 An audit & benchmark process in place	2.4.1	The accountable officers of the outsourcing organization and the IT suppliers periodically evaluate the cooperation and content of the IT outsourcing contract at predefined points in time and, if necessary, at any other point in time
	2.4.2	The IT outsourcing contracts clearly specify which process is to be followed should there be deviation from the agreements reached
	2.4.3	The audit process should be based on a general accepted standard

Table 10: Governance factor by Governance Factor for the IT Suppliers

Governance Factors for the IT Suppliers	Governance Indicators	
3.1 Adequate Contract and Account Management	3.1.1	Contract and account management tasks include: management of IT resources, overseeing the IT service provision, maintaining and increasing turnover, business development, optimization of the cooperation with other IT suppliers and ensuring internal priority for the IT outsourcing contract
	3.1.2	The responsibility of the contract and the account management for the outsourcing relationship is separated from the responsibility for the IT service provision and the IT business function
	3.1.3	The IT supplier's organization pays specific attention to IT outsourcing relationships and the IT outsourcing contracts are managed by a separate organizational unit: this unit is also responsible for the ongoing delivery of the IT services
	3.1.4	Contract and account management of the external IT supplier and the management of the IT business function posses business knowledge in addition to IT expertise
3.2 Adequate service delivery processes	3.2.1	The cost component required to deliver the IT services are well understood
	3.2.2	The available manpower capacity of an IT business function is used optimally
	3.2.3	The IT business functions include sufficient overhead costs to manage the IT business function
	3.2.4	IT employees spend an adequate number of training hours each year as part of an improvement-focused and concrete training plan to train themselves in new technologies and interpersonal and business skills
	3.2.5	Process govern service delivery
	3.2.6	Working with a limited group of dedicated resources to execute the IT service provision tasks
	3.2.7	The external IT supplier has alliances with specialized business organizations
3.3 The availability of human resources to the IT suppliers	3.3.1	IT employees are given the opportunity to gain experience with new technologies
	3.3.2	The IT supplier is able to attract a sufficient number of qualified new employees from the labor market
	3.3.3	The IT supplier has a high training budget and individual employees have a say in how their training budget is spend within the framework of an overall concrete and development-focused training plan
	3.3.4	The IT supplier has a low turnover percentage among the employees assigned to a specific IT outsourcing contract

SECTION IV:

IT GOVERNANCE IN ACTION

Chapter XIII

The Evolution of IT Governance at NB Power

Joanne Callahan
New Brunswick Power Corporation, Canada

Cassio Bastos
New Brunswick Power Corporation, Canada

Dwayne Keyes
New Brunswick Power Corporation, Canada

ABSTRACT

IT and business are inextricable linked. It is incomprehensible to think the two should ever again be independent. IT Governance will penetrate how IT is executed, as organizations like NB Power endeavor to become better at governing and managing their IT investments. NB Power, in 1998, implemented an IT Governance framework. Through IT Governance the organization was able to address the results of a diagnostic study on their internal IT service provider who was attempting to respond to a seemingly endless list of requests for IT support. Four years later, factors critical to the success of implementing an IT Governance framework are evident. The IT Governance framework is still evolving, but the organization is now well positioned to take advantage of its IT investment.

INTRODUCTION

The establishment and implementation of an Information Technology (IT) Governance framework to achieve business objectives through well managed and controlled IT investments is an increasing concern for enterprise. It is difficult to separate an organization's overall strategic mission from the underlying IT strategy that enables the mission. In a report on improving healthcare performance through information management, the Scottsdale Infomatics Institute (2001) suggests that having an IT committee is at least as important as having a finance committee. Finance, they say, is a matter of keeping score, but it is not playing the game. IT is driving the business.

Comments made by proponents of IT Governance at a roundtable discussion during the International Conference of the Information Systems Audit Control Association (ISACA), 2000, suggest that two of the most valuable aspects of organizations around the world are information and the technology that supports it. Often the success of an organization is extensively dependent on the effective management of information and IT systems. Organizations must satisfy the quality, fiduciary, and security requirements of their IT, as they do for other assets.

Williams (2001) contends that as critical business processes increasingly rely on IT, the benefits and risks grow exponentially. Senior executives and boards must actively address the governance of IT together with their other governance responsibilities.

IT Governance is needed because information is of critical importance to enterprise. Given the increasing volume and complexity of information it is difficult to measure value. It is necessary to align, integrate, and determine ownership of information and technology (Williams, 2000).

Prior to the implementation of IT Governance at NB Power, a publicly owned electric utility monopoly, the organization had difficulty maximizing its technology investments to gain economies of scale in IT investments and effectively manage IT resources. The benefits and value of IT investments were a challenge to communicate because the direction of IT in the organization was not clearly articulated. The organization should have been achieving greater value from its IT investment.

This chapter will discuss the business drivers that led one company to implement an IT Governance framework. The evolution of IT Governance over a four year period is illustrated and critical success factors and lessons learned are examined. Finally, trends in IT Governance are explored.

WHAT IS IT GOVERNANCE AND WHO SHOULD BE INVOLVED?

There is no definitive definition of IT Governance. Notable definitions that describe IT Governance include one provided by ISACA, 2000. They define IT Governance as a structure of relationships and processes to direct and control the enterprise in order to achieve the enterprise's goals by adding value while balancing risks versus return over IT and its processes.

IT Governance is also described as how those persons entrusted with governance of an entity consider IT in the supervising, monitoring, controlling, and direction of the entity. How IT is applied will have an immense impact on whether the entity will attain its vision, mission, or strategic goals (IT Governance Institute [IGI], 2001).

Implementing IT Governance is not always easy, nor is IT Governance readily embraced. Williams (2002) suggests that there is no clear governance in many organizations. Therefore, it is considered impossible to establish IT Governance. In the absence of a traditional IT Governance organizational structure, ideas differ when determining who should govern the corporate IT investment. A study from the Scottsdale Informatics Institute (2001) suggests that IT Governance has been lacking because IT has been seen as an operations matter best left to management. Board members often lacked interest or expertise in technology issues.

Contrary to this finding, some believe that IT Governance is the responsibility of the board of directors and, at the very least, executive management (Canadian Institute of Chartered Accountants [CICA], 2002; Board Briefing on IT Governance, 2001; Scottsdale Infomatics Institute, 2001).

Grembergen (n.d.) concurs. He recommends that senior managers should be questioning how they can get their Chief Information Officers (CIOs) and IT organization to return business value or ensure that they do not misappropriate capital funding or make bad investments. Furthermore, senior managers should provide mechanisms like an IT Balanced Scorecard to measure and manage the functions of the IT organization like IT-business alignment and the IT strategy development process.

Williams (2001) contends that boards and executives must become more concerned over IT investments because they ultimately affect shareholder value. He suggests a key question is whether the organization's IT investments are aligned with its strategic objectives, thus building the capability necessary to deliver long-term sustainable business value. IT Governance is a necessity, not a luxury. Organizations must bring IT Governance to the board level because it is not smart business to wait for an IT-related disaster before taking action.

The literature is very explicit in describing the roles and responsibilities of those who govern corporate IT investments, the board of directors or at the very least, the executive management team. The CICA (2002) in a recent report concludes that boards should review the strategic planning process, the approved plan, and performance against the plan. In so doing, they can ensure that the strategic planning process is structured so that IT investments are aligned with business objectives, and that the tactical plan to support the strategy is being executed.

The board must make certain there are policies and processes to ensure the integrity of internal controls and the information systems that are used for management. This function can be fulfilled by assigning the responsibility for the organization's use of and investment in IT to a board member, or IT subcommittee, and ensuring that the individual responsible for IT Governance holds a senior position. The board should be confident that management establishes programs to keep employees knowledgeable and compliant with information and security policies.

Finally, boards must exercise policies and processes that identify the business risks associated with IT, determine the level of risks the organization is willing to accept, and ensure that the risks are being monitored. In managing the risks of IT, boards will exercise due diligence of the IT systems and actively mitigate the risks of IT to the organization.

The IGI (2001) also suggests IT Governance is the responsibility of the board of directors and executive management. It is an integral part of enterprise governance and consists of the leadership, organizational structures, and processes that ensure IT sustains and extends the organization's strategies and objectives.

Boards and executives must provide the leadership, structures and processes that ensure that the organization's IT sustains and extends its strategies and objectives. They should keep themselves informed of the role and impact of IT on the enterprise, assign IT responsibilities, define constraints within which IT professionals operate, measure IT performance, manage IT risks, and ensure the compliance of IT Governance standards (Williams, 2001).

THE BENEFITS OF IT GOVERNANCE

When IT Governance is well deployed, business and IT benefits are numerous. Effective IT Governance helps ensure that IT supports business goals, maximizes IT investment, and appropriately manages IT-related risks. Good IT Governance helps achieve critical success factors by efficiently and effectively deploying secure, reliable information and applied technology (Lainhart, 2001).

Lainhart (2001) suggests that an appropriate IT Governance program helps organizations confidently address critical business issues such as addressing the risks of aging technologies or undertaking e-business. Through IT Governance an enterprise can protect its investment in IT and assure appropriate management of information assets, many of which are vital to the survival and growth of the enterprise itself.

The Scottsdale Infomatics Institute (2001) purports that effective IT Governance protects stakeholders, clarifies IT risks, directs and controls investments, opportunities, risks, and benefits, aligns IT and business, and sustains current operations. Given this mandate, IT Governance should be an integral part of the global governance structure.

Doughty (1998) contends that the effectiveness of the IT Steering Committee is of strategic importance to the overall success of the organization in achieving a competitive advantage from its IT investment. If the IT Steering Committee fails to monitor and supervise IT investment decisions and operations, then it could fail to achieve the corporation's strategic objectives.

Simply put, good governance, whether it is enterprise or IT, is good business (Lainhart, 2001).

A CASE TO IMPLEMENT IT GOVERNANCE

In 1998, NB Power endeavored to implement an IT Governance framework to address the results of a diagnostic study of their internal IT service provider, Business Information Systems (BIS). The results of the study indicated that the corporation had unachieved benefits from their technology investments. IT benefits that could have brought value to the organization were not being realized.

The study also revealed that operational inefficiencies prevented the corporation from gaining economies of scale in IT investments and human capital management. Planning and managing IT to support client requests was difficult because policies, standards, procedures, and methodologies were insufficient.

The morale among IT staff needed to be boosted. This was possibly the result of business units competing for IT resources in a process that was loosely structured and controlled. The requests for IT support and work were voluminous.

The study also suggested the imminent risk of not meeting future business needs because the direction of IT was unclear. Without knowing the direction of IT at NB Power, it was difficult to demonstrate how IT provided value and supported the achievement of business goals.

BIS had requests to support between 150 to 200 business initiatives. Prioritizing and scheduling requests to satisfy client needs was a monumental task. Based on previous experience, the IT senior management team realized the IT division and their clients were about to embark on another frustrating year because of the difficulty in scheduling the requests to deliver completed projects on time while satisfying client expectations. Planning and managing support for clients' requests was daunting. At the same time, the implementation of Y2K preparations required attention, and there was a major initiative to implement the SAP Customer Care and Service module to replace NB Power's legacy customer information system. More stressful, however, was the lack of articulation around business and IT alignment to ensure that IT resourcing would bring the best business value to the corporation. Armed with this knowledge, and knowing that additional requests for new business initiatives requiring IT support were inevitable, the corporate executive team endorsed the implementation of an IT Governance framework at NB Power.

STRUCTURES AND RELATIONSHIPS

The implementation of the IT Governance framework began with the establishment of two IT Governance committees: the IT Governance executive committee that included vice-presidents and general managers from each business unit of the corporation, and the IT Governance management committee, made up of senior managers to represent all business units and corporate services.

The IT Governance management committee was responsible for recommending an eighteen month plan that aligned business and IT initiatives for executive approval. The executive wanted assurances that the limited budget available for IT was being directed at initiatives which brought the most value to the corporation and helped mitigate corporate risks.

Table 1. Reasons for Implementing an IT Governance Framework at NB Power

	Business Drivers
Benefit Realization	Achieve maximum benefits from technology investments.
Economies of Scale	Gain economies of scale in IT investments.
Maximize Human Capital	Maximize human resource management.
Standardization	Standardize methodologies, policies, and procedures for IT.
Portfolio Management	Invest in tools for project prioritization and work request management.
Strategic Alignment	Set a clear direction for IT within the organization.
Communications	Communicate the benefits and value of IT to the organization.

Figure 1. Corporate IT Governance Organizational Structure and Supports

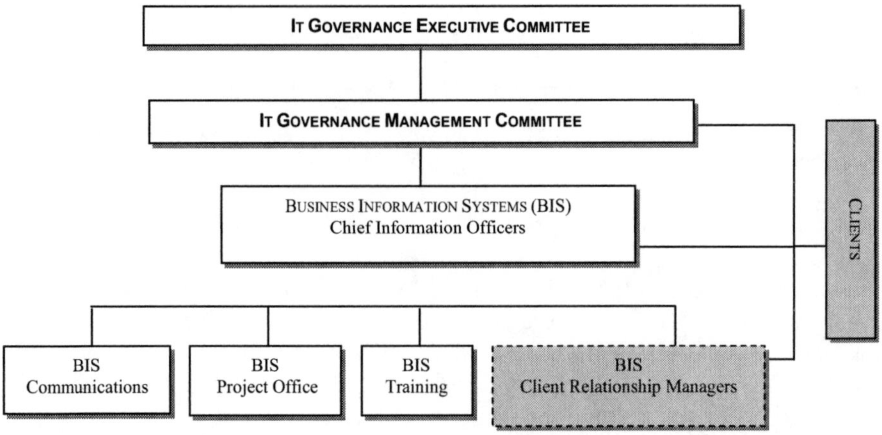

The management committee, in the absence of project prioritization tools, created a set of criteria to determine the value of each IT project in relation to how it supported corporate business objectives, and then ranked the projects according to implementation time and resource requirements. The exercise produced a corporate IT plan that garnered approval from the executive.

The IT Governance committees set about their business with a corporate IT plan in hand, a business depending on a predictable and reliable IT infrastructure, and an IT organization needing guidance.

Existing BIS organizational structures were modified to support IT Governance by overseeing the development of project initiation and delivery methodologies, quality assurance standards, resource planning tools and tools for reporting against the plan.

The role of the Client Relationship Managers became even more instrumental in establishing and building business relationships with clients, improving communications, and enhancing client service delivery. Another IT resource focused on providing an IT education and training to the newly formed governance committees.

A communications strategy to ensure an effective and targeted information flow through the organization was created.

The committees developed charters, established long-term goals around IT-business alignment, and defined their respective roles and responsibilities. The IT Governance Executive Committee was primarily responsible for setting IT direction and priorities consistent with strategic business direction, approving the corporate IT plan, ensuring IT projects achieved desired business results and providing the appropriate process to make the governance bodies effective.

The IT Governance Management Committee was responsible for managing barriers that jeopardized the completion of the IT plan. For instance, they confirmed that budget and human resources were available, and that unresolved cross-functional issues around IT projects were raised to the executive committee for resolution. They ensured that proposals for IT-related work followed defined business case criteria, monitored project implementation and results, and communicated and supported the IT direction of the organization.

The management committee met monthly. They met the executive IT Governance committee quarterly to review the IT plan and approve modifications.

IT GOVERNANCE AT NB POWER

The development of IT Governance at NB Power was marked by a series of events that contributed to its evolution, and ultimately changed its structure. For instance, the corporate IT plan, as approved by the executive committee, was quickly challenged. New business unit initiatives requiring IT resources were initiated without the governance committee's knowledge and the committee soon realized that the initial requests for IT support were underestimated. The prioritized plan of sequenced business unit initiatives requiring IT support began to change, as did the scope of the IT projects. The demands for IT work seemed mammoth.

The original members of the management committee began to change. Some members delegated their roles and responsibilities while others moved to new positions. This change in membership was perceived by some as a loss of key strategic thinkers. Although this perception has not been substantiated, it may possibly have affected the

Table 2. Roles and Responsibilities

	Roles & Responsibilities
IT Governance Executive Committee	1. Approve the 18-month IT plan. 2. Approve IT funding. 3. Approve IT policies. 4. Ensure alignment between the IT plan and the strategic direction of the organization. 5. Ensure the money spent on IT brought value to the organization. 6. Resolve issues beyond the mandate of the IT Governance Management committee.
IT Governance Management Committee	1. Prioritize IT projects on an 18-month plan. 2. Recommend an IT plan to the IT Governance Executive Committee. 3. Resolve issues (i.e., resourcing, systems integration, cross functional issues). 4. Ensure requests for IT projects have a business case. 5. Create criteria to determine the value of IT projects.
BIS - Project Office	1. Adopt project initiation and delivery methodologies. 2. Research and recommend resource-planning tools.
BIS - Client Relationship Mangers	1. Help defined client needs. 2. Ensured communications between BIS and the client. 3. Help resolve client issues. 4. Enhance client service delivery.
BIS - Training	1. Develop an education and training strategy for the Executive and Senior Managers around information systems and technology. 2. Implement the strategy.
BIS – Communications	1. Ensure timely communications between the governance committees and the organization.

committee's credibility. Some committee members believed they lacked the authority to make decisions. Others were not comfortable making decisions because of a perceived lack of knowledge about IT. Meetings between the two committees became less frequent.

Amid this seemingly tenuous existence, however, the committees met with success. Soon after the IT Governance framework was introduced, the concept of IT Governance and IT management began to infiltrate the organization, starting with the corporate executive, then the business units.

The executive endorsed several new corporate IT-related policies and standards around Internet and e-mail usage. Others followed. A corporate-wide project initiation and delivery methodology was created.

Executive members without significant IT experience participated in a year long education and training program that focused on addressing the needs of IT at NB Power. Regrettably, a similar plan was started for the management committee but did not achieve fruition.

Two of the committee's most significant contributions to IT at NB Power were the impetus to have project initiation and implementation methodologies, procedures, and templates developed and the creation of local IT Governance committees in each business unit. A team of middle managers and employees formed committees to assess the business value of IT-related projects for their respective businesses. They created tools to determine if requests for IT support aligned with business objectives and if the estimated costs for work provided a return on the investment. The local IT Governance committees made recommendations to the management committee as to whether or not a request for IT support should be considered for inclusion on the IT plan. The local governance committee prioritized the work for their own area based on business value and costs. Project prioritization and changes to the overall plan remained a responsibility of the management committees.

Ultimately though, the IT Governance management committee became frustrated. In fact, the committee stagnated. They believed they were doing little more than sequencing a list of business initiatives that required IT support rather than influencing corporate direction by aligning business and IT strategies. The committee faced a critical juncture: to dissolve their responsibilities or revitalize their original mandate to align business and IT. In 2001, the committee resolved their commitment to IT-business alignment and recommended to the executive that they absolve their current responsibilities and begin the development of a corporate IT strategy: a strategy that was business focused and business driven.

The committee presented the first drafts of a corporate IT strategy and tactical plan to the executive within six months. The strategy, approved autumn 2001, and the three year tactical plan, approved spring 2002, formed the foundation for all IT-related work at NB Power, unless otherwise mandated by government legislation.

The original two committee structures dissolved and one committee of executive members was formed, the Corporate IT Governance committee. Project managers for major capital projects, which are key to the future success of the corporation, participate on the committee until the project is complete. Based on Charles Popper's (2000) holistic framework of IT Governance, the committee accepted a new charter and mandate which elevated its responsibilities from IT project prioritization and management. Popper (2000) proposes three primary objectives for IT Governance and management: (1) fostering strategic and tactical alignment of IT with business; (2) relating the costs of IT to the value it brings to the business; (3) supporting a drive toward operational excellence. The mandate of this committee is not entirely unlike that of the previous committees, but instead of focusing only on IT-business alignment, the new committee will make a concerted effort around all three objectives. The committee is also committed to removing any barriers that jeopardize the completion of the corporate IT plan and strategy. The new committee meets every month.

FOUR YEARS LATER

As one reflects on the evolution of IT Governance at NB Power over the past four years, it is easy to identify factors that contributed positively to its initiation and growth. The executive support and sponsorship for the implementation of IT Governance at NB Power was critical to its success. The executive team provided corporate business direction and approved funding, and committed their time and that of management to manage the corporate IT plan.

The three Client Relationship Managers were instrumental in keeping the IT department abreast of client needs, educating the client about IT and educating the IT group about business. They were the conduits to effective communication between IT and business.

IT Governance at NB Power was successful because of the structures that supported the committee. Among other duties, resources in BIS, under the guidance of BIS senior managers and the IT Governance Management committee, created tools for project initiation that help capture the value of IT for business initiatives, methodologies for effective project delivery and integration, education strategies, programs, and prioritization tools. These governance mechanisms were critical in helping evolve the committee such that it could begin informal IT-business alignment exercises, make informed decisions regarding project prioritization, and understand how business decisions cannot be made in the absence of IT inputs. The idea that IT decisions must be made in the context of business objectives was also emphasized.

Having representation from all areas of the corporation has helped IT Governance succeed. The interests of every business unit and corporate service were represented, both by executive management and by senior managers. Not only did this structure ensure everyone's needs were being addressed, but it also provided an opportunity for others in the corporation to, either directly or indirectly, influence the way IT was governed and managed at NB Power.

Once the IT Governance executive and management committees were operational, local IT Governance committees were established in some parts of the organization. The local committees, in taking the responsibility to align their own IT-business interests at a local level, freed the other committees time to work on other tasks. One could argue that most importantly, it helped move the responsibility of initial IT-business alignment and determination of value/cost to yet another level of organization, thus permeating business with a greater awareness of IT-business relationships.

As the old adage, "hindsight is 20/20" implies, in looking back on the evolution of IT Governance one can derive many lessons.

A key lesson is that requests for changes to the IT plan or for IT funding must include input from the IT governing committees. This alone ensures that organizations can undertake IT-related initiatives that support organizational goals and strategies.

There is a difference between governance and management. Governance involves directing and controlling while management concerns planning, organizing, and directing. Governance is responsible for making sure the management framework is in place, not for executing it (Williams, 2000). IT Governance committees must resist the urge to manage IT initiatives or resolve issues better left for management teams and steering committees.

Besides ensuring that managers have the tools to manage IT, IT Governance committees need tools to perform their functions. Balanced Scorecards to assess the value of IT, IT auditing processes like CobiT to control IT investments, and best industry practices, like those from the IT Infrastructure Library (ITIL) are necessary for effective IT Governance.

Communications to the organization regarding IT must be effective and timely. Those responsible for governing IT, deploying IT, and requesting IT services must be informed about corporate IT decisions, goals, policies, and standards. Two-way communications between those responsible for governing and managing IT investments must remain open in order to make informed decisions about IT and business.

Business leaders must become educated about IT and those responsible for IT must develop business knowledge. An appreciation of how IT can enable or jeopardize business goals better positions everyone to make informed decisions.

FUTURE TRENDS

Predicting the future, especially for IT-related matters, is a precarious exercise at best. Researchers, however, make predictions about future trends, often with acute accuracy. Medina, Hollowell and Kline (2002) contend that the utility industry is pursuing a back to basics IT strategy. Senior managers asking to integrate the systems and applications they already have and own in order to get maximum benefits are driving this trend. Medina et al. (2002) say that many energy companies are leveraging their past IT investments rather than making new ones, limiting their technology risks, and insisting that new investments have a good chance of producing demonstrable improvements in operating efficiencies. Despite the challenges regarding how to invest in IT in order to

Table 3. Critical Success Factors and Lessons Learned

Critical Success Factors and Lessons Learned
1. Garner executive sponsorship and support.
2. Include representation from all areas of the corporation.
3. Involve the clients in prioritizing the IT support they want to receive.
4. Ensure that mangers have the tools required to successfully manage the organization's IT investments. For example, portfolio management tools, project methodologies, IT policies, and so forth.
5. Communicate the direction of IT to the organization.
6. Educate those making decisions regarding the IT investments about IT.
7. Create an IT strategy and work plan to execute the strategy.
8. Monitor and measure progress against the IT plan.
9. Have executive members approve requests for changes to the approved IT plan.
10. Have executive members approve IT funding.
11. Effectively management client relationships and expectations.

gain the greatest benefit, the authors suggest there are some fundamental strategies and tactical approaches to optimizing investments.

NB Power's own experience corroborates this suggestion. One of the six strategic imperatives of their Corporate IT Strategy is enabling realization of value from their SAP®™ and PeopleSoft®™ investments. Business unit initiatives to realize benefits from their enterprise resource planning (ERP) investments are currently underway and expected to continue for the next three years.

Since energy companies always assess their capital and operational spending because of regulatory, competitive and other economic factors, IT investments are among those items being scrutinized (Nicholson, 2002). Nicholson (2002) anticipates that the principle challenge facing IT organizations in the energy utilities industry is to ensure cost control actions are rational and targeted, and that IT is aligned with corporate strategies. Evidence to support this trend can be found at NB Power. Through the strategic IT planning exercise, the corporate executive team endorsed the strategic imperative of bringing consistent value to major capital projects through IT. This initiative was designed to ensure that technologies like SAP's Plant Maintenance module, being implemented to assist the refurbishment of one generating station, were done in such a manner that they could support other refurbishment projects.

An examination of several case studies that describe the implementation of IT Governance indicate that IT portfolio management is a critical factor for success (City of Mesa, Arizona, 2002; The North Carolina IRMC, 2001; Rosser, 2002). Those who are entrusted to govern the organization's IT investment must ensure that those managing IT have the proper tools and techniques to manage the IT portfolio. IT portfolio management is a disciplined approach for managing IT investments, including hardware, software, resources and projects. It allows organizations to balance risks, value, and costs. Companies can use this approach to identify and reduce redundant activities and then reprioritize their investment (Ward & Finnigan 2002).

Pucciarelli, Claps, Morello, and Magee (1999) predicted that by 2003 seventy percent of enterprises will institutionalize systematic assessment of IT-related projects, formalizing guidelines for project approval, continuation, suspension, and cancellation.

The META Group predicts that less than five percent of top-tier energy companies will begin to apply portfolio management to their IT budgeting process during 2002 and 2003. Between 2004 and 2006, however, adoption rates should increase rapidly and, by then twenty to thirty percent of energy IT organizations will likely be applying the principles of portfolio management in their budgeting. Another trend indicates that IT organizations are moving beyond efficiency to business effectiveness (Nicholson, 2002). In 2002, among CIOs' primary objectives will be the inculcation of value management into IT culture and the implementation of IT portfolio management as a communications and investment vehicle. By early 2004, 50 percent of IT groups will move to IT asset management (Boyle, 2002).

NB Power's own IT Governance experience has led the organization to implement portfolio management tools. Although the tool kit is incomplete, many initiatives are underway. The systems production and support teams are currently implementing ITIL best practices for some of their key processes, like change and incident management. The Business Relationship Mangers are working with the business units to sign service level agreements. A process for requesting IT work was recently implemented.

A pointed question often asked of an organization's IT group is, "Is IT bringing value to the organization and if so, how?" The stakeholders sometimes ask this out of genuine curiosity; other times they ask because of a perceived lack of creditability. When the IT organization finds itself lacking creditability, stakeholders begin to question whether IT investments and the organization itself are delivering appropriate, if any, business value (Gomoiski, Grigg, & Kirwin, 2002). Despite the reason for asking, the question needs to be addressed, though the answer can be nebulous. The definition of value, when there is one, can be subjective, or multiple meanings may exist. Value also depends on one's own perception (Brimson, 2002). Bannister and Remenyi (n.d.) maintain that the meaning of the term is assumed to be implicitly understood in IT literature. Consequently, a good definition to express the value of IT is not available.

Another difficulty with the IT-value question relates to identifying where value comes from. Each link in the business chain has a different concept of value. One of the difficulties in measuring added value is that many organizations do not understand the basic value of IT. Consequently, they cannot comprehend "added" value (Williams, 2002).

Despite the difficulties in defining value, Williams (2002) suggests a trend is emerging where board and executive members are asking, "How can I be sure that I'm getting value from my IT investments?"

Increasing the awareness of boards and executives is instrumental in answering this question. Poole (2001) says that one way to raise awareness of the value of IT is to demonstrate how to link business objectives to IT and then convince board directors why IT is crucial to sustain growth. Nothing that doesn't serve the ultimate goal of attaining business objectives is worth having unless it is driven by law or regulation.

Demonstrating how IT brings value to the organization remains a challenge for NB Power. There are initiatives in place like local IT Governance committees that assess the value of an IT request to ensure alignment with their own business goals before a formal request for support is made to the IT department. However, reputable governance mechanisms like the IT Balanced Scorecard and the CoBiT IT assessment tools are not yet employed.

CONCLUSION

IT Governance, though still an evolving idea, will continue to provide innumerable benefits to those who employ the framework and supporting mechanisms.

Boards and executives, as they continue to discern how IT provides value to the organization and ensure that IT and business are aligned, will provide the impetus to develop standard IT Governance frameworks, mechanisms and processes.

Presently, NB Power's corporate IT Governance committee has had to focus on ensuring alignment between IT and business while revisiting the corporate IT strategy and plan. In May 2002, the Provincial Government announced a major restructuring of NB Power into a Holding Company with four operating subsidiaries. The accomplishments of the former IT Governance committees and local governance groups has better positioned the new committee to meet the challenging needs of restructuring and strengthens NB Power's ability to benefit from further opportunities.

REFERENCES

Bannister, F., & Remenyi, D. (n.d.). *Value perception in IT investment decisions.* Retrieved December 14, 2002, from: http://is.twi.tudelft.nl/ejise/vol2/issue2/paper1.html.

Board Briefing on IT Governance. (2001). *Information Systems Audit and Control Foundation.* (Board Briefing on IT Governance) Illinois, USA: Author.

Boyle, L. (2002). Report: The business/IT dating game. *DATAMATION.* Abstract retrieved August 9, 2002 from database.

Brimson, J. (2002). Creating forward-looking value statements. *BetterManagement.com.* Retrieved June 27, 2002, from database.

Canadian Institute of Chartered Accountants. (2002). 20 Questions Directors Should Ask About IT. Toronto, Canada: Author.

Doughty, K. (1998). The myth or reality of Information Technology steering committees. *Information Systems Audit Control Association InfoByte,* (April). Retrieved April 7, 2000, from database.

Gomoiski, J., & Kirwin, G. (2002). The Business Value of IT and the IS Group Are Not the Same. *Gartner,* (March), SPA-15-6445. Retrieved April 28, 2002, from database.

Information Systems Audit and Control Association. (2001-2002). Standards Board. IS auditing guideline IT Governance *(Research Rep. No. 060.020.050).* Illinois, USA: Author.

IT Governance Case Study City of Mesa, Arizona, USA. (n.d.) Retrieved August 1, from: http://www.itgi.org/casestudy1.htm.

Keller, B., & Carr, J. (2001). Enterprisewide governance: The North Carolina IRMC. *Gartner,* CS-14-5938. Retrieved July 29, 2002, from Garter database.

Lainhart, J. (2001). *International standards provide guidance for IT Governance.* Retrieved May 3, 2002, from: http://www.ifac.org/News/SpeechArtical.tmpl?NID=99097368215895.

Medina, R., Hollowell, T., & Klima, C. (2002). How to optimize your IT investments. *Platts Global Energy,* (April), Feature. Retrieved July 3, 2002, from database.

Nicholson, R. (2002). Portfolio management coming to energy IT. *Platts Global Energy,* (April), Feature. Retrieved June 8, 2002, from database.

Poole, V. (2001). IT Governance roundtables spur discussion. *IT Governance*

Popper, C. (2000). *Holistic framework for IT Governance.* Cambridge, MA: Harvard University, Center for Information Policy Research.

Portal Resources, IT Governance Roundtables. Retrieved, May 10, 2002, from database.

Pucciarelli, J., Claps, C., Morello, D., & Magee, F. (1999). IT management scenario: Navigating uncertainty. *Gartner,* R-08-6153. Retrieved November 5, 2002, from Gartner database.

Rosser, B. (2002). AXA: From emotional appeal to economic model in IT. *Gartner,* CS-16-5555. Retrieved August 8, 2002, from Gartner database.

Staff. (2001). Closing the governance gap: Bringing boards into the IT equation. *Scottsdale Institute Information Edge.*

Van Grembergen, W.. (n.d.). *The balanced scorecard and IT Governance.* Retrieved June 8, 2001, from: http://www.itgovernance.org/resources.html.

Ward, P., & Finnegan, S. (2002). Proven methodology is key to balancing risks with ROI — Aligning IT and business expectations. *META Group*, Press Room. Retrieved July 18, 2002, from META Group database.

Williams, P. (2000). IT Governance roundtable — Sponsored by the governance institute. *IT Governance Portal Resources*, IT Governance Roundtables. Retrieved, May 10, 2001, from database.

Williams, P. (2000). Transcript of IT Governance roundtable. *IT Governance Portal Resources*, IT Governance Roundtables. Retrieved, June 6, 2001, from database.

Williams, P. (2002). Bringing the board online with IT Governance. *ComputerWeekly CW360Ú, Strategy, Governance and Organizational Structure.* Retrieved June 5, 2002, from *ComputerWeekly* CW*360°* database.

Williams, P. (2002). Value versus costs: Governing IT on a reduced budget. *ComputerWeekly CW360Ú, Strategy, Governance and Organizational Structure.* Retrieved January 18, 2002, from *ComputerWeekly* CW*360°* database.

Chapter XIV

Governance Structures for IT in the Health Care Industry

Reima Suomi
Turku School of Economics and Business Administration, Finland

Jarmo Tähkäpää
Turku School of Economics and Business Administration, Finland

ABSTRACT

In this chapter we bind together three elements: governance structures, the health care industry and modern information and communication technology (ICT). Our hypothesis is that modern ICT has even more than before made the concept and operation of the governance structures important. ICT supports some governance structures in health care better than others, and ICT itself needs governing. Our research question also is: which kinds of governance structures in health care are supported and needed by modern ICT?

Our chapter should be of primary interest for Health Care professionals. They should be given a new, partly revolutionary point of view to their own industry. For parties discussing governance structure issues in Health Care, the chapter should give a lot of support for argumentation and thinking. The models and conclusions should be extendable to other industries too. For academic researchers in Governance Structure and IT issues, the chapter should contain an interesting industry case.

INTRODUCTION

The pressures for the Health Care industry are well known and very similar in all developed countries: altering population, shortage of resources as it comes both to staff and financial resources from the taxpayers, higher sensitivity of the population for health issues, and new and emerging diseases, just to name a few. Underdeveloped countries dwell with different problems, but have the advantage of being able to learn from the lessons and actions the developed countries made already maybe decades ago. On the other hand, many solutions also exist, but they all make the environment even more difficult to manage: possibilities of networking, booming medical and health-related research and knowledge produced by it, alternative care-taking solutions, new and expensive treats and medicines, promises of the biotechnology…you name it (Suomi, 2000).

From the public authorities' point of view the solution might be easy: outsource as much as you can out of this mess. Usually the first ones to go are marginal operational activities, such as laundry, cleaning and catering services. It is easy to add information systems to this list, but we believe this is often done without a careful enough consideration. Outsourcing is too often seen as a trendy, obvious and easy solution, which has been supported by financial facts on the short run. Many examples, however, show that even in basic operations support outsourcing can become a costly option, not to speak of lost possibilities for organizational learning and competitive positioning through mastering of information technology.

In our chapter, we discuss the role of IT in Health Care, and focus on the question, "Which governance structure(s) are best suited for managing IT within the Health Care industry?" Our basic hypothesis is that information is a key resource for Health Care and that the managing of it is a core competence for the industry (Suomi, 2001).

Our analysis is restricted to public primary Health Care. We maintain that the governance problems are most acute there, because of several reasons. First, here the total spectrum of different customers and diseases is met. No customers can be selected or neglected, but all are entitled to some basic level of care. Possibilities to collect more resources, say through customer fees or through financial market operations, are limited (Suomi & Kastu-Häikiö, 1998). As compared to special care units, the activities are fragmented and performed in smaller, less well-equipped units. As compared to the private sector, commercial thinking is less mature: outsourcing for commercial actors in the Health Care industry is a natural topic, as the private companies themselves are just the ones to whom activities are outsourced.

Our chapter shortly discusses two case examples — the primary Health Care in a small and in a middle-sized city — that are reported in closer detail elsewhere (Holm, Tähkäpää, & Suomi, 2000; Suomi, Tähkäpää, & Holm, 2001; Tähkäpää, Suomi, & Holm, 2001; Tähkäpää, Turunen, & Kangas, 1999). Here we aim at showing how the concepts we introduce are reflected in the reality. Picking out two different cases also serves the goal of discussing the competences and resources needed for outsourcing IT activities. Our message is that keeping the IT activities in-house demands certain resources and skills, but similarly outsourcing them cannot happen without any own resources and skills dedicated for this purpose. We discuss the situation of small entities that seem to be stuck in a dead-end situation: not enough expertise and resources either for in- or outsourcing.

THE CONCEPT OF
GOVERNANCE STRUCTURE

The concept of a governance structure is by no means settled or well defined. We define a governance structure here as:

"A structure giving meaning and rules to an exchange relationship."

Lately we have seen many writings simply stating that governance structure is the same thing as management. Governance issues would be those of management issues. We strongly believe and stress that management is a different concept from governance structures, when of course management is needed in the case of governance structures, and governance structures also exist in management.

Our definition above conveys many details. First, the term *structure* refers to something stable that will last over a period. Governance structures are sure to change over time, but economizing exchange relationships necessitates that governance structures are of lasting nature. Should governance structures change all the time, no exchange relationship would be on a permanent basis.

With *meaning* and *rules* we refer both to the motivation and *guidance functions* of exchange relationships. As governance structures are there to guide exchange relationships, it is natural to expect that they try to foster them. As meaning refers to something meaningful, governance structures of course try to eliminate negative behavioral effects in exchange relationships, such as opportunism (Conner & Prahalad, 1996; Dickerson, 1998; Genefke & Bukh, 1997; Lyons & Mehta, 1997; Nooteboom, 1996), bounded rationality (Simon, 1991) and information asymmetry (Seidmann & Sundararajan, 1997; Wang & Barron, 1995; Xiao, Powell, & Dodgson, 1998), moral hazard (Jeon, 1996), and small numbers bargaining or negative network externalities (Kauffman, McAndrews, & Wang, 2000; Koski, 1999; Shapiro & Varian, 1999).

The *rules* or *guidance functions* contain three types of entities:

- rules on how an exchange relationship can be entered,
- rules on how to perform an exchange relationship,
- rules on how to control and follow-up an exchange relationship.

For example, certain exchange relationships can be reserved just for qualified partners. Just certified partners are entitled to run many transactions, say buy and sell options in stock exchanges or sell medicines. Rules on how to perform exchange relationships can be many and detailed. As a control mechanism should be seen as a permanent entity, it should have some control mechanisms that will foster successful exchange relationships and eliminate bad exchange relationships on the long run. To take a fanciful example, the www-site looking for flawed health care information Quackwatch (see *Quackwatch www-site*, 2003) is an example or guidance functions of an exchange relationship.

Finally, the *exchange relationship* can be understood in many ways. First, an exchange relationship can be seen as a transaction, where the relationship between the transaction partners is usually both short and well defined. The other end of the continuum is a long-term relationship. Further, the exchange relationship can be onerous, or happen without any visible or instant payment. The object of the exchange relationship can be any, including information, which role of course gains in importance in the information society.

Missing governance structures can be a major problem. Take for example the Eastern European transition economics. Missing regulations and infrastructure elements severely hamper trade and any exchange relationships, as for example documented in Kangas (1999). The same is true for the new e-business, where many environmental factors for exchange relationships still are underdeveloped, making actions in the new economy risky ones.

This leads us to ask who is responsible for the governance structures. We must differentiate between *obligatory* and *selected* governance structures. *Obligatory* governance structures are something that must be taken into account and cannot be omitted. Binding legislation is a good example. *Selected* governance structures are picked up by the exchange relationship parties themselves. For example, exchange partners can select which channel (for example an information system) to use in the exchange, and for example in which currency the transaction will be settled.

Governance structures exist in many levels. Any exchange relation between two parties needs some kind of governance structure. Even when organizing their own personal activities, people naturally use some kind of reference frame, a governance structure. A governance structure has a relation to the roles individuals carry. The same person is sure to behave differently, also to have different governance structures for exchange relationships, say in the roles of a husband, father or boss. Running the analysis on an organizational level, we can differentiate between *intra- and inter-organizational* governance structures.

Governance structures are used in many disciplines. Naturally the topic suits to *political science*, where government of the citizens and organizations is a central topic. *Economics* is a natural place to study governance structures, and here it is very much about economizing exchange transactions (Williamson, 1985; Williamson, 1989); typical examples can be seen in Madhok (1996) and Mylonopoulos and Ormerod (1995). The most diverse and recent application area is *management*, including information systems management, where governance structures can be applied in many ways. As we discuss contracts, *law* is a natural background too. On the contrary to the tradition in economics, in management governance structures are not just used for economising purposes, but exchange relationships can be seen as tools for many organizational goals. Typical managerial problems are resource allocation (Loh & Venkatraman, 1993), eliminating risks in exchange relationships (Nooteboom, Berger, & Noorderhaven, 1997), motivating organizational stakeholders (Hambrick & Jackson, 2000; Peterson, 2000) or building trust in exchange relationships (Calton & Lad, 1995; Nooteboom, 1996).

There are some major tools to establish governance structures. At the national and international level, *legislation* is the key concept. At an individual exchange relationship level, there is usually a *contract* defining the exchange relationship. *Organizations* are strong mechanisms to cover exchange relationships inside them. Fluent exchange relationships are made possible by *information* (and information systems), *money*, *trust* and *cultural* customs, such as organizational culture or trading conventions within an industry. These are also all elements of governance structures.

Using the term "governance structure" is by no means new in the health care sector. Pelletier-Fleury and Fargeon have used the concept in connection with the process of

Figure 1. The Main Tools to Affect Governance Structures and Disciplines Providing the Conceptual Basis for Studying Them

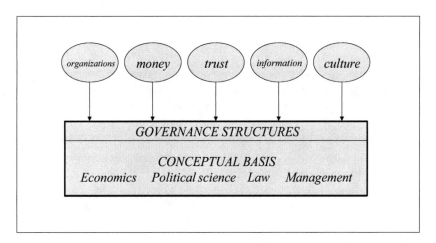

diffusion of telemedicine (Pelletier-Fleury & Fargeon, 1997). Spanjers et al. have studied the general networking or governance structure strategies of hospitals (Spanjers, Smiths, & Hasselbring, 2001). Further, Donaldson and Gray (1998) discuss how quality can be maintained through the smart design of governance structures. There is a rich research tradition on the new opportunities modern ICT offers for organizing health care; see for example Suomi (2001).

TOOLS FOR ANALYZING GOVERNANCE STRUCTURES

In this section, we shortly discuss three disciplines to study governance structures. They are handpicked and surely do not cater for all the possibilities of analyzing governance structures. The first and most classic is that of transaction costs. Agency cost concepts are closely linked to those in the transaction cost analysis. An established concept is also that of a value chain. Finally, trust as an element in governance structures is shortly touched upon.

Transaction Costs

The transaction cost approach (TCA) is founded upon the following assumptions (Williamson, 1985):

- The transaction is the basic unit of analysis.
- Any problem that can be posed directly or indirectly as a contracting problem is usefully investigated in transaction cost economizing terms.
- Transaction cost economics are realized by assigning transactions (which differ in their attributes) to governance structures (which are the organizational frameworks

within which the integrity of contractual relation is decided) in a discriminating way. Accordingly:

- the defining attributes of transactions need to be identified,

- the incentive and adaptive attributes of alternative governance structures need to be described.

- Although marginal analysis is sometimes employed, implementing transaction cost economics mainly involves a comparative institutional assessment of discrete institutional alternatives — of which classical market contracting is located at one extreme; centralized, hierarchical organization is located at the other; and mixed modes of firm and market organization are located in between.

- Any attempt to deal seriously with the study of economic organization must come to terms with the combined ramifications of bounded rationality and conjunction with a condition of asset specificity.

A very central concept is that of a transaction cost. Transaction is a difficult concept that materializes in several levels (Figure 2). First, each transaction has its exchange object(s), actors performing the transaction, and some channel(s) through which the transaction is performed. These offer the basic ramifications for any transaction and its associated transaction costs. In general, transactions tend to be more fluent the better the channel for them and the more voluminous they are. In literature, the main conceptual reasons for transaction costs are those of asset specificity, complexity of product description, bounded rationality and opportunistic behaviour. From the concepts, there is still a long way to the actual measurement of transaction costs, which is a difficult task.

The basic distinction of TCA among different organizational forms is the distinction between markets and hierarchies (Coase, 1937), which are forms of economic organizations. Given the division of labor, economic organizations control and coordinate human activities.

A market is an assemblage of persons which tries to arrange the exchange of property, where prices serve as both coordinating guides and incentives to producers in affecting what and how much they produce — as well as the amount they demand. At the equilibrium free-market price, the amounts produced equal the amounts demanded — without a central omniscient authority (Alchian & Allen, 1977).

In a hierarchy (firm) market transactions are eliminated and in place of the market structure with exchange transactions we find the entrepreneur-coordinator, the authority who directs production (Coase, 1937).

In addition to these two basic forms of organizational design, research on the subject has produced several sub-forms of organizations.

In the early days of transaction cost approach, the focus was mostly on hierarchies, as this was the dominant governance structure. An example of this focus is A.D. Chandler's division of hierarchies into multidivisional and unidivisional structures (Chandler, 1966).

The most important of the current developments of organization forms is the concept of groups or clans by Ouchi (1980). He breaks down hierarchies into bureaucracies and clans. These two organizational forms differ in their congruence of goals. Clans have a higher goal congruence than bureaucracies, and thus are further along in their attempt to eliminate transaction costs.

Figure 2. A Tri-Level Transaction Cost Framework

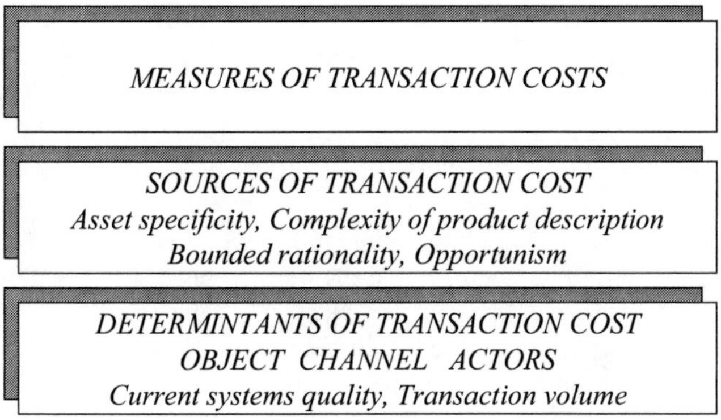

MEASURES OF TRANSACTION COSTS

SOURCES OF TRANSACTION COST
*Asset specificity, Complexity of product description
Bounded rationality, Opportunism*

DETERMINTANTS OF TRANSACTION COST
OBJECT CHANNEL ACTORS
Current systems quality, Transaction volume

Cooperative behavior among firms is the root of many success stories of today's management (Jarillo, 1988). Like many other authors, he calls for a generally accepted framework for the study of inter-organizational systems. His contribution to the framework formulation is the concept of a strategic network. In discussing markets and hierarchies, he further divides markets into two segments, the segments of "classic market" and "strategic network". The difference between these two concepts lies in how transactions are organized: they can be based on competition or on cooperation, respectively.

Thomas Malone introduces several other organizational designs. He studies organizational forms and their effects on production, coordination, and vulnerability costs. Focusing on the internal organization of a firm, he introduces the following organizational designs (Malone, 1987):

- product hierarchy,
- decentralized market,
- centralized market,
- functional hierarchy.

In a product hierarchy, divisions are formed along product lines. In a functional hierarchy, similar processors are pooled in functional departments and shared among participants. In the realm of decentralized markets, different kinds of processors can be freely acquired from the market: processors supplied by different organizational units can be freely interchanged. In the case of centralized markets, freedom to choose remains but all processors must be collected from the same place.

We can further differentiate between six types of transaction costs (Casson, 1982):

- information costs,
- costs caused by requirement analysis,
- costs caused by negotiating,
- costs caused by initiating the transaction,
- costs caused by monitoring the transaction,
- costs caused by making the transaction legal.

The Value Chain

One of the most established governance structure concepts is that of the value chain as presented in Porter (1985). Since then the concept has been widely used, but has also awakened a lot of critique for its simplicity. The basic idea of the value chain is a one-directional flow of material and information in a production process. The value chain emphasizes the resources needed for production, but does not mention information or information systems, at least not explicitly. Analysis of the *flows of information, money and physical goods* is a key task for understanding any exchange transaction.

The strength of the value chain is its simplicity. It paved the way to the thinking that organizations should concentrate on the main value-adding activities, later called core competencies. The weakness of the value chain lies in its one-direction flow of activities. The value chain is unable to explain complicated market-based interactions, not to speak of modern virtual organizations.

The value chain helps individual participants in exchange relationships to understand their place in the totality. It is too strong in focusing attention to the *value-adding elements* of any exchange relationship, calling for less attention to those traits that do not add value to the exchange relationship.

Trust

Trust is a general concept usable in all human activity. It is present in some way, most visibly when absent, in all exchange relationships. We can define it as a one- or two-direction relationship between a human and a system, which according to Checkland (1981) can be one of the following:

- Natural system, including human,
- Designed activity system,
- Designed abstract system,
- Designed technical system,
- Transcendental system.

With the two first ones, the Trust relationship can materialize in two directions. You can trust a natural system and a designed activity system, and that one can trust you. Trust might be defined as an individualistic feature of human relations. Even in case of Trust existence as interorganizational Trust, de facto it is Trust between those organizations' managers and their staff consultants. Here we would like to refer to Berger (1991): *"The most important experience of others takes place in the face-to-face situation, which is the prototypical case of social interaction. All the other cases are derivatives of it."*

In transaction cost economics, Trust is not a key concept. However, the discipline puts emphasis on at least two dysfunctional phenomena that exist in a transaction if Trust is absent: Opportunism, Moral hazard.

In this connection, Trust is a key element in the fight against transaction costs. As Thompson (1967) cites: *"Information technology belongs to those technologies, like the telephone and money itself, which reduce the cost of organizing by making exchanges more efficient."* We might add, *"Trust belongs to those technologies, like the telephone, information technology and the money itself, which reduce the cost of organizing by making exchanges more efficient."*

We summarize the basic conceptual tools usable for studying governance structures in Table 1.

TRENDS AND PRESSURES IN THE HEALTH CARE INDUSTRY IN EUROPE

In the European region, as also in the rest of the world, the health systems and services are undergoing a major transformation as a consequence of changes in age structures, social imbalance, increase in unhealthy lifestyles and new diseases. Fundamental differences between countries and especially between Western and Eastern European countries and their health problems have emerged. EU published a report (*The State of Health in the European Community*, 1996) about the state of the health in 1996 in which a few main features in development of population in EU region were discovered: fewer children, more older people, people live longer and differences persist between countries and regions. Especially the difference in life expectancy between the Western and Eastern European countries was highlighted in the report, as well as in the European Health report 2002 by World Health Organization (WHO) (*The European Health Report*, 2002). The difference of the share of population living below the poverty line is considerable between these countries. Poverty has a clear effect for the upward amount of illnesses due to communicable diseases (HIV/AIDS, tuberculosis) in the Eastern Europe. In the Western Europe the non-communicable diseases (e.g., cardiovascular diseases, cancer, neuropsychiatric disorders, overweight) account for about 75% of the burden of ill health and constitute a "pan-European epidemic". In addition to the concept of Digital Divide (Compaine, 2001; Norris, 2001) we can also well establish the concept of "Health Divide".

In the Communication on the development of Public Health Policy (OECD, 2003) EU has defined the following challenges facing the Member States: *"Health care systems in the Member States are subject to conflicting pressures. Rising costs due to demographic*

Table 1. Conceptual Tools Offered by Different Disciplines to the Analysis of Governance Structures

---

The transaction cost approach
- Transaction as a unit of analysis
- Basic governance structures: markets and hierarchies
- Transaction costs and economising them

The value chain model
- The value chain, the individual exchange relationship as a part of a totality
- Flows of information, money and physical goods
- Value-adding activities

Trust
- One- or two-directional relationship
- Inter-personal and inter-organizational trust
- Trust as an eliminator of transaction costs

---

factors, new technologies and increased public expectations are pulling in one direction. System reforms, greater efficiencies and increased competition are pulling in another. Member States must manage these conflicting pressures without losing sight of the importance of health to people's well-being and the economic importance of the health systems."

This Communication points out several important challenges, but in economic perspective, rising costs and the economic importance of the health system are especially interesting. Further, in discussion about technological and other supply-driven developments the Communication brings out the management issues. *"Computerisation and networking, including the implementation of health care telematics, may help reduce health costs, particularly in relation to the management of health care."*

The increasing costs have driven countries to develop and find new solutions in organizing their health care but still retaining the high standards and availability. Reorganizing functions and processes, new strategies, information systems and management issues play an important role in this effort. WHO has found four trends in organizing health services in Europe (*The European Health Report*, 2002):

- Countries are striving for better balance sustainability and solidarity in financing. Especially in the Western countries the solidarity is kept at a relatively high level.

- There is an increasing trend towards strategic purchasing as a way of allocating resources to providers to maximize health gain. Those are e.g.:
  - Separating provider and purchaser functions,
  - Moving from passive reimbursement to proactive purchasing,
  - Selecting providers according their cost-effectiveness,
  - Effective purchasing is based on contracting mechanism and performance-based payment.

- Countries are adopting more aggressively updated or new strategies to improve efficiency in health service delivery.

- Effective stewardship is proving central to the success of health system reform. This role (health policy, leadership, appropriate regulation, effective intelligence) is usually played by governments but it can also involve other bodies such as professional organizations.

Issues like financing problems, management strategies, provider-purchaser models and professional economical management are not the concepts which have been under very close attention in health care. The late adoption of these concepts have resulted that health care organizations are still in relatively early stages in learning to internalize them. The boost for the adoption is mostly the result of a serious recession in economics in the 1990s, which forced also public organizations — including health care — to consider issues like effectiveness. Before that, health care did by no means waste money or resources but those were not as scarce as today.

In the business environment effectiveness has been a central mantra for decades; mostly, therefore, these concepts were adopted from there. However, business environment is quite different in many ways, which don't make the adoption any easier. The differences can be seen to originate already from research paradigms, which are different, e.g., in medicine and business economics. The paradigms have influence on education and through that also to the professions and are therefore deep in organizations and difficult to change (Turunen, 2001).

Although it is also a consequence of the effectiveness demand, increased use of information and communication technology can be seen as a trend as well. In health care there can be seen several trends and visions about the increased use of ICT. After the use of ICT started almost from scratch in the beginning 1990s it has increased exponentially.

One of the most visible and effective trends has been the introduction of electronic patient record (EPR) systems, which are widely in use today. For example in Finland about 63% of public health care organizations use EPRs and the number is increasing rapidly. Use of electronic records has several advantages like easier and faster access to customer information, and information is in real time and thus the reliability of information is better. However, most of the advantages are still most likely not yet achieved. E.g., though implementation of systems is successful, organizations have mostly failed in renewing their processes which new systems enable and require to become effective. Another nation-wide attempt in Finland is to integrate the local systems to one nation-wide system where the information of patient is available for the clinician regardless of time or place.

Another interesting trend caused by increased electronic health information is the use of Internet. Health information is one of the most frequently sought topics on the Internet, with more than 40% of all Internet users. It is second in popularity after pornography (Nicholas, Huntington, Williams, & Jordan, 2002). Increased use of the Internet has surfaced also the question about reliability of the information acquired. Reliability is also concerning electronic systems in health care organizations. Along with the increased use of electronic health information, also organizations assessing that information have increased. Organizations like Health on the Net Foundation (HON) assess information on the Internet. National and international example is Finnish organization FinOHTA, which supports and coordinates health care technology assessment and distributes both national and international assessment results within the health care system (Järvelin, 2002).

Internet has advanced also the use of different types of call, contact and communication centres. Those are established in an attempt to concentrate some of the services on one place including usually phone or Internet contacts. Centralizing certain services in one place should bring the advance for both to the patient when she/he can contact one place to get service, and to the organization that can offer information through phone or Internet and thus avoid unnecessary visits to doctors or nurses. One of the best-known examples of these services is in UK offered by NHS (NHS Direct Online). In Figure 3 we present a conceptual model of a call centre. Without going into details, in the figure we can see that the usage pattern of a call centre is dictated by the users (actor domain) and by the functionalities built in the system (system domain). For more information on the model, see Suomi and Tähkäpää (2003).

A current trend in the health care technology can be seen in the use of mobile technology in communication between the patient and the clinician. Sending and receiving information through mobile phone is however still quite clumsy and the benefits of the use of it are not yet fully proven. Wireless communication is however in use, e.g., in hospitals and health centres where doctors use portable computers in their daily visits to the wards to get access in patient health records online.

Figure 3. A Conceptual Model of a Call Centre in Health Care (Adopted from Suomi
& Tähkäpää, 2003)

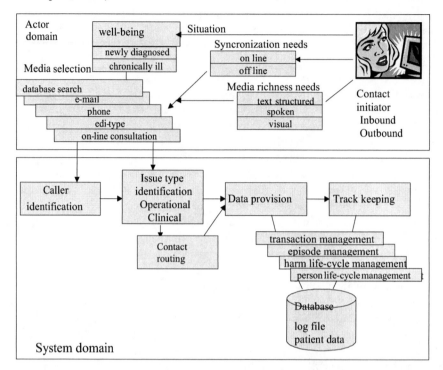

There are several other trends and visions about use of ICT in health care (like use of digital images) which are not mentioned here (Table 2). However, the basis of the use of ICT seems to be the use of an integrated database, like EPR, in which all the information about a single patient is integrated so that it is available for the clinicians in their decision-

Table 2. Summary of Trends in the Health Care Industry

- The gap between the rich and poor countries, the "Health Divide" even growing
- Increasing costs and increased economic importance of the health care system
- Demographic pressures to the health care system
- Increased expectations of the public
- New technologies offering solutions and pressures too
- Health care must happen in an environment valuing solidarity and sustainable environment
- Outsourcing health care functions from the public sector might be a solution
- Strategic planning is increasingly taking place
- Effective stewardship is of central importance
- Thinking patterns and paradigms in business management and health care are integrating in the health care sector
- Application of modern ICT is increasing fast
- The Internet offers new functionalities for the health care sector
- Several electronic means are used for communication between patients and health care staff

making despite the time or location. The EPRs in use today are not yet able to offer all the required information from one application.

However, two trends or visions are certain: First, the pressure for more effective organizations is not going to diminish in health care for the next few decades and new solutions, forms of organizations and methods have to be searched for. Second, once the health care industry got off the ground in using ICT, it is not going to diminish its use, and therefore the development of new technology for health care is going to increase ever faster (as in most of the other industries too) and governance structures and management issues are going to play ever-increasing roles in trying to get all the benefits from it.

THE PUBLIC-PRIVATE HEALTH CARE GOVERNANCE ISSUE

In this section we discuss the different role of private and public health care and the differences in their management and governance structures. The aim is to sketch the complexity of public sector health care governance issues. This issue is made even more complex through the recent adoption of market thinking, business originated strategies and competition discussion in the sector.

When organizing the services, public and private health care use quite similar processes at the operational level. The visit to a nurse or a doctor because of flu or fracture generates very much similar processes and transactions both in the public and in the private sector. The same information and materials are needed in both organizations to cure the illness so you could claim that the value chain is similar and produces the same value to the organization and to the patient. The differences appear of course in money flows, but the basic cure process is very much the same.

However, when the organizations are studied at the upper, strategic level, the difference of the governance structures starts to appear. The public and private sectors differ from each other in several ways in terms of goals, decision-making, fund allocations, job satisfaction, accountability and performance evaluation. Typically public organizations have little flexibility in terms of fund allocations and very little incentive to be innovative. Rigid procedures, structured decision making, dependence on politics, high accountability by public and administration, and temporary and politically dependent appointments are features connected to public sector organizations and employees (Aggarwal & Mirani, 1999).

The private sector has different goals as they seek to enhance stakeholders' value and maximize profits. They are more flexible than public organizations in terms of budget allocation, personnel decisions and organizational procedures. Merit and award systems are mostly well defined and new ideas that maximize firms' value are encouraged (Aggarwal & Rezaee, 1995).

As these definitions show, the difference is high, especially in organizing activities. While the public sector has to follow strict rules, private companies can organise their activities according to the market situation. When we look at the definition of governance structure and the words *meaning* and *rules* in it, the importance of the latter is very big in the public sector. Most exchange relations have to follow strict rules, say especially in purchasing: the public organizations have to organise a public competitive bidding when purchasing services or goods above a certain value.

On the other hand the word *meaning* has probably a stronger emphasis in the private sector. Public health care organizations are guided by national politics and political decisions, which may be well thought of, but which anyway are given from above and which are thus more distant and abstract than strategies built by the organizations themselves.

However, because of several changes in political and economic environments as well as the changes in technology, public sector is facing the same uncertain and turbulent environment as the private sector has always faced. In this new environment, public sector organizations are expected to exhibit many features usually seen in the private sector, including some scope of entrepreneurial behaviour. This shift has not been totally accepted in the public sector and there is a concern that the application of the language of consumerism, the contract culture, excess performance management and the use of quasi markets might create problems. It is argued that all these need to be balanced by approaches that recognize the value of the public sector (White, 2000). The complexity is of course dependent also on the size of the organization. The larger the organization is, the more administrative information is included (Spil, 1998).

Increased complexity and turbulent environment refers to the changing *structures* of the public sector. Until the last decade the structures of the public sector have remained quite stabile because of governments' strong role in steering them. Starting from the 1980s, however, decentralization and local empowerment have also invaded the public health care sector. Therefore one could say that at the moment the structures in the public sector health care are not on a permanent basis — rather they are in a turbulent phase. It might be that effectiveness cannot be achieved in the public sector because of the ongoing turnover phase of the industry.

One distinctive difference between the public and private sector which cannot be bypassed since it greatly affects governance structures through management is the group of stakeholders the sectors have to satisfy. While the private sector is to maximize the profits of the owners (to use rough generalization), the public sector has more critical stakeholders. Of course neither in the private sector is this so simple as, e.g., employees are a strong stakeholder group with its own interests inside the organization. Employee demands cannot be set aside. And despite the differences managers must work in both organizations to find a point where most of the stakeholders are satisfied most of the time. In many cases increasing the satisfaction of one group of stakeholders decreases the satisfaction of others (Dolan, 1998). This affects the structure of exchange relationships as the stakeholders eventually decide (consciously or unconsciously) whether the relationships the organization maintains are in accordance with their demands.

Another view to the public-private sector governance structure is to discuss the issue from a national perspective. The private and public sectors share the health care markets and the national government and legislation have a great effect on those shares. The obligatory governance structures play an essential, role especially in the public sector: they have many responsibilities that they cannot escape. Next we will describe some features of the roles and market shares of public and private health care using Finland as an example.

All health care services are financed mostly (60%-80%) from the state or municipal taxation and the remainder from the National Insurance Scheme (10%-20%) and co-payments. The private health care sector is seen more or less as complementing rather than competing against the public health care. The markets for the private sector have established themselves slowly, mainly because of the extensive role of the public services. By 1996 the share of private doctor consultation in Finland was 16% of all doctor consultations and the share of doctors who practice solely in the private sector was 5%. The total share of private health care services was 22%. The private sector has the strongest market share in general practice visits, dentist and physiotherapist services and in employee health services.

In Finland health care authorities at the local level have gained more independence in organizing their governance structures since the state subsidiary system changed in 1997. Earlier the rule was that local public health care should produce primary care as an internal service and that the state subsidy was granted on the bases of population, morbidity, population density and land area. Since 1997 the criteria were changed and local authorities gained more independence in organizing the services according to the local needs. They were encouraged to use methods and approaches familiar from the private sector business environment.

Some opposite developments have been seen at the international level. In European countries the need to strengthen the stewardship role of the state appeared with the introduction of new market mechanisms and the new balance between the state and market in health systems. Thus, policy makers have sought to steer these market incentives to achievement of social objectives (*The European Health Report*, 2002).

The government's target is therefore both to increase the independence but at the same time to steer the development. This is a hybrid form of market and hierarchies where the market works with the rules set by an entrepreneur-coordinator (government) who directs the production (transactions).

IT SOURCING GOVERNANCE ISSUES

Outsourcing is seen in many organizations as a means to get rid of everything unpleasant or unknown. The solution is in some cases good but in some cases it is also a way to lose even the last understanding and control of the outsourced activity. The solution is not as simple as it might sound. Many firms have made sourcing decisions based just on anticipated cost saving without further consideration of its effects on strategies of technological issues (Kern & Wilcocks, 2000).

IT outsourcing can be defined in the following way (Kern & Wilcocks, 2000): *"IT outsourcing can be defined as a decision taken by an organization to contract out or sell the organization's IT assets, people, and/or activities to a third-party vendor, who in exchange provides and manages assets and services for monetary returns over an agreed time period."* Kern and Willcocks stress the significance of contract and control in their article. They further underline that *"the client-vendor relationship is indeed more complex than a mere contractual transaction-based relationship."* Also control is seen as an essential but complex issue in these relationships. IT outsourcing tends to be more complex than many other forms of outsourcing such as cleaning, catering or calculation of salaries because IT pervades, affects and shapes most organizational processes in some way (Kern & Willcocks, 2002).

The tools for analyzing the governance structures described earlier included the element of trust. Contracts are the ones with which the mistrust is tried to reduce. With a comprehensive and mutually agreed upon contract the trust can be improved but no matter how comprehensive it is, there is always a possibility to understand it in a different way than a partner. As trust is an individualistic feature of human relation, contract and especially its reading are a result of this feature: You (the individual) can interpret it to your own benefit if needed.

Thus, outsourcing needs first of all a good contract. Another main issue in outsourcing is control. As long as you have your information systems insourced the biggest concern is to control the professionals in the department. They have to be skilled, motivated and they have to have enough resources to perform their work. However, it is mostly the matter of technology and personnel management. Of course you have to also to ensure the quality of the outcome in a way that it serves the organization most effectively.

When outsourcing, the management loses a part of this control. Also new issues connected with trust like opportunism and moral hazard are emerging and those are a lot more difficult to control than just internal resources. In outsourcing the contract is juridical, done between two organizations, but the trust is always emerging (or not) between two individuals. Therefore choosing persons to negotiate and consummate an agreement is not a trivial case. In addition to that those persons have to be familiar with the technical and juridical issues of the contract, they have to have also some knowledge about human nature and knowledge about the meeting technique.

GOVERNANCE STRUCTURES IN ICT SOURCING SOLUTIONS

In this chapter we discuss two research cases that we conducted in the Finnish health care industry. The first case is about development of information management strategy for a small health care federation of municipalities. Creating a strategy for an organization is a typical internal project and suggests that there is a need to study the goals of the organization and structures with which it can achieve those goals. Internal governance structures are an essential part of this development. The second case is an extensive evaluation of a large information systems implementation project with outsourcing solutions. Outsourcing activities suggest that there is a need for developing and managing external governance structures. We also have two different types of cases to discuss governance structures. The purpose of this is to find out differences between medium and small size organizations and between internal and external governance structures and needs.

Case 1: Organizing the Internal Governance Structures of a Small Health Care Unit

In the first case two small municipalities compose a health care federation running a health centre to provide health care services for the population of some 13,000 people in the area. In the Finnish scale it presents a small organization, though a fairly typical one. The organization had used a new system for over a year including hardware, EPR and network. The patient records were earlier in manual form and the new EPR is the heart of the new information system of the organization. The implementation had gone well and the staff began to get acquainted with the system. However, problems started to appear after some time. They were not yet massive, but the management noticed that they did not have enough expertise to develop the system towards the goal of better supporting organization goals and strategies.

The organization had created a comprehensive business strategy, which the new system was supposed to support. One of the most important issues in the strategy was that the organization should be able to support the provider-purchaser model. The purchaser (the municipalities) expected to get a fixed per unit price for the services they buy. To be able to satisfy the purchasers the health care federation needed a system from which they could dig the required information to set the prices. In addition, the health care centre is expected to give an assessment of the level of demand for different health services by the population.

Our research team developed an information management strategy in co-operation with the management and the staff of the federation. The ultimate goal was to solve how the federation could manage its ICT development effectively to support and also to steer business strategy. The strategy development team included our research team (three researchers) and members from the management and from the staff (three to five persons) of the federation. The development team had meetings weekly. We carried out some 60 interviews including staff and management of the federation, management of municipalities, and vendors.

The starting point was that the governance structures to steer ICT functions were not clear and therefore one of the main goals was to develop those structures through creating a strategy for the management of ICT. The development needs were mostly intra-organizational (Figure 4). Although there were also problems in inter-organizational relationships, say in connections to the municipalities and to the system vendors, the problems in those relationships were mostly due to lack of clear internal organization to handle them.

One of the first tasks was to solve the problem of responsibilities for different components of the system. The system had five clear areas (see Figure 4) and each of them was allocated an owner. Earlier the manager of the federation had that responsibility of all areas almost totally himself. Owners were selected from the management of the organization, and they had total responsibility over their areas. Main users who were links to the practice in different departments supported them. Main users were selected from the staff on the basis of their position, ICT knowledge and interest. The structure was primarily defined to clarify the organization of ICT management, but since ICT is the heart of the value chain, the work naturally served also the management of organization's value chain. In an information-intensive industry like health care the fluent flow of information can be "dead serious".

With defined responsibilities the organization had a clearer structure to manage ICT, which is crucial in any governance structure attempting to economize the exchange relationships. A structure should be lasting, so in our model the ownership is not bound to a person but to a position. The persons may change but the position is most likely going to remain. With the clear structure of system owners the information flows more efficiently also to the management. Achieving a lasting structure gave possibilities to define rules to manage it — a stable structure supports our main goal, setting up a decent information management strategy.

Exchange relationships with external stakeholders became also clearer. The organization had now a structure with rules with which it could handle relationships more

Figure 4. The Governance Structure in the Insourcing Case

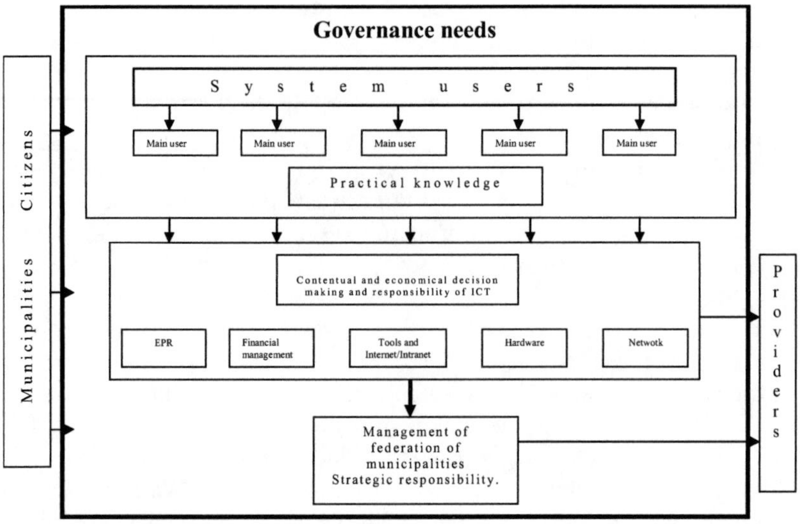

effectively. Especially, relations to system vendors which were earlier haphazard, became now clearer. There was now a defined contact person for each part of the system and the contacts could be conducted through them. The structure also serves the guidance functions mentioned in our definition of governance structure earlier.

The case pointed out that even or better yet especially a small organization needs clear governance structures to handle various functions. In the transaction costs point of view the health care is moving from a hierarchy where market transactions are eliminated towards a market. This change has been fast and health care organizations have faced difficulties in adopting new ways of operating. Industry has not had enough time to adjust to new situation and organizations have tried to cope with old structures. The governance structures should be emphasized especially in the case of functions with high effect on processes and activities but with little expertise to handle them in the organization. ICT is naturally not a core business in any health care organization but affects almost every process and activity when implemented.

Case 2: Organizing the External Governance Structures of a Medium-Sized Health Care Unit

In the second case the authors conducted evaluation research of the large information system implementation project called Primus during the year 2000. The project was executed in the public sector health care department in the fifth largest city of Finland. The Primus included four subprojects: EPR, telecommunications network, process development and three smaller development projects. The first two subprojects included the basic infrastructure solutions and were quite successfully implemented. The last two subprojects were more connected with the exchange relationships in daily health care, and were considerably more difficult to master. During the project, 800 users in 440 workstations began using the new patient record systems in about a hundred different units around the city's health care department. As in our previous case, also here the infrastructure and especially the EPR were in key roles. Before the project the patient records were in manual form.

Our evaluation research was divided in two parts. In the first part we evaluated the process, which led to the implementation of new ICT. We focused on management and strategic issues, negotiations, sourcing decisions, supportive issues (training, help desk) and technical solutions: aspects that were decided in the planning process of the project. The outsourcing solutions were also in close examination by the public. The network and maintenance of hardware was outsourced to a large Finnish teleoperator, which was a solution that did not satisfy everybody.

In the second part we evaluated the results of the project. The evaluation included the cost-benefit aspects and end user and patient satisfaction in order to assess how the project has influenced the activities of the organization. The research included 90 interviews, two questionnaires (staff and customers), two group interviews and one half-day seminar for interest groups. We had also meetings in research steering group at least once in a month.

During the evaluation project we also did some comparative research with a local private health care clinic. One of the main interests of comparative research was to find out which kind of organization held responsibility about ICT in the private sector and the governance structure they used.

Although there were several functions in the implementation project, probably the most critical parts of it were outsourcing network, hardware maintenance and acquiring EPR (Figure 5). As said earlier, outsourcing does not mean that outsourced functions can be forgotten and left without attention. In our case organization there had not been earlier experience about outsourcing of such large technological solutions, and even nation-wide the solution was a new in public health care environment. So there was no baseline determined from which the organization could have sought examples about structure or guidance for the solution. Outsourcing contract was agreed for a five-year period after which the outsourced services were set under public competition, so it can be considered as a long-term exchange relationship. New situations set high requirements for managing the relationships and new governance structures had to be created.

In such contracts, the rules and guidance functions play extremely important roles. From our determination earlier the rules on how to perform an exchange relationship and especially how to control and follow-up an exchange relationship are most vital. Without strict rules the transaction cost can rise unexpectedly high. In Figure 2, the sources of transaction cost are mentioned: complexity of product, opportunism and bounded rationality. Opportunism is related also to trust. With contracts these elements can be eliminated to some extent, but not all of them, as we noted few times in our case. There were misunderstandings about maintenance level and responsibilities and especially with the software provider about corrections and new features in EPR. Problems and limitations in EPR caused problems in practical work.

Although these two cases are different in size and in research focus, they give an excellent opportunity to study the governance structures in public health care. Public health care organizations operate basically with the same rules and procedures. However, like was said earlier, the organization's size affects its complexity in administration. Complexity affects naturally also governance structures and the management. On the other hand, small organizations have less expertise to execute, e.g., ICT projects, and that lack makes the project complex even in that environment. Although the large organiza-

Figure 5. The Governance Structure in the Outsourcing Case

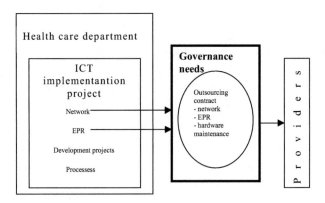

tions have more complex projects they also have more resources to solve them. We mentioned earlier also that the IT-outsourcing is more complex than many other forms of outsourcing since it pervades, affects and shapes most organization processes in some way and this is the case whether the organization is small or large.

The above is maybe a rough generalization, but comparing these two cases, in the smaller case the management was many times totally desperate since they had no expertise to solve problems and they were too small to put enough pressure for providers to have more attention. The larger organization was the largest customer to the system provider so the situation was completely different (not meaning that they did not have any problems). In these situations the contracts (rules) and trust become an essential in the governance structure perspective.

Since health care as industry has long traditions in organization and governance structures it is not easy to create new structures. However, also, health care has to learn to create and use new governance structures if it wants to keep up with the technological development. ICT facilitates and forces organizations to consider new governance structures, so in that way ICT is acting as a catalyst in renewing health care structures even deeper and wider than just ICT requires. Old structures are challenged and their existence is put under close examination.

In both cases, introduction of ICT challenged old governance structures and made it possible to introduce new structures. The smaller organization fought to find governance for the ICT internally, whereas the bigger organization made a risky outsourcing decision. The bigger organization had an opportunity to change general governance structures because of modern ICT, but this proved to be a difficult road to go.

CONCLUSION

A carefully and intelligently defined and implemented governance structure relieves organizations and their management of a lot of stress. However, in a constantly turbulent environment, no organization is a running machine that would not need maintenance. Governance structures are means to control exchange relationships, but in rare cases means to totally automate them.   Especially in the popular outsourcing literature and marketing, outsourcing is pictured as a panacea of getting rid of management troubles — say in a complicated case of ICT management. As we however know, this responsibility of management cannot be escaped.

Exchange relationships are the value-adding activities of organizations. Governance structures give meaning and rules to them. Exchange relationships are there to add value to all partners, but unfortunately also contain costs and risks. For an exchange relationship to take place, the added value produced must be bigger than those of the transaction costs for each partner, taking into account also his/her/its risk profiles. In just a few simple transactions, the total outcome of the exchange can be counted beforehand, if even afterwards.

Value chains are a typical tool to analyze governance structures and exchange relationships happening in them. Their task is to show the total flow of activities, whether it is either the case of information, material or money. With a value chain, each individual actor can understand its place in the totality. Value chains are too valuable in their focus towards value-adding elements. Should an exchange relationship not add value to anybody, it should not happen. Governance structures should be there to eliminate non-value-adding elements from value chains. This is too often forgotten.

Trust is a concept that is currently heavily studied, but too seldom in the field of governance structures research and practice. For us to produce a metaphor, trust is the glue that keeps governance structures together and the crease that makes them fluently serve exchange transactions.

The pressures towards health care are many. Many of these pressures would have emerged despite ICT, but also in many cases ICT has even emphasized those pressures. On the other hand, ICT offers many new possibilities to build governance structures. Finally, ICT is a resource and exchange relationship field to be governed. ICT has also three roles in their relationship to governance structures:

- Emphasizer and visualizer of pressures towards governance structures in an industry, say health care,
- Facilitator of new governance structures to handle exchange relationships,
- An exchange relationship field to be governed.

Many see privatization as a panacea for all the problems in most industries, including health care. We do not however share this view. In many aspects private organizations are faster and more flexible in their activities, but health care too has many characteristics that cause that activities there cannot be solely left to market forces. This is not to say that public organizations should not try to learn the best practices from the private sector.

A whole book could be devoted to the discussion of health care ICT sourcing decisions. A fact seems to be that all health care organizations are compelled to outsource at least some ICT functions. They simply lack the expertise. In this sector, fine-tuned thinking is however needed, so that the core competencies of health care are not outsourced. Modern health care is anyway much about management of information about the patient, diseases and their cures. Infrastructure-oriented ICT tasks could and should be outsourced, but not the whole task of knowledge management.

Table 3. Summary of Conclusions

- Governance structure is not a means to get rid of management responsibilities, but to handle them
- Governance structures give meaning and rules to exchange transactions
- In exchange transactions, risks and transaction costs are lowered
- Value chains are useful in defining exchange relationships: added value is emphasized
- Trust is a central resource in any governance structure: trust as glue and crease of governance structures
- ICT acts both as a challenger and solution in the health care governance structures
- Public health care organizations are in a pressure to adopt market-like governance structures from private institutions
- IT tends to be outsourced in modern health care
- The bigger the health care organization, the more diversified the possibilities for outsourcing

---

Our cases handled two organizations of different sizes. The smaller one had even difficulties in finding resources to outsource basic ICT — even this process needs careful management. The bigger organization outsourced ICT infrastructure management and searched for possibilities to redesign governance structures in other fields. To some extent success was visible, but the number of available options seemed to be out of scope for the organization. Again, management time and energy was a scarce resource.

REFERENCES

Aggarwal, A. K., & Mirani, R. (1999). DSS model usage in public and private sectors: Differences and implications. *Journal of End User Computing, 11*(3), 20-28.

Aggarwal, R., & Rezaee, Z. (1995). Internal control structure in telecommuting. *Internal Auditing, 11*(1), 16-23.

Alchian, A. A., & Allen, W. R. (1977). *Exchange and production: Competition, coordination and control.* Wadsworth.

Berger, P., & Luckmann, T. (1991). *The social construction of reality.* Penguin Group.

Calton, J. M., & Lad, L. J. (1995). Social contracting as a trust-building process of network governance. *Business Ethics Quarterly, 5*(2), 271-295.

Casson, M. (1982). *The entrepreneur. An economic theory.* Oxford.

Chandler, A. D. J. (1966). *Strategy and structure: Chapters in the history of the industrial enterprise.* Doubleday.

Checkland, P. (1981). *Systems thinking, systems practice.* Chichester, UK: John Wiley & Sons.

Coase, R. H. (1937). The nature of the firm. *Economica N.S., (4),* 386-405.

Compaine, B. M. (Ed.). (2001). *The digital divide: Facing a crisis or creating a myth?* Cambridge, MA: MIT Press.

Conner, K. R., & Prahalad, C. K. (1996). A resource-based theory of the firm: Knowledge versus opportunism. *Organization Science, 7*(5), 477-501.

Dickerson, C. M. (1998). Virtual organizations: From dominance to opportunism. *New Zealand Journal of Industrial Relations, 23*(2), 35-46.

Dolan, T. C. (1998). Balancing stakeholder satisfaction. *Healthcare Executive,* (November/December).

Donaldson, E. J., & Gray, J. (1998). Clinical governance as a quality duty for health organizations. *Quality in Health Care,* (7), 37-44.

The European Health Report (2002). Copenhagen: World Health Organization, Regional Office for Europe.

Genefke, J., & Bukh, P. N. D. (1997). On hikers, tigers, trust and opportunism. *Academy of Management Review, 22*(1), 11-13.

Hambrick, D. C., & Jackson, E. M. (2000). Outside directors with a stake: The linchpin in improving governance. *California Management Review, 42,* 108-127.

Holm, J., Tähkäpää, J., & Suomi, R. (2000). *Primus-hankkeen arviointi. Prosessinäkökulma.* Turku: Turun kaupungin terveystoimen julkaisuja no. 4:2000.

Jarillo, J. C. (1988). On strategic networks. *Strategic Management Journal, 9*(1), 31-41.

Järvelin, J. (2002). *Health care systems in transition - Finland, 4*(1). European Observatory on Health Care Systems- WHO Regional Office for Europe.

Jeon, S. (1996). Moral hazard and reputational concerns in teams: Implications for organizational choice. *International Journal of Industrial Organization, 14*(3), 297-315.

Kangas, K. (1999). Competency & capabilities based competition and the role of Information Technology: The case of trading by a Finland-based firm to Russia. *Journal of Information Technology. Cases and Applications, 1*(2), 4-22.

Kauffman, R. J., McAndrews, J., & Wang, Y. M. (2000). Opening the "black box" of network externalities in network adoption. *Information Systems Research, 11*(1), 61-82.

Kern, T., & Wilcocks, L. (2000). Contracts, control and "presentation" in IT outsourcing: Research in thirteen UK Organizations. *Journal of Global Information Management, 8*(4).

Kern, T., & Willcocks, L. (2002). Exploring relationships in information technology outsourcing: the interaction approach. *European Journal of Information Systems, 11,* 3-19.

Koski, H. (1999). The implications of network use, production network externalities and public networking programmes for firm's productivity. *Research Policy, 28*(4), 423-439.

Loh, L., & Venkatraman, N. (1993). Corporate governance and strategic resource allocation. The case of Information Technology investments. *Accounting, Management and Information Technologies, 3*(4).

Lyons, B., & Mehta, J. (1997). Contracts, opportunism and trust: Self-interest and social orientation. *Cambridge Journal of Economics, 21*(2), 239-257.

Madhok, A. (1996). The organization of economic activity: Transaction costs, firm capabilities, and the nature of governance. *Organization Science, 7*(5), 577-590.

Malone, T. W. (1987). Modeling coordination in organizations and markets. *Management Science, 33*(October 10), 1317-1332.

Mylonopoulos, N. A., & Ormerod, R. J. (1995). A microanalytical approach to the efficient governance of IT service provision: The case of Outsourcing. Paper presented at the *Proceedings of the Third European Conference on Information Systems, Athens, Greece.*

Nicholas, D., Huntington, P., Williams, P., & Jordan, M. (2002). NHS Direct Online: Its users and their concerns. *Journal of Information Science, 28*(4), 305-319.

Nooteboom, B. (1996). Trust, opportunism and governance: A process and control model. *Organization Studies,* (6), 985-1010.

Nooteboom, B., Berger, H., & Noorderhaven, N. G. (1997). Effects of trust and governance on relational risk. *Academy of Management Journal, 40*(2), 308-338.

Norris, P. (2001). *Digital divide: Civic engagement, information poverty, and the Internet worldwide.* Cambridge: Cambridge University Press.

OECD. (2003). *Communication on the development of Public Health policy.* OECD. Available online: http://europa.eu.int/comm/health/ph/general/phpolicy2.htm [2003, 31.1.].

Ouchi, W. G. (1980). Markets, bureaucracies, and clans. *Administrative Science Quarterly, 25*(1).

Pelletier-Fleury, N., & Fargeon, V. (1997). Transaction cost economics as a conceptual framework for the analysis of barriers to the diffusion of telemedicine. *Health Policy, 42*(1), 1-14.

Peterson, R. R. (2000). Emerging capabilities of Information Technology governance: Exploring stakeholder perspectives in financial services. Paper presented at the *Proceedings of the Eighth European Conference on Information Systems, Vienna.*

Porter, M. E. (1985). *Competitive advantage: Creating and sustaining superior performance.* New York, London: Free Press, Collier Macmillan.

Quackwatch www-site (2003). Available: http://www.quackwatch.org/) [2003, 31.1.].

Seidmann, A., & Sundararajan, A. (1997). The effects of task and information asymmetry on business process redesign. *International Journal of Production Economics, 50*(2-3), 117-128.

Shapiro, C., & Varian, H. R. (1999). *Information rules: A strategic guide to the network economy.* Boston, MA: Harvard Business School Press.

Simon, H. A. (1991). Bounded rationality and organizational learning. *Organization Science, 2*(1), 125-134.

Spanjers, R., Smiths, M., & Hasselbring, W. (2001). Exploring ICT-enabled networking in hospital organizations. In R. A. Stegwee & T. A. Spil (Eds.), *Strategies for Healthcare Information Systems* (pp. 164-180). Hershey, PA: Idea Group Publisher.

Spil, T. A. M. (1998). *From professional healthcare to where? A healthcare information management reference model.* Paper presented at the IRMA International Conference.

The State of Health in the European Community (1996). Brussels: European Union, Office for Official Publications of the European Communities.

Suomi, R. (2000). Leapfrogging for modern ICT usage in the health care sector. Paper presented at the *Proceedings of the Eighth European Conference on Information Systems, Vienna.*

Suomi, R. (2001). Streamlining operations in health care with ICT. In T. A. Spil & R. A. Stegwee (Eds.), *Strategies for Healthcare Information Systems* (pp. 31-44). Hershey, PA: Idea Group Publishing.

Suomi, R., & Kastu-Häikiö, M. (1998). Cost- and service effective solutions for local administration - the Finnish case. *Total Quality Management, 9*(2-3), 335-346.

Suomi, R., & Tähkäpää, J. (2003). Establishing a contact centre for public health care. Paper presented at *The 36th Hawaii International Conference on System Sciences (HICSS-36), Big Island, Hawaii*.

Suomi, R., Tähkäpää, J., & Holm, J. (2001). Outsourcing of health care information systems - why and why not. In R. Suomi & J. Tähkäpää (Eds.), *Health and Wealth Through Knowledge. Information System Solutions in the Health Care Sector* (pp. 65-75). Turku: Turku Centre for Computer Science.

Tähkäpää, J., Suomi, R., & Holm, J. (2001). *Primus-hankkeen arviointi. Vaikuttavuusnäkökulma*. Turku: Turun kaupungin terveystoimen julkaisuja no. 2:2001.

Tähkäpää, J., Turunen, P., & Kangas, K. (1999). Information management in public healthcare; A case of small municipality federation. Paper presented at the *IRMA (Information Resources Management Association), Hershey, PA, USA*.

Thompson, J. D. (1967). *Organizations in action*. Russell: Sage.

Turunen, P. (2001). *Tietojärjestelmien arviointimenetelmien valinta terveydenhuolto-organisaatiossa- sidosryhmänäkökulma*. Unpublished Dissertation, Turku School of Economics and Business Administration, Turku.

Wang, E. T. G., & Barron, T. (1995). Controlling information-system departments in the presence of cost information asymmetry. *Information Systems Research, 6*(1), 24-50.

White, L. (2000). Changing the whole system in the public sector. *Journal of Organizational Change Management, 13*(2), 162-177.

Williamson, O. E. (1985). *The economic institutions of capitalism. Firms, markets, relational Constructing*. The Free Press.

Williamson, O. E. (1989). Transaction cost economics. In R. Schmalensee & R. D. Willig (Eds.), *Handbook of Industrial Organization* (Vol. 1, pp. 137-182). New York: North Holland.

Xiao, Z., Powell, P. L., & Dodgson, J. H. (1998). The impact of information technology on information asymmetry. *European Journal of Information Systems, 7*(2), 77-89.

About the Authors

Wim Van Grembergen is a professor and chair of the MIS Department at the Business Faculty of UFSIA-RUCA (University of Antwerp) and executive professor at the University of Antwerp Management School (UAMS), Belgium. He is a guest professor at the Free University of Amsterdam, has been a guest professor at the University of Leuven (KUL), and had teaching assignments at the University of Stellenbosch in South Africa and the Institute of Business Studies in Moscow. He teaches information systems at the undergraduate and executive level, and researches in business transformations through information technology, audit of information systems, IT Balanced Scorecard and IT Governance. He served as academic director of the MBA Program of UFSIA (1989-1995) and presently he is academic coordinator of an IT-audit master program and an e-business master program. Dr. Van Grembergen presented at the European Conference on Information Systems (ECIS) in 1997 and 1998, at the Information Resources Management Association (IRMA) Conferences in 1998, 1999 and 2000, and at the Hawaii International Conference on Systems Sciences (HICSS) in 2001 and 2002. He has been Track Chair on IT Evaluation for the IRMA-conferences since 2000 and Minitrack Chair "IT Governance and its Mechanisms" for the HICSS-conferences since 2002. He published articles in journals such as *Journal of Strategic Information Systems, Journal of Corporate Transformation, Journal of Information on Technology Cases and Applications, IS Audit & Control Journal* and *EDP Auditing (Auerbach)*. He also has several publications in leading Belgian and Dutch journals and, in 1997, published a book on business process reengineering in Belgian organizations and in 1998, a book on the IT Balanced Scorecard. He edited a book entitled *IT Evaluation Methods and Management* which was published in 2001 by Idea Group Publishing, and in 2002, *Information Systems Evaluation Management*, also published by IGP. Currently, he is editing a book entitled *Strategies for Information Technology Governance*. Professor Van Grembergen serves on the editorial boards of *Journal of Global Information Technology Management*

(JGITM), Journal of Information Technology Cases and Applications (JITCA) and *Annals of Cases on Information Technology and Management in Organizations.* Professor Van Grembergen was engaged in the development of CobiT 3rd Edition and has presented papers and conducted workshops at ISACA (Information Systems Audit and Control Association) international conferences. He is also a member of the Academic Relations Task Force of ISACA and is currently conducting a research project for ISACA on IT Governance. Dr. Van Grembergen is a frequent speaker at academic and professional meetings and conferences and has served in a consulting capacity to a number of firms. He is a member of the Board of Directors of an IT company servicing a Belgian financial group. His e-mail address is: wim.vangrembergen@ufsia.ac.be.

* * *

Isabelle Amelinckx is a research assistant in management information systems at the University of Antwerp (UFSIA) in Belgium. She is engaged in research in the domain of the application of the concept of the Balanced Scorecard and business-to-business e-business. Currently, she is preparing a PhD on the changes in organizations and industry structure resulting from B2B e-business initiatives.

Cassio Bastos is a Business Process Solutions Systems analyst for the New Brunswick Power Corporation Business Information Systems division (Canada). His special area of interest is the configuration and use of enterprise resource planning systems in business analysis and reporting and strategic management. His business and technical knowledge lies in business planning and systems design. Recent contributions include the formulation of corporate and divisional business planning processes and mechanics of IT Governance. Bastos has more than 10 years of progressive experience in mechanical industrial engineering, business planning and information technology. He has a master's degree in Business Administration from the University of New Brunswick, Fredericton, New Brunswick, a bachelor's degree in Mechanical Industrial Engineering from the School of Industrial Engineering, São José dos Campos, Brazil, and an Applied Information Technology diploma from the Information Technology Institute, Moncton, New Brunswick.

Erik Beulen, PhD, is associated with Tilburg University, The Netherlands, and is employed as an international business development manager with Atos Origin. He obtained his PhD from Tilburg University. His research concentrates on outsourcing and the management of outsourcing relationships. His papers have been published in the proceedings of the HICSS and ICIS. He is the author and coauthor of various Dutch language books on the subject of IT outsourcing.

Joanne Callahan lives in New Brunswick, Canada, and works for the provincial utility company NB Power in the Business Information Systems division. Her initial involvement in IT governance was through the Executive and Senior Management Education Information Technology Program. She is now responsible for Corporate IT Governance at NB Power and the divisional business planning process for Business Information Systems. She has worked in IT for eight years in her current capacity and in the e-learning industry.

Steven De Haes is responsible for the Information Systems Management executive programs at the University of Antwerp Management School (UAMS), Belgium. He is engaged in research in the domain of IT Governance and in this capacity performing research for ISACA (Information Systems Audit and Control Association). Currently, he is preparing a PhD on the practices and mechanisms of IT Governance. He has several publications on IT Governance, primarily in the *Information Systems Control Journal* and in the *Journal for Information Technology Case Studies and Applications (JITCA)*.

Alea Fairchild is the director of Greiner International, Inc. Her special area of interest is the development and use of intelligence and technology in product planning and business strategy. Her technical expertise lies in open architectures and interoperability. Recent areas of research have included knowledge management and productivity metrics for technology in financial institutions. Dr. Fairchild has more than 13 years of experience in global IT market analysis, and has worked for many of the major market research agencies as both an analyst and as a consultant. She has also performed as a consultant for the European Commission, as well as major multinational IT companies throughout Europe. Her forthcoming book, *Technological Aspects of Virtual Organizations*, will be published by Kluwer in 2003. As an acknowledged IT market expert, Dr. Fairchild has been quoted by many of the leading business and technology journals. She is a member of both the American Economic Association (AEA) and The Strategic Planning Society (SPS). She is the author of three books: *Interoperability for Enterprise Information Systems*, published in September 1996, *Year 2000 Compliance: The Guide to Successful Implementation*, published in May 1997, and *Reengineering and Restructuring the Enterprise*, published in February 1998, all published by CTR. Dr. Fairchild received her Doctorate in Applied Economics from Limburgs Universitair Centrum in Belgium, in the area of banking and technology. She has a master's degree in International Management from Boston University/Vrije Universiteit Brussel, Brussels, Belgium, and a bachelor's degree in Business Management and Marketing from Cornell University, Ithaca, New York.

Petter Gottschalk is professor of Information Management in the Department of Leadership and Organizational Management at the Norwegian School of Management, Norway. He received his MBA in Germany, MSc in the USA and DBA in the UK. His executive experience includes CIO at ABB Norway, CEO at ABB Datacables and CEO at the Norwegian Computing Center. Professor Gottschalk's research has been published in journals such as *Information & Management, Long Range Planning* and *European Journal of Information Systems*.

Erik Guldentops, CISA, was until recently security advisor for the Society of Worldwide Interbank Financial Telecommunication (SWIFTsc) in Brussels, Belgium, where he previously held the positions of chief inspector and director of information security. SWIFT provides secure global communication to more than 7,000 financial institutions in more than 190 countries. More than five million messages valued in trillions of dollars are sent over SWIFT's network every business day. Guldentops is advisor to the board of the IT Governance Institute, USA, and an executive professor in the management school of the University of Antwerp, Belgium, where he teaches on the subjects of IT security and control, IT governance and risk management. He initiated and has headed up the developments of Control Objectives for Information and related Technology (CobiT) since the early nineties and is currently the chair of Information Systems Audit and Control Association's CobiT Steering Committee.

Dwayne Keyes is manager of Business Planning and IT Governance and has been involved in the Utilities and IT industry at New Brunswick Power, Canada, for 22 years. During these years, he has had various Information Technology assignments including: programmer analyst; project manager; customer relationship manager; operations manager; project office manager; manager, Business Planning & IT Governance. Dwayne has been involved in most aspects of the applications life cycle, and business planning and management processes. He was responsible for the management and evolution of the IT Governance process at NB power. He is currently coordinating all programs supporting a major corporate restructuring initiative.

Jerry Luftman is the executive director and Distinguished Service professor for the graduate information systems programs at Stevens Institute of Technology, USA. His 22 year career with IBM prior to his appointment at Stevens included strategic positions in management (IT and consulting), management consulting, Information Systems, marketing, and executive education. He played a leading role in defining and introducing IBM's Consulting Group. As a practitioner he held several positions in IT, including a CIO. Dr. Luftman's research papers have appeared in leading professional journals and he has presented at many executive and professional conferences. His new book, *Competing in the Information Age: Align in the Sand*, published by Oxford University Press, is one of the bases for this chapter. His PhD in Information Management is from Stevens Institute of Technology.

Manuel Mogollon is senior manager of Global Pricing at Nortel Networks, USA, where he provides pricing analysis and pricing strategies to ensure competitiveness and profitability based on corporate pricing guidelines. He has developed financial analysis, customer value analysis, and price modeling tools, based on knowledge of competitors' pricing strategies and discounting practices. He also developed Return-on-Investment tools for different Nortel Networks products and designed an Internet online rebate calculator that allowed end-users to calculate their rebate for Nortel Networks marketing programs.

Nandish V. Patel has published research on deferred system's design, deferred design decisions and tailorable information systems for which he was invited to prepare a position paper for the American Association for Computing Machinery (ACM) Special Interest Group on Computer Supported Co-Operative Work. This work is collected together in an international book entitled *Adaptive Evolutionary Information Systems*. He initiated a studentship [Ref: 00302238] on deferred system's design offered by British research agency EPSRC. Dr. Patel has written nationally and internationally refereed journal and conference papers and book chapters, and has spoken at, and been a member of programme committees for conferences on information systems in the USA and Europe. He has refereed national and international journal papers for the *European*

Journal of Information Systems and *International Journal of Human Computer Studies* and international conferences like the highly prestigious Hawaiian International Conference on System Sciences and European Conference on Information Systems. His work on evaluating evolutionary IS has received a citation of excellence for practical implications. He was entered for the recent British Research Assessment Exercise for his work on deferred system's design and tailorable IS. The submission was awarded five (World Class research). He is currently working with a data exchange company and potentially the British Ministry of Defense to investigate emergent information needs and networked organisations. He is the deputy director of Studies for The Brunel MBA, UK.

Ryan R. Peterson is professor of Information Systems Management at the Instituto de Empresa (Madrid, Spain). He holds an MSc in Organization Science, and a PhD in Information Systems. His research focuses on the governance, organization and management of IT and e-business, with a keen interest and research experience in financial services, health care, educational institutions, and small and medium enterprises. He has been involved in various programs on supply chain management, customer relationship management, application service provisioning, e-health, and e-business innovation. His publications have appeared in international journals and magazines, and he is a regular speaker at international conferences and management workshops. Professor Peterson was previously associate researcher at the Telematics Research Institute and the Center for Research on Information Systems and Management, and formerly MBA instructor at the TIAS Business School, where he lectured on IT Governance, strategic information management, and IT auditing.

Mahesh S. Raisinghani, is a program director of E-Business and a faculty member at the Graduate School of Management, University of Dallas, USA, where he teaches MBA courses in Information Systems and E-Business. Dr. Raisinghani was the recipient of the 1999 UD Presidential Award and the 2001 King Hagar Award for excellence in teaching, research and service. His previous publications have appeared in *Information and Management, Journal of Global IT Management, Journal of E-Commerce Research, Information Strategy: An Executive's Journal, Information Resources Management Journal, Journal of IT Theory and Applications, Enterprise Systems Journal, Journal of Computer Information Systems,* and *International Journal of Information Management*, among others. He serves as an associate editor and on the editorial review board of leading information systems/e-commerce journals and on the board of directors of Sequoia, Inc. Dr. Raisinghani is included in the millennium edition of Who's Who in the World, Who's Who Among America's Teachers and Who's Who in Information Technology.

Ronald Saull, MBA, CSP is the senior vice-president and CIO of the Information Services Division of Great-West Life, London Life (Canada) and Investors Group headquartered in Winnipeg, Canada. Mr. Saull is responsible for the strategic application of information technology and for the integration of information services to each of the Power Financial Group of Companies. He has more than 25 years' experience as an information systems professional and manager in both the public and private sectors. Mr. Saull is a vice-president of the International Board of Directors of ISACA (Information Systems Audit

and Control Association) consisting of more than 25,000 information systems professionals worldwide. He is the past chairman of ISACA's Research Board and is currently
active as a member of the Board of Directors, the IT Governance Board and the CobiT
Steering Committee. In 1999 he was appointed to the CobiT (Control Objectives for
Information and related Technology) Steering Committee. This Committee is charged
with the continuing development of CobiT, an emerging global best practice for
governance and control of the IT function. He has published an article on the IT Balanced
Scorecard in the *Information Systems Control Journal* and presented on this topic at
ISACA conferences.

Reima Suomi is a professor of Information Systems Science at Turku School of
Economics and Business Administration, Finland, since 1994. He is a docent for the
universities of Turku and Oulu. He concentrates on topics around management of
telecommunications, including issues such as management of networks, electronic and
mobile commerce, virtual organizations, telework and competitive advantage through
telecommunication-based information systems. His current research interests focus on
the tourism industry and health care industry. Reima Suomi has altogether more than 200
publications, and has published in journals such as *Information & Management,
Information Services & Use, Technology Analysis & Strategic Management, The
Journal of Strategic Information Systems* and *Behaviour & Information Technology*.
For the academic year 2001-2002 he was a senior researcher, "varttunut tutkija" for the
Academy of Finland.

Jarmo Tähkäpää is a PhD student and a researcher in Institute of Information Systems
Science at Turku School of Economics and Business Administration, in Finland, since
1998. During 1999-2002 he has held positions of research associate and senior researcher
associate and at the moment he is finalizing his dissertation. His research topics are from
the area of health care industry and include topics like planning, managing and evaluating
health care information systems. Together he has some 30 publications in conferences
and professional magazines.

Michalis Xenos is a lecturer in the Informatics Department of the School of Science and
Technology of the Hellenic Open University, Greece. Dr. Xenos also works as a
researcher in the Computer Technology Institute of Patras and has participated in more
than 15 research and development projects in the area of software engineering and IT
development management. His research interests include, inter alia, IT Management and
Software Quality. He is the author of five books in Greek and more than 30 papers in
international journals and conferences.

Index

E

e-business initiatives 152, 160
e-business IT governance 81
e-business models 83
e-business projects 188
e-commerce 83
e-commerce applications 187
e-mall 93
e-shop 93
e-wakening 39
economic viability 88
employee retention 194
enabler 131
entrepreneur 260
exchange relationship 359
external metrics 227
extranets 84

F

figurehead 259
functional hierarchy 363
future orientation 130

G

generic e-business balanced scorecard 157
generic IT balanced scorecard 157
global governance structure 271
governance maturity 102
governance structures 357

H

health care industry 357
human resource consideration 107
human resources 329

I

information and communication technology
    (ICT) 357
information economics 28
information management 322
information systems (IS) 82
information technology (IT) 82
information technology (IT) budgets 188
information technology governance 37

insourcing 311
integration strategies 37
intellectual property 174
internal metrics 223
internal rate of return (IRR) 154
Internet marketing 162
intranets 84
IT balanced scorecard 129
IT-business alignment exercises 351
IT function 247
IT governance 1, 343
IT governance committees 348
IT governance framework 344
IT governance solutions 40
IT governance tactics 216
IT Infrastructure Library (ITIL) 352
IT investment 175
IT investment 271
IT management 4
IT manager 247
IT organization 247
IT outsourcing partnerships 310
IT risks 271
IT steering committees 22
IT strategy 85, 169, 321
IT strategy committee 22
IT supplier 313

K

key goal indicators 270
key performance indicators 270
knowledge management 169
knowledge metrics 169
knowledge sharing 28
knowledge workers 175

L

liaison 260
line management 247

M

management guidelines 270
managerial know-how 195
maturity models 270
measurability of IT 270

NEW from Idea Group Publishing

- The Enterprise Resources Planning Decade: Lessons Learned and Issues for the Future, Frederic Adam and David Sammon/ ISBN:1-59140-188-7; eISBN 1-59140-189-5, © 2004
- Electronic Commerce in Small to Medium-Sized Enterprises, Nabeel A. Y. Al-Qirim/ ISBN: 1-59140-146-1; eISBN 1-59140-147-X, © 2004
- e-Business, e-Government & Small and Medium-Size Enterprises: Opportunities & Challenges, Brian J. Corbitt & Nabeel A. Y. Al-Qirim/ ISBN: 1-59140-202-6; eISBN 1-59140-203-4, © 2004
- Multimedia Systems and Content-Based Image Retrieval, Sagarmay Deb
  ISBN: 1-59140-156-9; eISBN 1-59140-157-7, © 2004
- Computer Graphics and Multimedia: Applications, Problems and Solutions, John DiMarco/ ISBN: 1-59140-196-86; eISBN 1-59140-197-6, © 2004
- Social and Economic Transformation in the Digital Era, Georgios Doukidis, Nikolaos Mylonopoulos & Nancy Pouloudi/ ISBN: 1-59140-158-5; eISBN 1-59140-159-3, © 2004
- Information Security Policies and Actions in Modern Integrated Systems, Mariagrazia Fugini & Carlo Bellettini/ ISBN: 1-59140-186-0; eISBN 1-59140-187-9, © 2004
- Digital Government: Principles and Best Practices, Alexei Pavlichev & G. David Garson/ISBN: 1-59140-122-4; eISBN 1-59140-123-2, © 2004
- Virtual and Collaborative Teams: Process, Technologies and Practice, Susan H. Godar & Sharmila Pixy Ferris/ ISBN: 1-59140-204-2; eISBN 1-59140-205-0, © 2004
- Intelligent Enterprises of the 21st Century, Jatinder Gupta & Sushil Sharma/ ISBN: 1-59140-160-7; eISBN 1-59140-161-5, © 2004
- Creating Knowledge Based Organizations, Jatinder Gupta & Sushil Sharma/ ISBN: 1-59140-162-3; eISBN 1-59140-163-1, © 2004
- Knowledge Networks: Innovation through Communities of Practice, Paul Hildreth & Chris Kimble/ISBN: 1-59140-200-X; eISBN 1-59140-201-8, © 2004
- Going Virtual: Distributed Communities of Practice, Paul Hildreth/ISBN: 1-59140-164-X; eISBN 1-59140-165-8, © 2004
- Trust in Knowledge Management and Systems in Organizations, Maija-Leena Huotari & Mirja Iivonen/ ISBN: 1-59140-126-7; eISBN 1-59140-127-5, © 2004
- Strategies for Managing IS/IT Personnel, Magid Igbaria & Conrad Shayo/ISBN: 1-59140-128-3; eISBN 1-59140-129-1, © 2004
- Information Technology and Customer Relationship Management Strategies, Vince Kellen, Andy Drefahl & Susy Chan/ ISBN: 1-59140-170-4; eISBN 1-59140-171-2, © 2004
- Beyond Knowledge Management, Brian Lehaney, Steve Clarke, Elayne Coakes & Gillian Jack/ ISBN: 1-59140-180-1; eISBN 1-59140-181-X, © 2004
- Multimedia Security: Steganography and Digital Watermarking Techniques for Protection of Intellectual Property, Chun-Shien Lu/ ISBN: 1-59140-192-5; eISBN 1-59140-193-3, © 2004
- eTransformation in Governance: New Directions in Government and Politics, Matti Mälkiä, Ari Veikko Anttiroiko & Reijo Savolainen/ISBN: 1-59140-130-5; eISBN 1-59140-131-3, © 2004
- Intelligent Agents for Data Mining and Information Retrieval, Masoud Mohammadian/ISBN: 1-59140-194-1; eISBN 1-59140-195-X, © 2004
- Using Community Informatics to Transform Regions, Stewart Marshall, Wal Taylor & Xinghuo Yu/ISBN: 1-59140-132-1; eISBN 1-59140-133-X, © 2004
- Wireless Communications and Mobile Commerce, Nan Si Shi/ ISBN: 1-59140-184-4; eISBN 1-59140-185-2, © 2004
- Organizational Data Mining: Leveraging Enterprise Data Resources for Optimal Performance, Hamid R. Nemati & Christopher D. Barko/ ISBN: 1-59140-134-8; eISBN 1-59140-135-6, © 2004
- Virtual Teams: Projects, Protocols and Processes, David J. Pauleen/ISBN: 1-59140-166-6; eISBN 1-59140-167-4, © 2004
- Business Intelligence in the Digital Economy: Opportunities, Limitations and Risks, Mahesh Raisinghani/ ISBN: 1-59140-206-9; eISBN 1-59140-207-7, © 2004
- E-Business Innovation and Change Management, Mohini Singh & Di Waddell/ISBN: 1-59140-138-0; eISBN 1-59140-139-9, © 2004
- Responsible Management of Information Systems, Bernd Stahl/ISBN: 1-59140-172-0; eISBN 1-59140-173-9, © 2004
- Web Information Systems, David Taniar/ISBN: 1-59140-208-5; eISBN 1-59140-209-3, © 2004
- Strategies for Information Technology Governance, Wim van Grembergen/ISBN: 1-59140-140-2; eISBN 1-59140-141-0, © 2004
- Information and Communication Technology for Competitive Intelligence, Dirk Vriens/ISBN: 1-59140-142-9; eISBN 1-59140-143-7, © 2004
- The Handbook of Information Systems Research, Michael E. Whitman & Amy B. Woszczynski/ISBN: 1-59140-144-5; eISBN 1-59140-145-3, © 2004
- Neural Networks in Business Forecasting, G. Peter Zhang/ISBN: 1-59140-176-3; eISBN 1-59140-177-1, © 2004

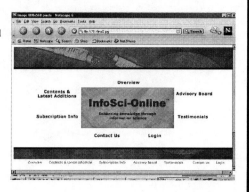

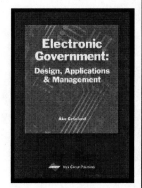